T0172580

HISTORIC ENGLAND
PRACTICAL BUILDING CONSERVATION

CONSERVATION BASICS

Series Editors: Bill Martin and Chris Wood

Volume Editor: Iain McCaig

HISTORIC ENGLAND
PRACTICAL BUILDING CONSERVATION

CONSERVATION
BASICS

Historic England

Routledge
Taylor & Francis Group

LONDON AND NEW YORK

First published in paperback 2024

First published 2013 by Ashgate Publishing Limited

Published 2021
by Routledge
4 Park Square, Milton Park, Abingdon, Oxon OX14 4RN

and by Routledge
605 Third Avenue, New York, NY 10158

Routledge is an imprint of the Taylor & Francis Group, an informa business

British Library Cataloguing-in-Publication Data

Practical building conservation.

    Conservation basics.

    1.   Historic buildings – Conservation and restoration.

    2.   Historic buildings – Law and legislation.

    I.   Historic England.

    363.6'9-dc22

Library of Congress Control Number: 2009928383

ISBN: 978-0-754-64551-1 (hbk)
ISBN: 978-1-032-609157- (pbk)

To the memory of John Ashurst (1937–2008), an inspiration and friend to all the editors, whose encouragement and support was a great motivation for this new series of Practical Building Conservation.

NOTES ON VOLUME EDITORS & CONTRIBUTORS

The professional profiles of editors and contributors described below were those at the time of initial publication.

Volume Editor: Iain McCaig

Iain McCaig is a Senior Architectural Conservator in the Building Conservation and Research Team at Historic England. He studied architecture before specialising in building conservation and has many years of experience, working in local authority, private practice and within Historic England.

Principal Contributors: Graham Abrey, Paul Bryan, Alan Cathersides, Paul Drury, Steve Emery, Emily Gee, Anna McPherson

Graham Abrey is a chartered building surveyor specialising in the maintenance and repair of historic buildings. He has extensive practical experience in project design, documentation, management and supervision. **Paul Bryan** is the Geospatial Imaging Manager in Historic England's Remote Sensing Team in York, specialising in photogrammetry, laser scanning and other metric survey techniques. **Alan Cathersides** is a Senior Landscape Manager at Historic England. He is particularly interested in methods of management that benefit both historic and natural environments. **Paul Drury** is a chartered surveyor, archaeologist, architectural historian and heritage consultant, with a particular interest in the concept of conservation and the development of historic environment policy. **Steve Emery** is Fire Safety Adviser at Historic England. He has wide-ranging experience in fire safety engineering in historic buildings and is particularly interested in emergency planning for the heritage sector. **Emily Gee**, Head of Designation at Historic England, is an architectural historian who also studied building conservation at the Architectural Association. **Anna McPherson** is a chartered architect and heritage consultant, with extensive experience of advising on historic environment policy and practice and the repair of historic buildings.

The volume editor would like to take this opportunity to thank all the contributors for their help and support with this book, as well as the information and images they have so generously provided.

Other Contributors:

Michael Copeman, Kathryn Davies, Keith Emerick, Sophie Godfraind, Kate Gunthorpe, Karen Gwilliams, Alison Henry, Hugh Johnson, Jon Livesey, Duncan McCallum, Adam Menuge, Richard Morrice, John Neale, Clare Parfitt, David Pickles, Sarah Pinchin, Brian Ridout, Malcolm Starr, John Stewart, Julie Swann, Martyn West, Richard Whittaker, Chris Wood, John Yates

CONTENTS

This series of *Practical Building Conservation* technical handbooks supersedes the original five volumes written by John and Nicola Ashurst, and published in 1988. This work was published by English Heritage in 2013. In 2015, the organisation changed its name to Historic England. The two names are reflected appropriately in this reprinting.

The series is aimed primarily at those who look after historic buildings, or who work on them. The ten volumes should be useful to architects, surveyors, engineers, conservators, contractors and conservation officers, but also of interest to owners, curators, students and researchers.

The contents reflect the work of the Building Conservation and Research Team, their colleagues at Historic England, and their consultants and researchers, who together have many decades of accumulated experience in dealing with deteriorating building materials and systems of all types. The aim has been to provide practical advice by advocating a common approach of firstly understanding the material or building element and why it is deteriorating, and then dealing with the causes. The books do not include detailed specifications for remedial work, neither do they include a comprehensive coverage of each subject. They concentrate on those aspects which are significant in conservation terms and reflect the requests for information received by Historic England.

Building conservation draws on evidence and lessons from the past to help understand the building, its deterioration and potential remedies; this encourages a cautious approach. New techniques, materials and treatments often seem promising, but can prove disappointing and sometimes disastrous. It takes many years before there is sufficient experience of their use to be able to promote them confidently. Nonetheless, understanding increases with experience and building conservation is a progressive discipline, to which these books aim to contribute.

The volumes also establish continual care and maintenance as an integral part of any conservation programme. Maintenance of all buildings, even of those that have deteriorated, must be a priority: it is a means of maximising preservation and minimising costs.

Most of the examples shown in the books are from England: however, Historic England maintains good relations with conservation bodies around the world, and even where materials and techniques differ, the approach is usually consistent. We therefore hope the series will have a wider appeal.

Duncan Wilson
Chief Executive, Historic England

"The past is never dead. In fact, it's not even past" (William Faulkner 1879–1962). The past, in other words, lives in the present. Historic buildings and places are an integral part of our collective heritage or patrimony. As Sir Bernard Feilden once observed: *"they bring us messages from the past"*; they tell stories of how and why we came to the places we are now. But they are also very much a part of the present and are, for the most part, adaptable to current expectations of commodity, firmness and delight. The choices we make about what, and how, to conserve – and our changing opinions about conservation – reflect current attitudes to the past and the values we ascribe to its material remains. These choices, in their turn, become part of the story that we pass on to future generations who will reinterpret it from their perspective, as we have done in our time, and add further chapters of their own.

In recent years there have been developments in building conservation theory and principles that have led to the adoption of a 'values-based' approach. This is based on the idea that heritage values are ascribed to places by people, rather than being inherent or intrinsic characteristics. Understanding the reasons why people value a place is therefore fundamental to the process of conservation. There have also been some changes in the language used to express these ideas, with terms such as heritage assets, significance and conservation planning entering the lexicon.

This book is about managing the maintenance and repair of historic buildings and places: what we may now refer to as 'heritage assets'. It explains the theories and principles that underpin building conservation in the 21st century, and shows how they may be applied in practical terms.

The book begins with a chapter tracing **The Evolving Concept of Building Conservation** in England from its origins up to the present day, and examines the ways in which attitudes to conservation have changed – and will continue to change – under the influence of developing ideas, beliefs, and wider political, social, economic and cultural factors. **Current Law, Policy & Guidance** reviews planning legislation, guidance, codes and standards concerning building conservation in England. **Conservation Planning for Maintenance & Repair** explains how a values-based method for safeguarding and sustaining the significance of heritage assets translates into practice. The wide range of **Survey & Investigation Methods** for understanding the history and evolution of a building, its construction and condition, are reviewed. **Ecological Considerations** examines the relationship between the natural and historic environments, and emphasises the importance of understanding the ecological interest of historic places, and how to reconcile conflicts between fauna and flora and historic buildings. **Managing Maintenance & Repair** describes the processes involved in devising, planning and managing programmes of maintenance and repairs for buildings of all types and sizes. Finally, the chapter **Planning for Emergencies** deals with risk assessment, risk management and preparing for unforeseen events such as fire and flood. Practical advice is given on management and measures to mitigate threats to significance.

USING THESE BOOKS

For accessibility and ease of use, the information given in the text has not been footnoted, and rather than references, short lists of further reading are given at the end of the appropriate chapters. References to other sections within the text are given in **bold**, and references to other publications in ***bold italics***.

Links to other books in the *Practical Building Conservation* series are indicated throughout the text by the relevant volume symbol, showing that more information on the topic will be found in that volume.

The other volumes in the series are:

- Building Environment   ⟳ENVIRONMENT
- Concrete   ⟳CONCRETE
- Earth, Brick & Terracotta   ⟳E B & T
- Glass & Glazing   ⟳GLASS
- Metals   ⟳METALS
- Mortars, Renders & Plasters   ⟳MORTAR
- Roofing   ⟳ROOFING
- Stone   ⟳STONE
- Timber   ⟳TIMBER

Although every attempt has been made to explain terms as they first occur in the text, a glossary has also been included, and will be found just before the index.

THE EVOLVING CONCEPT OF BUILDING CONSERVATION

This chapter discusses the origins and evolution of conservation in England.

The recognition of conservation as a distinct philosophical approach to inherited fabric has its origins in the Italian Renaissance. The idea that physical remains – as well as written texts – could provide a unique insight into the human past stemmed from the development of scientific methods of inquiry, coupled with an increasing awareness of the process of historical change. In parts of Italy, the remains of Classical Antiquity began, by the 15th century, to be recognised as worthy of public protection. The conflict between restoring the perceived aesthetic values of buildings and sculpture as works of art, and retaining their authenticity as true remains of the past, quickly became a subject for discourse that has continued ever since.

These ideas surfaced in England in the middle of the 16th century, but only really began to flourish more than a century later, in the context of a growing awareness of the value of the medieval inheritance. From the end of the 18th century, English thought and practice in building conservation developed through sometimes bitter contention, becoming widely influential through the ideas and writings of its leading figures, particularly John Ruskin and William Morris. Evolving ideas about the cultural heritage values that people attach to some buildings and places, and the role of the state in sustaining them, are central to the history of conservation. From these ideas stem public decisions about what should be protected and how the public interest in heritage should be managed. Technical skill and innovation (including the rediscovery of lost arts) tend to evolve in parallel to provide the means of implementation.

During the 20th century, more diverse elements of the historic environment were recognised as having public value. Approaches to their conservation developed in parallel, championed by various public and professional groups, and underpinned by different legislation. These different strands, which this chapter attempts to explore, continue to influence thought and practice at a time when government is seeking to establish common principles, policies and processes for the conservation of all significant elements of our historic environment.

BUILDING CONSERVATION: THE CONCEPT

Building conservation is distinctly different from the physical processes of repair and adaptation. It is an attitude of mind, a philosophical approach, that seeks first to understand what people value about a historic building or place beyond its practical utility, and then to use that understanding to ensure that any work undertaken does as little harm as possible to the characteristics that hold or express those values. Conservation in the 21st century needs to be explained in such terms, rather than by technical directives (that is to say, to be operative rather than prescriptive), simply because of the diversity of the buildings and places that people now value and wish to hand on to future generations.

Valuing the remains of Classical Antiquity in the Italian Renaissance

Detail from *The Arrival of Cardinal Francesco Gonzaga*, by Andrea Mantegna 1465–74, from the *Camera degli Sposi* in the Palazzo Ducale, Mantua.

The background to this fresco shows an ideal city, in which the ruins of a Roman aqueduct and Classical sculpture are incorporated into what appears to be a landscape garden.

The practice of conservation involves judgement guided by professional ethics and public policy. It is based on an understanding of the relative importance of the heritage values attached to a building or structure, how they are represented in its fabric, and the effects on them of different approaches to repair. The intellectual arguments for conservation originally put forward by antiquaries and critics, often prompted by the threatened destruction of valued buildings, have gradually developed into professional statements of ethics and good practice.

The concept has evolved over a long time, but the language used to articulate it is changing. As conservation becomes a more complex and public activity, approaches to the conservation of buildings are seen as being closely linked not only to the conservation of objects, but also to sustaining cultural values in the historic environment as a whole. In that context, English Heritage, in its *Conservation Principles, Policies and Guidance* (2008), adopted a formal definition of conservation as *"The process of managing change to a significant place in its setting in ways that will best sustain its heritage values, while recognising opportunities to reveal or reinforce those values for present and future generations"*.

Destroying a fallen stone by fire and water at Avebury, 20th of May, 1724, by William Stukeley (1687–1765). Stukeley wrote: *"And this stupendous fabric, which for some thousands of years, had brav'd the continual assaults of weather, and by the nature of it, when left to itself, like the Pyramids of Egypt, would have lasted as long as the globe, hath fallen a sacrifice to the wretched ignorance and avarice of a little village unluckily plac'd within it."*

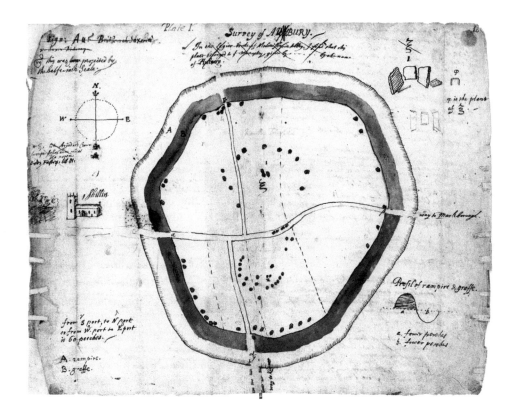

Plan of Avebury in Wiltshire, by John Aubrey (1626–97), from his *Monumenta Britannica*. Architects and antiquaries were drawn to the awe-inspiring Neolithic monuments of Stonehenge and Avebury from the 17th century onwards. The first systematic accounts of them were published by William Stukeley in 1740 and 1743, leading, as with historic buildings, to pressure to preserve them from harm.

Throughout Europe, the cultural significance of historic buildings and places is generally recognised as a public interest in property, regardless of who owns it, justifying the use of law, public policy, and public investment to protect that interest. There are differences, however, about what buildings and areas are valued sufficiently to warrant legal protection, both quantitatively (the number of buildings and areas) and qualitatively (the values ascribed to them). Although the values of some places have long been recognised and tend to become more clearly established through time, attitudes to others (often of more recent date) may change, sometimes quite rapidly, within an evolving culture. Conservation therefore requires an awareness of the mutability of heritage values. Current policies and good practice about what should be conserved and how that should be done represent a snapshot in time, rather than unchangeable truths.

THE ORIGINS OF BUILDING CONSERVATION IN ENGLAND

AN EMERGING RESPECT FOR THE MEDIEVAL INHERITANCE

An appreciation of the value of historic buildings, particularly the medieval inheritance of Romanesque and Gothic architecture, was a necessary precursor to the development of the concept of building conservation. In the summer of 1709, Sir John Vanbrugh (1664 [baptised]–1726) – adventurer, playwright and, ultimately, architect – was overseeing the construction of Blenheim Palace and lodging in the semi-ruinous remains of the medieval royal manor of Woodstock. He advanced his *Reasons Offer'd for Preserving some Part of the Old Manor at Blenheim* as an ornament to the landscape park to the Duchess of Marlborough thus:

"There is perhaps no one thing, which the most Polite part of Mankind have more universally agreed in; than the Value they have ever set upon the Remains of distant Times. Nor amongst the Several kinds of those Antiquities, are there any so much regarded, as those of Buildings; Some for the Magnificence, or Curious Workmanship; And others; as they move more lively and pleasing Reflections (than History without their Aid can do) on the Persons who have Inhabited them; On the Remarkable things which have been transacted in them, Or the extraordinary Occasions of Erecting them" (*Dobrée, Webb 1927*).

This page: Sir John Vanbrugh, engraved by John Faber the Younger, 1727.

Facing page: *The Old Manor House at Woodstock*, late 18th-century watercolour and chalk by George Marquis of Blandford after an engraving of 1711.

In 1709, Vanbrugh urged the Duchess of Marlborough to retain the ruins of the Old Manor as an ornament in the landscape park of the new palace of Blenheim (but the Duchess refused).

Despite being written three centuries ago, Vanbrugh's reasoning seems remarkably modern: people value the physical legacy of the past, particularly buildings, for a wide variety of reasons. In the event, the imperious Duchess rejected Vanbrugh's plea, and the remains of Woodstock Manor were destroyed. He was not alone, however, in advocating that the *"Remains of distant Times"* should be respected. In the same year, 1709, the Society of Antiquaries was founded, with a *Royal Charter* following in 1751. Over time, the Society would add to Vanbrugh's case an archaeological emphasis on the value of physical objects, including buildings and places, as primary documents of the human past.

THE SOUTH-VIEW OF WIGMORE-CASTLE, IN THE COUNTY OF HEREFORD.

The South-View of Wigmore-Castle in the County of Hereford, 18th-century engraving by Nathaniel and Samuel Buck.

During the 18th century, there was an explosion of topographic drawings and prints, many of them illustrating ruined monuments, symptomatic of an increasing *"awe, wonder and respect for medieval ruins and buildings"* (*Jokilehto 1999*). More artless mending and preservation of functionally useless but old structures were undertaken than is often realised, decisions having been made or influenced by instinctive perceptions of heritage value. Close study of individual monuments increasingly brings examples of this to light, like Castle Arch at Guildford, repaired by John Carter as early as 1669.

1596	1620	**1700**	1707
Robert Redhead attempts to dismantle Clifford's Tower in York, resulting in protests that it was *"an ornament to the City and a landmark along with the Minster"* (**Cooper 1911**).	Inigo Jones undertakes the first survey of Stonehenge.		Society of Antiquaries of London founded for the *"encouragement, advancement and furtherance of the study and knowledge of the antiquities and history of this and other countries"* (**Charter of George II, 1751**).

The repair of the Castle Arch at Guildford in 1669 by its then owner, John Carter, was commemorated by a date stone (which is now in Guildford Museum).

The emergence of this growing respect for, and study of, the ruins of the medieval past in England during the 18th century is in some ways similar to the rediscovery of the remains of Classical Antiquity in the Italian Renaissance. Yet the differences are equally significant. English Gothic was not quite dead, rather than long dead, and, after a slow start, it was re-established as one, rather than the standard, idiom of contemporary building. Through a century of competition, Classical styles had become the norm in polite English circles by the middle of the 17th century. Yet Gothic lived on into the beginning of the 18th century, particularly as an occasional architectural expression of the national church.

Even as late as the 1720s, when Christopher Wren (1632–1723) and Nicholas Hawksmoor (1662–1736) added the western towers to Westminster Abbey, it was still just possible in that context to design from tradition in a Gothic idiom. In the 18th century, it was primarily the ongoing study of Antique and Renaissance buildings in distant lands that influenced contemporary (overwhelmingly classical) architecture in England, not the study and reinvigoration of its own Gothic past. 'Gothick' was occasionally used as a paper-thin domestic ornamental style, as novel as *chinoiserie*, or as a lightly-applied mantle over essentially Classical buildings, for picturesque effect. Scholarly study as a basis for the revival of Gothic architecture lay in the future.

1709

Sir John Vanbrugh proposes that the ruins of Woodstock Manor should be preserved, as *"the Remains of distant Times"* (**Dobrée, Webb 1927**).

1751

Society of Antiquaries receives Royal Charter.

1782

Start of James Wyatt's work to 'restore' Salisbury Cathedral.

1800

1804

Criticism of Wyatt's approach to stylistic 'restoration' culminates in an essay by John Carter in the *Gentleman's Magazine*.

CONSERVATION BASICS
THE EVOLVING CONCEPT OF BUILDING CONSERVATION

CONSERVATION PHILOSOPHY DEVELOPING THROUGH CONTENTION

THE ANTIQUARIAN REACTION

While, in Vanbrugh's words, *"the most Polite part of Mankind"* (**Dobrée, Webb 1927**) might value medieval buildings, until the late 19th century there was no sense that those values should be recognised as a public interest in private property, justifying public constraints on the rights of owners. They could rework or demolish medieval houses without public censure. But the most conspicuous medieval monuments, the cathedrals and churches of the Church of England, were a different matter, since they were held by the established national Church. What happened to them was seen as a valid public interest, and late in the 18th century became a matter of public controversy. In consequence, one hugely influential strand of conservation thinking was established by contention, addressing its objectives, methods and weight as a public interest. To generalise, the arguments were between antiquaries and later art critics and social reformers on one side, and architects and churchmen on the other, at least until the Arts and Crafts movement gained ground towards the end of the 19th century.

In the 1780s, James Wyatt (1746–1813), a classically-trained architect, was commissioned to 'restore' Lichfield, Salisbury, Hereford and Durham Cathedrals. He was driven by the idea that medieval churches should be presented unencumbered by later additions or insertions, in a state of *"beautiful simplicity"* (**Gough 1789**). At Salisbury, between 1782 and 1791, he demolished porches, chapels and a bell tower to emphasise the unity of the 13th-century church. Wyatt gave more weight to architectural qualities, interpreted in the light of contemporary aesthetic values that he believed were universally valid, than to the buildings themselves as authentic historic documents, illustrative of the way in which they had been built, evolved and used through time.

1817	**1830**	1834	1836
Thomas Rickman publishes *An Attempt to Discriminate the Styles of Architecture in England from the Conquest to the Reformation.*		Institute of British Architects founded in London; later to become the Royal Institute of British Architects [RIBA].	Publication of *Contrasts*, by A. W. N. Pugin.

James Wyatt Esq RA, engraved by Joseph Singleton, c.1795.

This imposition of a contemporary rational aesthetic approach to the remodelling of cathedrals essentially resulted in a clash of ideology. The Society of Antiquaries was provoked to come to the defence of ancient buildings: bitter criticism by Richard Gough (1735–1809) and John Carter (1748–1817), two members of the Society, mobilised the antiquarian lobby. Through public contention, the importance of age, authenticity and integrity over transient architectural fashion was established, at least for a relatively small group of buildings clearly recognised at the time as being of national significance. In consequence, the proposed 'improvements' to Durham Cathedral were greatly reduced.

Out of this clash came a specific condemnation of 'restoration', in the sense of repair that went beyond the minimum necessary to keep a building wind and weathertight. As John Carter put it in 1804, in relation to Henry VII's Chapel at Westminster Abbey: *"when Restoration comes – why then the original will be no more. For my part, I am for no restoration of the building; I am content with it even as it is. For repair indeed, I am ready enough to agree to that; such as carefully stopping open joints, making good some of the mullions of the windows, putting the glazing of some of the windows in proper conditions; but no further would I go"* (**Carter 1804**).

1839	**1840**	1843	1849
Cambridge Camden Society (later the Ecclesiological Society) founded to promote the *"study of Gothic architecture and of Ecclesiastical Antiques"*. The first issue of its journal, *The Ecclesiologist*, is published in 1841.		British Archaeological Association [BAA] founded to promote the study of archaeology, art and architecture, and the preservation of national antiquities in Britain.	John Ruskin publishes *The Lamp of Memory*.

Salisbury Cathedral, engraved by
Wenceslaus Hollar, 1671, showing
the detached bell tower demolished
by James Wyatt.

Thus the philosophy of 'minimum intervention' emerged as the desirable approach
to conservation. Buildings that were, or looked, 'ancient' were valued as the tangible
expression of a different distant age and outlook, and historic fabric was absolutely
privileged over design values. The cultural context was one of rapid social, economic
and physical change as England industrialised, its burgeoning prosperity providing
the means to fund major works to historic buildings. By contrast, there was a
growing counter-cultural desire to prevent the legacy of the past from being entirely
overwhelmed, to confront the increasingly prevalent excesses of speculative or
stylistic restoration and crude intervention.

1860

1860

Union of Benefices Act passed,
empowering the Church of England to
demolish redundant churches in the City
of London and sell the sites to finance
church expansion into suburban areas.

1864

The RIBA establishes a sub-committee on the Conservation
of Ancient Architectural Monuments and Remains,
and publishes *Conservation of Ancient Architectural
Monuments and Remains: General Advice to the
Promoters of the Restoration of Ancient Buildings*.

A composition showing a decorated interior, original drawing for *An Attempt to Discriminate the Styles of English Architecture...*, by Thomas Rickman, *c.* 1817.

By providing an archaeological analysis of multi-period buildings, Rickman brought scientific research about the past to bear on current architectural issues.

As Gothic as a living tradition receded, Gothic scholarship began, especially amongst a group of Cambridge antiquarians who sought to understand, rather than simply observe and record, the Gothic style. They included James Essex (1722–84) and the Rev G. D. Whittington (1781–1807). Essex was arguably the first architect to take a scholarly interest in Gothic and apply it as a system of design. Whittington, although little known today, studied the origins and evolution of Gothic in England and France. His overview was published posthumously in 1809, ahead of any comparable French scholarship (*Watkin 1980*).

1873	**1874**	**1875**	**1877**
First Ancient Monuments Bill is introduced into Parliament, but is never enacted.	Hampstead's Georgian Parish Church is saved after a campaign led by George Gilbert Scott, Jr and G. F. Bodley.	Society for Photographing Relics of Old London is founded.	The Society for the Protection of Ancient Buildings [SPAB] founded by William Morris, who drafts its *Manifesto*.

1870

As time went on, seeking to understand Gothic architecture simply through observation and recording was not enough. *An Attempt to Discriminate the Styles of Architecture in England from the Conquest to the Reformation*, published by Thomas Rickman (1776–1841) between 1815 and 1817, brought scientific research about the past to bear on current architectural issues. Rickman based his book on an archaeological analysis of multi-period buildings, in the belief that understanding the fabric of churches was a prerequisite to answering what was (to him) the most important question of church builders and restorers: 'How should we build today?' *"Independently of the value of such investigations to the history of the science of construction, the knowledge of the methods actually employed would greatly assist us in the imitation of the works of each period"* (*Willis 1842*). Slightly later, Robert Willis (1800–75) stands out for his study and publication of Gothic architecture, and his 'modern' approach of comparing documentary evidence with that of the fabric itself.

Willis' approach has been characterised as *"keep[ing] religion out of architecture"* (*Watkin 1980*), but for A. W. N. Pugin (1812–52), Roman Catholic architect, designer and polemicist, the two were inseparable. For him, Gothic was the only true Christian form of architecture. He was forthright in his criticism of those who failed to adopt its 'true principles', but instead used it in an essentially decorative fashion or, worse, adopted another style. His was an imperative not to conserve, but to restore churches to their medieval Catholic form. On the Anglican side, the Cambridge Camden Society, founded in 1839 by J. Mason Neale (1818–66) and Benjamin Webb (1819–85), proved hugely influential through their monthly journal, *The Ecclesiologist*, which publicised both new design and the restoration and repair of historic parish churches. The Society believed that recovering the setting of medieval piety could lead to the recovery of that piety itself. It promoted both the Decorated Style (*c.*1260–1360) as the purest and 'most pious' architecture, and the adoption of the essentially medieval re-ordering of church interiors that still dominates today. One result of this injection of religious and moral fervour into architecture was the typical High Victorian drastic reconstruction and re-ordering of medieval churches, making all new, sharp and correct.

1880

1882

First *Act for the Better Protection of Ancient Monuments* passed, giving limited protection to specific prehistoric monuments.

1884

The City of Chester obtains powers to protect its medieval walls by an Act of Parliament, followed a few years later by similar action in Colchester and Newcastle.

1890s

Archaeologist General Pitt-Rivers appointed as Inspector of Ancient Monuments.

At the end of the 18th century, Wyatt had sought to impose alien Classical ideas on medieval buildings. Growing knowledge about the evolution of the Gothic style now led to an equally misplaced confidence to restore, based on perceptions of understanding, not so much archaeological, as stylistic. The French architect Eugène Viollet-le-Duc (1814–79) was perhaps the strongest 19th-century advocate of 'stylistic restoration'. In his *Dictionary* of 1866, he acknowledged that *"The term Restoration and the thing itself are both modern. To restore a building is not to preserve it, to repair or to rebuild it; it is to reinstate it in a condition of completeness which may never have existed at any given time"* (*Jokilehto 1999*). Although his text goes on to promote a more subtle approach, it did lead to a movement which was to favour design, or design intent, over material or evolution.

William Morris, photographed by Frederick Hollyer in 1874.

In reaction to the 'scholarly' restoration movement, the conservation case for 'minimum intervention' was widened and elaborated beyond the antiquarian argument that the building should be preserved as a document. In *The Lamp of Memory* (1849), for example, the art critic John Ruskin (1819–1900) emphasised the value of the marks of the craftsmen and age on a building, concluding that *"Restoration, so-called, is worse than Destruction"*. His disquiet went beyond the romantic and picturesque values of what would later sometimes be called the 'patina of age'. He was concerned about loss of authenticity, believing that ancient buildings stood proxy for his imagined lost ideals of the medieval past, in which the role of the craftsman was emphasised above that of the designer. Criticism of repairs and restorations, especially of churches, grew steadily more strident through the 1850s, 1860s and 1870s, particularly as national and local architectural and archaeological societies took root and flourished.

1894	1896	1897	1898
The Survey of London is established to create a systematic inventory of its buildings and monuments. The National Trust founded by Octavia Hill, Sir Robert Hunter and Canon Hardwicke Rawnsley.	The National Trust acquires its first building, Alfriston Clergy House.	*Country Life* first published, featuring one country house in each issue.	*London County Council Act* gives the Council the power to acquire monuments and buildings of 'architectural and historic interest' by compulsory purchase.

The east front of Kelmscott Manor, frontispiece of *News From Nowhere*, by William Morris, 1892.

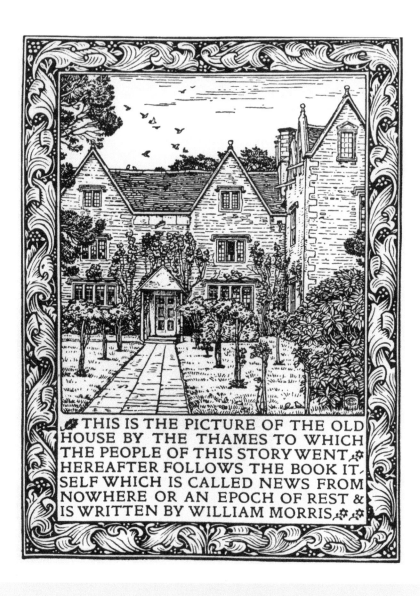

THIS IS THE PICTURE OF THE OLD HOUSE BY THE THAMES TO WHICH THE PEOPLE OF THIS STORY WENT ✿ HEREAFTER FOLLOWS THE BOOK IT SELF WHICH IS CALLED NEWS FROM NOWHERE OR AN EPOCH OF REST & IS WRITTEN BY WILLIAM MORRIS ✿ ✿

1900

1900

First volume of the **Survey of London** published, covering Bromley-by-Bow, edited by C. R. Ashbee.

Ancient Monuments Protection Act permits inclusion of any structure of 'archaeological or historic interest' on the schedule.

1901

First volume of the **Victoria County History** published.

Tintern Abbey purchased on behalf of the nation.

The arguments of William Morris (1834–96) went even farther. A progressive socialist, Morris reacted against the dehumanising consequences of the Machine Age and a rigid, highly unequal society by promoting radical libertarian socialism. He contrasted what he imagined to be the life and sensibility of the medieval free craftsman with the regimented life of the 19th-century factory worker, an argument already made by Pugin in *Contrasts* (1836). Morris emphasised the inherent value of craftsmanship, contemporary as well as historic, as it came to be manifested in the Arts and Crafts movement.

In his utopian vision *News from Nowhere* (1890), Morris imagined the Thames Valley as a pseudo-medieval pastoral idyll, inspired by art and beauty rather than by Pugin's Catholic Church. The transformation was achieved after a massacre of Combined Workers by the army in Trafalgar Square, followed by a civil war in which the workers finally triumphed.

The established intellectual position of English building conservation towards the end of the 19th century thus remained 'minimum intervention', poetically, if uncompromisingly, restated in Morris' 1877 *Manifesto for the Society for the Protection of Ancient Buildings*. Morris initially founded the Society for the Protection of Ancient Buildings [SPAB] in reaction to George Gilbert Scott's imminent restoration of Tewkesbury Abbey, but the Society was concerned with secular and vernacular buildings, as well as ecclesiastical ones and major monuments. In the *Manifesto*, Morris asserted that ancient buildings should be preserved with the minimum of intervention, staving off repair through daily care. Repairs which were unavoidable should not attempt to match the original, since any attempt to restore or copy would result in the loss of authenticity, the creation of a fake. But it was still mostly buildings, intact or ruined, which were, or appeared to be, 'ancient', usually medieval and so at least three centuries old, that were considered worthy of conservation. They were visibly of another age, and most were ecclesiastical.

1905	1906	1908
The Care of Ancient Monuments is published by Gerard Baldwin Brown, surveying the conservation of historic buildings across Europe.	*London Squares and Enclosures (Preservation) Act* protects 64 squares.	Royal Commission on the Historical Monuments of England [RCHME] established to prepare inventories of pre-1700 structures and sites.

THE ANCIENT MONUMENTS ACTS & THE OFFICE OF WORKS

The first *Act for the Better Protection of Ancient Monuments*, passed in 1882, was primarily concerned with specific prehistoric sites. Its passing into law nonetheless established, after a long parliamentary campaign led by Sir John Lubbock (1834–1913), the principle that their care was a public interest that should override private property rights. A second *Ancient Monuments Act*, in 1900, widened the definition of monument to include structures of architectural or historic interest, allowing them to be taken into guardianship; that is, into the care of the Commissioners of Works on behalf of the state. Responsibility for managing and maintaining these sites came to rest with the Office of Works, later to become the Ministry of Works.

An attempt to incorporate cathedrals into the *Ancient Monuments Consolidation and Amendment Act 1913* failed mainly because of resistance from the Bishops. This nonetheless prompted the Church of England to improve its own faculty jurisdiction governing work to historic churches, with the first Diocesan Advisory Committee established in 1916. But the *1913 Act* did introduce the concept of a Preservation Order and, in 1914, the Office of Works served an order on an empty, largely Georgian house at 75 Dean Street in Soho, London, to prevent its demolition. Success would have extended the practical scope of the *1913 Act* (which, as its name suggests, was focused on monuments) to more recent buildings of 'special historic, architectural, traditional, artistic or archaeological interest' not in ecclesiastical use. But such orders, which required all works to the structure concerned to be approved by the Commissioners of Works, had to be confirmed by Parliament. A Select Committee of the House of Lords refused to confirm the order for 75 Dean Street, and awarded costs to the owner. The Office made no further attempt to use such compromised powers, withdrawing from involvement in historic buildings to concentrate almost exclusively on ruined monuments and government-owned buildings. In consequence, its approach to their conservation developed in isolation from that of historic buildings in use.

Facing page: 75 Dean Street, Soho, London, photographed in 1912.

The house was the subject of the first and only Preservation Order under the 1913 Ancient Monuments Act, served by the Commissioners of Works in 1914. It was traditionally identified as the home of Sir James Thornhill (1675/6–1734), painter of the dome of St Paul's Cathedral, to whom the painted stair hall was attributed; but it was actually built *c.*1735, as emerged during the owner's successful appeal against the Order.

1910

1911

Tattershall Castle, Lincolnshire, saved from dismantling and removal to the United States by Lord Curzon.

1913

Ancient Monuments Consolidation and Amendment Act supersedes earlier legislation, and creates the Ancient Monuments Board.

1914

75 Dean Street demolished after courts fail
to uphold a Preservation Order.

1916

Establishment of first Church of England Diocesan Advisory
Committee to control works to historic churches.

CONSERVATION BASICS
THE EVOLVING CONCEPT OF BUILDING CONSERVATION

During the early 20th century an increasing number of 'ancient monuments' came under state guardianship, or, like Tintern Abbey in 1901, were purchased on behalf of the nation. The Office of Works adopted minimum intervention ('repair as found') as its approach to these mostly ruined structures, whose 'use' was entirely didactic. The concept came to be narrowly interpreted. Nothing was visibly added to the inherited structure or (normally) restored. Vegetation, accumulated destruction deposits and post-medieval fabric tended to be cleared away so that sites could be displayed simply, surrounded by immaculate green lawns, as the fabric that was considered to be 'historic' had survived.

Such was the fate of Tintern Abbey, once a favourite destination for artists (including J. M. W. Turner) and early tourists in search of the 'picturesque': the first guidebook to the Wye Valley, by William Gilpin (1724–1804), appeared in 1782. Stripped of the vegetation that had made it appear so 'romantic', but which was concealing and contributing to its decay, Tintern was presented, once repaired, like an object in an invisible museum case.

Certain values – those that provided evidence and illustration of aspects of the past – came to be prized absolutely over others, especially artistic and aesthetic values. Over the course of the 20th century, the philosophy of 'conserve as found' established by Sir Charles Peers (1868–1952), Chief Inspector of Ancient Monuments in the Office of Works, and its chief architect Sir Frank Baines (1877–1933), often proved woefully inadequate as a basis for managing the increasingly complex places taken into state care and the wide range of heritage values they represented. Some buildings in guardianship were deliberately unroofed; others were archaeologically 'unpicked' to remove even 16th-century work. 'Conserve as found' had become a formulaic response to exposing and consolidating medieval work. The normal use of strong mortars, including those gauged with cement, was creating problems for the future, the scale of which would only become apparent a few decades later (for example, at the White Tower in the Tower of London).

Tintern Abbey

The Wye Valley, Monmouthshire, was a favourite destination for late 18th-century travellers in search of the 'picturesque'.

Facing page: An 18th-century watercolour of Tintern Abbey, by Thomas Girtin (*left*). The ruins were controversially stripped of vegetation and consolidated by the then Office of Works when the monument was purchased on behalf of the nation in 1901. The site remains in this state today (*right*).

1920

1924

Ancient Monuments Society founded.

Royal Fine Art Commission [RFAC] established.

1925

Council for the Protection of Rural England founded.

1926

Bishop of London attempts to demolish 19 city churches, but is prevented from doing so by the joint efforts of the City Corporation, the London County Council [LCC] and SPAB.

1927

On the advice of the Norwich Society, the Norwich Corporation purchases and renovates houses in Elm Hill; this becomes the first example of a townscape (rather than individual buildings) being protected.

CONSERVATION BASICS
THE EVOLVING CONCEPT OF BUILDING CONSERVATION

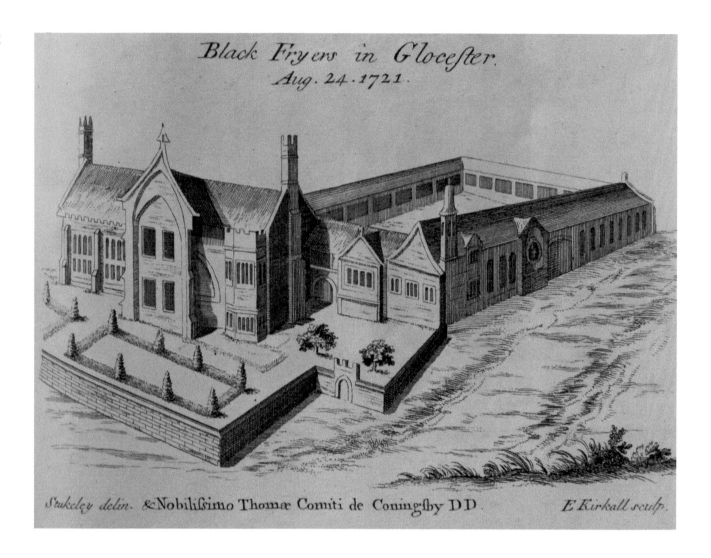

Black Fryers in Glocester.
Aug. 24. 1721.

Stukeley delin. &Nobilissimo Thomæ Comiti de Coningsby DD. *E Kirkall sculp.*

1928

Publication of Clough Williams-Ellis' *England and the Octopus*, the first popular book wholly about the preservation of historic buildings and areas.

1929

A. R. Powys' *Repair of Ancient Buildings* first published by SPAB.

Montacute House, Somerset, becomes the first country house to be given to the National Trust.

The Landscape Institute founded as the Institute of Landscape Architects [ILA].

Blackfriars, Gloucester

Founded in the mid-13th century, Gloucester Blackfriars is one of the most complete surviving Dominican Friaries in England. After the dissolution of the monasteries, it was acquired in 1539 by Sir Thomas Bell, then mayor of Gloucester, and used as his residence. The monastic buildings became workshops for his cloth and cap manufactory. Now an English Heritage property, the Friary remained in private ownership with a miscellany of uses until the 20th century.

Facing page: Gloucester Blackfriars before removal of later fabric.
This page: Gloucester Blackfriars came under the guardianship of the Ministry of Works in 1960. Most of the historically important fabric of Thomas Bell's house was subsequently removed to present the medieval fabric in isolation. A similar approach to other parts of the complex was later halted by English Heritage and the embedded Georgian houses were returned to use.

1930

1931

Ancient Monuments Act empowers local authorities to preserve monuments and their settings.

Athens Charter for the Restoration of Historic Monuments adopted.

1931

The *London Squares Preservation Act* protects 461 squares and other green spaces.

International conservators agree the *Athens Charter*.

The sad fact is that by the 1960s the Ministry of Works (as the Office of Works had then become) was no longer always in the forefront of good conservation practice; private sector specialists, both professionals and contractors, had increasingly taken the lead. Things began to change, however, when the professional Research and Technical Advisory Service [RTAS] was established. Headed from 1968 by Principal Architect John Ashurst (1937–2008), its main role was researching the reasons for the deterioration of building materials, and devising conservation treatments to address them. Repair techniques were increasingly based on scientific understanding of the composition and performance of existing fabric.

John Ashurst (1937–2008)

The field experience and casework notes compiled by John Ashurst over 25 years formed the basis for the first series of *Practical Building Conservation*, published in 1988.

The clarity and succinctness of the series made them standard references on building conservation, both in the UK and abroad. John Ashurst taught the value of methodical and analytical problem solving: an approach involving rigorous observation, asking the right questions, making an accurate diagnosis, selecting appropriate remedies and reviewing the results.

John Ashurst played a key role in planning the new series, and was a lead contributor to the *Mortars, Renders & Plasters* volume.

1932

Shell Guides first published, edited by John Betjeman.

The *Town and Country Planning Act* gives local authorities power to make Building Preservation Orders.

1936

The *City of Bath Act* is passed to help preserve the character of the 18th-century city.

THE ORIGINS OF CONSERVATION AS A DISTINCT TECHNICAL SKILL

Contemporary and traditional building practices began to diverge at around the end of the 18th century, and the rate of change has increased ever since. Problems of intent in repair were increasingly compounded by problems in execution, as innovations in building technology began to be used in the repair of historic buildings and workmanship evolved in parallel. By the middle of the 19th century, heated debate emphasised the need to establish how good conservation practice differed from contemporary building. In 1865, the Royal Institute of British Architects [RIBA] published the first conservation guidance promulgated by a professional body in England, *Conservation of Ancient Monuments and Remains*. It was drafted by a committee and sought to establish a generally acceptable line.

The most problematic introduction would turn out to be Portland cement, increasingly used in mortars for building and repointing masonry walls, and in renders to cover them. Hard and impervious in nature, it would accelerate the decay of much 'soft' historic brick and stonework, as well as proving to be an inherent problem for a good deal of new building in stone.

The increasing use of machinery to cut stone, and better steel tools to work it by hand, producing sharp arrises and fine joints, resulted in a growing contrast between historic fabric and repair work: this was reinforced by the introduction of harder and sharper machine-made bricks and tiles. Even if the aim was minimal repair, the results were often visually and technically problematic. Conservation increasingly needed to embrace both intent and the means of execution.

1937

The Georgian Group founded to advocate the preservation of buildings dating from 1700 up to the 1830s.

National Trust launches the 'National Trust Country House Scheme' to acquire large country houses, often allowing their original owners to remain as tenants.

Country Life publishes a list of 639 large country houses worth preserving.

1938

The first open-air museum in Britain opens at Cregneash, Isle of Man.

Diocesan Advisory Committees become statutory.

The most enduring and positive legacy of the first fundamental strand of conservation thinking, embodied in the SPAB *Manifesto*, has been the promotion of regular maintenance, gentle repair and, if necessary, incremental change to overcome problems. SPAB has increasingly encouraged the use of materials and techniques close to the original through a body of technical literature and practical training. A. R. Powys' *Repair of Ancient Buildings*, produced in 1929 for the Society (but still in print), and John and Nicola Ashurst's first edition of this *Practical Building Conservation* series, published in 1988 for English Heritage, are widely separated in time, but both are very much in the tradition of passing on understanding gained through practical experience, increasingly strengthened by science-based investigation.

General building construction continued to move ever further from the use of traditional materials and practice during the course of the 20th century. Cement-based mortars and gypsum plasters became all but universal after the Second World War. New materials displaced old, and traditional craft skills were lost. But as the often undesirable longer-term effects of using substitutes in the repair of historic buildings became clear, serious initiatives were taken to maintain, or re-establish, supplies of traditional materials and the skills to use them. The use of lime mortars, renders and plasters for historic buildings became accepted practice. Training courses were established for both professionals and craftspeople. At the same time, as more recent buildings have come to be valued, the problems of conserving modern materials have had to be addressed, often, as with reinforced concrete, on a very considerable scale.

1940

1940

The National Buildings Record founded (later becoming the National Monuments Record).

The first national lists of historic buildings developed by the Ministry of Works in cooperation with the National Buildings Record, in response to the threat of wartime damage and destruction.

A NATIONAL COLLECTION OF BUILDINGS

ORIGINS & EVOLUTION

A second distinct strand in English conservation emerged around the beginning of the 20th century, focused on the large number of post-medieval secular buildings mostly in everyday use. It ultimately led first to the identification and then the protection of what is now a 'national collection' of buildings, valued principally for their contribution to the present and future, as well as their recollection of the past. The Royal Commission on the Historical Monuments of England [RCHME] was established by government in 1908 to document this heritage, as well as more ancient structures, with a terminal date of 1700, subsequently extended to 1714.

During the first half of the 20th century the National Trust set the example of charitable acquisition in the public interest.

The Trust's first purchase was the Clergy House at Alfriston (*right*), a 14th-century Wealden hall house, which was bought in 1896 for £10.

1944

The ***Town and Country Planning Act*** introduces the (then theoretical) concept of listed buildings.

The Council for British Archaeology [CBA] founded, partly to oversee the repair of war-damaged towns.

These 16th-century timber-framed houses at Elm Hill, Norwich, were acquired by Norwich Corporation in 1927. Their subsequent restoration was perhaps the first example of conservation of a townscape, rather than of an individual building.

1946

'Instructions to investigators for the listing of buildings of special architectural interest' published under section 42 of the *Town and Country Planning Act 1944*, drafted by John Summerson.

National Land Fund (later the National Heritage Memorial Fund) is established to purchase objects and sites in memory of the war dead.

On 1st August, Gosfield Hall in Essex becomes the first building to be spot-listed.

Public interest in buildings of the 16th and 17th centuries was now acknowledged, embracing the early history of English classicism, and narrowing the time lag between their construction and recognition of them as part of the national heritage. In a sense, this was a 'new renaissance': the rediscovery of the value of historic (specifically English) classical design, as well as historic fabric. The large number of these buildings meant that only a few could ever be taken into public or charitable care: their future survival clearly depended on them remaining in beneficial use.

Prompted particularly by urban redevelopment and expansion, calls soon began for legislative protection of buildings identified by the Commission as 'worthy of preservation'. Nationally, these expressions of public interest in private property were frustrated by landed interests, as the Dean Street case illustrates. Locally, however, a few buildings in areas of 'special architectural, historic or artistic' interest were protected in local town-planning schemes made under early legislation, including the *Town and Country Planning Act 1932*. The most notable example was probably New Romney, Kent, which accounted for about one third of the small number of the 20 or so Preservation Orders made in the inter-war years. The *1932 Act* nonetheless set a precedent for the direction of post-war legislation, which would include national protection for historic areas as well as for buildings. Otherwise, through the first half of the 20th century, survival of historic buildings depended upon sympathetic ownership, with the National Trust, founded in 1895, setting the example of charitable acquisition in the public interest.

Slum clearance and commercial development (especially in and around London) drove the redevelopment of historic buildings and areas between the wars. Despite protests, and the emergence of the Georgian Group as the first of the 'period societies' in 1937, it took dramatic destruction through bombing in the Second World War to focus attention on the value of built heritage. The new world of the post-war Labour Government also led to a fundamental, if temporary, shift in the balance of power between public and private interests in land.

1947

The *Town and Country Planning Act* obliges the relevant Minister to compile lists of historic buildings, but protection still requires a local authority to serve a Building Preservation Order.

First general list is published for five parishes in the Rural District of Blofield and Flegg, Norfolk.

The *Town and Country Planning Act 1947* (developing its 1944 precursor) introduced a comprehensive national system for regulating the use of land in the public interest. It also allowed for Preservation Orders to be made on historic buildings in use, and required the relevant minister to produce a national list of buildings of 'special architectural or historic interest', the start of statutory listing. The emerging list owed much to pre-war unofficial local surveys, which had helped to inform lists of buildings compiled in 1940–44 as priorities for first aid repair in the event of bomb damage.

Protection inevitably demanded selection. The initiative was now with architectural historians in the Ministry of Housing, rather than archaeologists in the Ministry of Works. Age was, and remains, key to selection for protection, with inclusion in the statutory lists becoming progressively more selective through time. The lists soon included a small number of Victorian buildings as the principal works of the leading architects of the time, so that, albeit very selectively, the potential interval between construction and recognition of cultural significance narrowed to half a century.

As the listing criteria gradually embraced first 'Edwardian', then inter-war, and finally post-war, buildings, all initially with very restrictive selection criteria, so the gap closed further. A '30-year rule' was set, and soon was reduced to 10 years for buildings deemed to be of 'outstanding' importance and under threat. Architects have had to come to terms not only with their buildings being listed during their lifetime – an honour – but also to working on them for a second time, like Norman Foster at the Willis (originally Willis Faber & Dumas) Building in Ipswich, under the constraints of listing .

Facing page: When this glass-clad building designed by Foster Associates for the Willis Faber & Dumas Company in 1973–75 was given Grade I status in 1991, it became the first structure less than 30 years old to be listed.

In 1992, it was also the subject of the first Listed Building Management Agreement, a formal agreement between the owner and the local planning authority providing *"clarification as to what proposals for the building may not require Listed Building Consent and/or planning permission."*

1950

1950

The Gowers Report recommends that owners of outstanding country houses should be eligible for tax and death duty relief (although this does not happen until 1980).

1951

Founding of the International Institute for Conservation of Historic and Artistic Works [IIC].

Creation of the first National Park (in the Peak District).

1951

Publication of the first volume of
Nikolaus Pevsner's *Buildings of
England* series, covering Cornwall.

1953

Historic Buildings and Ancient Monuments Act
introduces national grant schemes for the repair of
historic buildings and ancient monuments, and creates
the Historic Buildings Council to advise government.

1954

Howard Colvin's *Biographical
Dictionary of British Architects
1600–1840* first published.

The closing of the gap between (present) cultural value and (past) cultural heritage was perhaps an inevitable consequence of the emphasis from the outset (in the selection criteria for listing) on architectural, rather than historic, interest, and thus on design value as well as fabric. The incremental shift could not have been achieved without public support. When, in the 1980s, systematic listing of post-war buildings was mooted, English Heritage asked whether these buildings really were 'Something worth keeping?': the public's answer was 'yes'. Professional support is evident from campaigns by the architectural press to list threatened modern buildings.

The statutory 'list' now represents the nation's selection of architecture collectively deemed of sufficient cultural value to be secured for the benefit of future generations as part of their heritage. The scope of the list demonstrates that, while age is important, national significance is judged on how built fabric reflects a wide range of cultural heritage values. For more recent buildings, these tend to emphasise the value of their design expressed through the essentially complete survival of their key elements. While the legislation allows for selection on the basis of architectural or historic interest, buildings and structures associated with historically significant people or events have not normally been listed unless they also have discernible architectural quality. However, some recent decisions including the listing of a zebra crossing in Abbey Road in London, famous for its association with The Beatles; Brixton Market, also in London; and the childhood homes of John Lennon and Paul McCartney in Liverpool, mark the beginning of a departure from this convention, and illustrate the celebration of places for primarily historical associations.

The parallel with other visual arts, where public collections are built up and actively curated, is nonetheless becoming more valid. There is, of course, an important difference; most listed buildings are privately owned. Rather than buying them in the market, the state pays nothing to 'designate' new additions to the collection. In that sense, listing can be seen as the last vestige of the post-war Labour Government's short-lived nationalisation, in the *1947 Act*, of the development value of land. Listed buildings are sustained (or occasionally not) by their owners primarily for their usefulness and market value, rather than their cultural value.

Facing page: In April 2010, the Secretary of State listed the arcade buildings of Brixton Market, London, at Grade II. The listing recognised the market's cultural importance to an Afro-Caribbean community, and its contribution to the social and economic history of Brixton, particularly since the 1950s, as well as its architectural importance: such arcades, once common, are now rare.

1954

Forty-eight country houses demolished; more than in any other single year.

1957

Civic Trust founded; the Civic Trust Awards scheme is launched two years later.

Friends of Friendless Churches founded to campaign for and rescue redundant churches.

1958

Victorian Society founded to promote the awareness and preservation of 19th- and early 20th-century buildings.

1960

1962

Euston Arch demolished.

Local Authorities (Historic Buildings) Act enables local authorities to offer grants towards the repairs of listed or unlisted buildings, including churches.

1964

Venice Charter drafted as the successor to the 1931 *Athens Charter*.

Foundation of ICOMOS (International Council on Monuments and Sites).

CONSERVATION BASICS
THE EVOLVING CONCEPT OF BUILDING CONSERVATION

CURATING THE COLLECTION

Creating a list is one thing; retaining, let alone conserving, the buildings included on it is a different matter. While the statutory list is compiled and maintained nationally (since 1984, with the advice of English Heritage, now Historic England), responsibility for acting upon it is essentially local. This is an inevitable consequence of its origins in planning legislation, which is administered by local planning authorities. It is also a practical necessity, given the large number of listed buildings compared with ancient monuments, which are still administered centrally. Initially, listing was not particularly effective in securing preservation. A local authority had to serve a building preservation notice if it wished to control the alteration or demolition of a listed building, and some authorities were inevitably more enthusiastic than others in doing so, guided by local priorities.

Grants from central government for the repair of buildings of 'outstanding' importance were introduced by the *Historic Buildings and Ancient Monuments Act 1953* (which also created the Historic Buildings Council to advise the Minister on grant applications), but the post-war *zeitgeist* tended to favour comprehensive urban redevelopment over conservation. The Macmillan Government arguably saw the destruction of the Euston Arch, London, in 1961–62 as a symbolic act of national modernisation. It actually marked, insofar as any one event can, a turning point in post-war attitudes to heritage conservation. Protest was not confined to 'heritage' interests, but came from radical modern architects like Peter and Alison Smithson (1928–93 and 1923–2003) as well, providing a reflection of the emerging nature of the statutory list as a national architectural collection.

1966	1967	1968
The Garden History Society founded.	*Civic Amenities Act* introduces conservation areas.	*Town and Country Planning Act* introduces the need for Listed Building Consent, the notification of demolitions to the amenity societies (SPAB, Georgian Group, Ancient Monuments Society, Victorian Society, Council for British Archaeology and the RCHME), and repair notices for neglected buildings. First official conservation advice in *Ministry of Housing and Local Government Circular 61/68*.

Demolition of Euston Arch

The 1837 Euston Arch in London, designed by Philip Hardwick (1792–1870), was the first monument of the railway age, and the largest Greek Revival propylaeum ever built. Its vigorously contested and controversial demolition in 1961–62, to allow the redevelopment of Euston Station and electrification of the West Coast mainline, proved to be the catalyst for the subsequent preservation of many Victorian buildings.

1968

Conservation reports on York, Chester, Lincoln and Bath are commissioned by the Ministry of Housing and Local Government, published as *A Study in Conservation*.

Redundant Churches Fund (later the Churches Conservation Trust) is founded.

Nationwide resurvey of listed buildings begun.

1969

The first 37 inter-war Modern Movement buildings listed.

In 1968, a general requirement to obtain Listed Building Consent for works affecting the character of listed buildings was introduced. Decisions were still made locally but, over time, steadily became more conservative. This was due to more effective national overview through both policy and the requirement to refer important cases to central government. From 1980, the resurvey of the original highly-selective lists was accelerated, championed by Michael Heseltine as then Secretary of State for the Environment. This resulted in a more-or-less comprehensive review by 1993.

Conservation philosophy had to develop to meet new challenges. The *Athens Charter*, the result of an international conference held in 1931, reflected many of the principles already well-established in England through SPAB. It also recommended that *"the occupation of buildings, which ensures the continuity of their life, should be maintained, but that they should be used for a purpose which respects their historic or artistic character"* (*Article 1*). The use of modern technology in repair, especially reinforced concrete, preferably concealed, was endorsed more enthusiastically than would later prove wise. The Charter's avowed successor, the *Venice Charter* (1964) was more tightly worded: *"The conservation of monuments is always facilitated by making use of them for some socially useful purpose. Such use is therefore desirable, but it must not change the lay-out or decoration of the building. It is within these limits only that modifications demanded by a change of function should be envisaged and may be permitted"* (*Article 5*).

These charters were not inter-governmental conventions, but statements of best professional practice, drafted by leading practitioners. The *Venice Charter* still forms the guiding principles of the International Council on Monuments and Sites [ICOMOS], founded at the time of its drafting. Primarily, the authors of the charters had in mind monuments whose exceptional significance was evident in national, and often international, contexts. Their ideal of limited adaptation to changing needs, however, proved difficult to apply to the growing range and number of protected buildings and structures of more modest significance, like the majority of those included in the statutory list of England. For these, sustaining ongoing use was not just a desirable option, but essential for their survival.

1970

1971

RESCUE, the British Archaeological Trust, founded.

Covent Garden market buildings saved from demolition after a public campaign.

1972

Field Monuments Act introduces a system of payments to landowners with scheduled monuments on their land.

While these international charters were influential in professional thinking in England, national guidance was a government initiative. When the requirement to obtain Listed Building Consent was introduced, ministers thought that *"it may be helpful to local planning authorities to have some broad guidance on criteria in dealing with such applications"*. This was set out in the **Ministry of Housing and Local Government Circular 61/1968**, whose content would remain recognisable in its successor circulars up to 2010. It also introduced a *"presumption ... in favour of preservation"*, since, unless the rate of demolition of listed buildings *"is, if not completely arrested, at least slowed down, the virtually complete disappearance of all listed buildings can be predicted within a calculable time"*. Appendix IV of this circular's successor, **Department of the Environment Circular 23/77**, was presented as a *"Technical Digest on Alterations to Listed Buildings... compiled by Historic Buildings Investigators of the Department to assist local authorities in deciding on the suitability of alterations to listed buildings"*. Beyond the preamble, which stated that *"alterations... should be kept to a minimum wherever possible"*, much of the advice under *"alterations in detail"* emphasised the need for appropriate repair rather than replacement. Traditional or matching design was generally seen as the most suitable for extensions, *"in keeping with other parts of the building... great care should be taken to follow the fenestration and detailing of the original"*.

This circular advocated an essentially pragmatic approach. It was based on practical advice (much of which seems obvious now, but was not so at the time), rooted in an arts and crafts preference for blending new work seamlessly with the inherited structure, matched with a realistic concern to ensure that historic buildings could remain in use. The guidance was repeated almost identically in **DoE Circular 8/87** (save that the Investigators had now become the Historic Buildings Inspectors of English Heritage). Its long-lived successor, **Planning Policy Guidance Note (PPG) 15: Planning and the Historic Environment** (1994–2010), set out more general principles for alterations. These were based on the need to understand and justify the potential impact of the changes on the 'special architectural or historic interest' of the building which had given rise to its listing, rather than the predominantly visual concerns of earlier guidance. The technical guidance was rewritten, embracing advice on elements of 20th-century buildings, and the preference that extensions should be 'in keeping' was dropped.

1973

Adam Fergusson's *The Sack of Bath: A Record and an Indictment* published.

The UK joins the European Economic Community, and so becomes obliged to conform to *EC Regulations and Directives*.

UK ratifies *European Convention on the Protection of the Archaeological Heritage*.

1974

'Destruction of the Country House' exhibition at the Victoria and Albert Museum.

Town and Country Amenities Act requires planning permission for full or partial demolition of buildings within conservation areas.

Early policy guidance that extensions or alterations to historic buildings should be in traditional or matching design, *"in keeping with other parts of the building"* (*DoE Circular 8/87*), gave way to more general principles based on the need to understand the impact of the changes on the 'special character' of the building (*PPG15*), allowing contemporary interventions such as this addition to the courtyard and crypt of Lambeth Palace, London, designed by Richard Griffiths Architects.

1975

European Architectural Heritage Year held under the auspices of the Council of Europe.

SAVE Britain's Heritage founded.

1976

Architectural Heritage Fund established.

1977

Updated conservation policy guidance issued as **Department of the Environment [DoE] Circular 23/77.**

Although the legal controls applied only to works of alteration perceived to affect a building's character, official guidance was always strongly oriented towards encouraging like-for-like repair, which did not normally need consent. Later, *PPG15* was explicitly supported by Christopher Brereton's *The Repair of Historic Buildings; Advice on Principles and Methods* (1995), which grew out of the experience of English Heritage architects responsible for monitoring grant-aided works to listed buildings. A succinct book of great good sense, Brereton's definition of the primary purpose of repair, "*...to restrain the process of decay without damaging the character of buildings and monuments, altering the features which give them their historic or architectural importance, or unnecessarily disturbing or destroying historic fabric*", remains entirely valid.

As well as stemming the rate of demolition, the circulars and their successors were influential in helping to define building conservation practice in England in the late 20th century. Ultimately, however, the role of these planning policy documents was to balance the public interest in the conservation of historic buildings with other public (and private) interests that might champion the benefits of greater levels of intervention, including demolition. In contrast to ethical documents such as the SPAB *Manifesto* or the *Venice Charter*, the role and status of the circulars as government guidance precluded anything more absolute than a "*presumption in favour of preservation*" (*PPG15 1994*) (subsequently 'conservation' in *PPS5: Planning for the Historic Environment, 2010*). In parallel, a wide range of textbooks began to appear, including the magisterial *Conservation of Historic Buildings* (1982) by Sir Bernard Feilden (1919–2008).

The trend through the later 20th century was steadily more conservationist. The focus of concern moved from preventing demolition to achieving conservative repair and alteration, and, increasingly, to protecting the setting of listed buildings. A normative statement of best conservation practice specific to the UK context, written by a group of professionals who were not constrained by official policy, was produced (after an arduous 15-year gestation) in 1998. This was the *British Standard BS 7913:1998 Guide to the principles of the conservation of historic buildings*. Revised in 2013, it carries weight as a BS Guide, and remains current. Unfortunately, the high price and restrictive copyright conditions attached to BS publications have limited its adoption within the sector. Otherwise, modern practice in built heritage conservation in England has been codified primarily through public policy and guidance, rather than standards produced by professional bodies.

1979

Ancient Monuments and Archaeological Areas Act introduces the 'Scheduled Monument Consent' regime.

United Kingdom Institute for Conservation of Historic and Artistic Works [UKIC] becomes autonomous from the International Institute for Conservation of Historic and Artistic Works [IIC].

The Thirties Society (later the Twentieth Century Society) founded.

ICOMOS Australia agrees the Burra Charter, which introduces the concept of conservation management.

BUILDING ARCHAEOLOGY

In the 19th century, the application of archaeological techniques of analysis to historic buildings was fundamental to establishing the chronology of Gothic architecture. *Windsor Castle: An Architectural History* (1913), by W. H. St John Hope (1854–1919), stands out as a commanding study of a single site. The RCHME always maintained an analytical approach to understanding building fabric as well as earthworks. From the late 1960s, archaeologists whose background lay primarily in excavation below ground began applying their techniques of highly detailed recording and analysis to standing buildings. At first, efforts were focused on churches, particularly to address the issue of defining Anglo-Saxon architecture embedded within later structure. Stone-by-stone recording and analysis began to yield some quite surprising results about the complex evolution of key buildings, such as Deerhurst Church in Gloucestershire.

Soon this approach spread to other complex standing structures, such as the medieval and Tudor house at Acton Court, South Gloucestershire, and was often combined with below-ground excavation. Radiocarbon dating began to confirm the antiquity of some medieval carpentry, which thus became a subject worthy of study, initiated by Cecil Hewett (1926–1998). Precision came through dendrochronology, which, given the right conditions, can date the year, indeed the season, of felling of timber. The value of vernacular architecture as a source of information about the past – and the revelation that some of it is of considerable antiquity – is all part of the extension of 'archaeological' interest to a wider range of standing material remains of the past. The development of historic paint analysis – science-based archaeology on a very small scale – meant that, if the paint layers survived, a virtually complete sequence of decorative schemes could be recovered, transforming understanding of the historic use of colour and our ability to restore historic decorative schemes.

1980

1980

The demolition of Wallis Gilbert's Firestone Factory without consent, resulting in pressure to extend protection of inter-war buildings.

National Heritage Act establishes the National Heritage Memorial Fund in place of the Land Fund, and provides for the donation of property in lieu of death duties.

Building archaeology

Top: Rivenhall Church, Essex. Removal of cement render from the north side of the chancel of a church then thought to have been completely rebuilt *c.*1838–39 revealed a long sequence of medieval work, from the 11th-century chancel, to which an apse was later added, in turn replaced by a chancel extension, and the remodelling of the earlier structure early in the 14th century.

Bottom: Osterley Park Stable block. The use of rectified photographs of the elevations of the stables enabled the complex phases of construction and history of alterations to be rapidly and extensively recorded and interpreted.

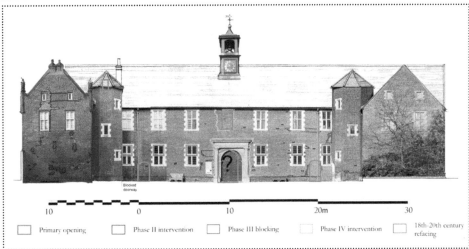

| Primary opening | Phase II intervention | Phase III blocking | Phase IV intervention | 18th-20th century refacing |

1981

The Historic Buildings Council recommends a further 150 inter-war buildings for listing.

Accelerated Resurvey of listed buildings begun.

1982

Association of Conservation Officers founded.

Institute of Field Archaeologists [IFA] founded.

ICOMOS' *Florence Charter* on the protection of historic gardens.

First edition of Bernard Feilden's *Conservation of Historic Buildings* is published.

CONSERVATION BASICS
THE EVOLVING CONCEPT OF BUILDING CONSERVATION

The consequences of these developments (and many more) for building conservation have been profound. On one level, it has become the norm to include building archaeologists in a team working on major repair projects to significant buildings. The results can add to (and sometimes transform) academic understanding. They can also provide a more detailed understanding of the evolution and significance of parts of a building, and so inform conservation decisions, ranging from structural repair to reinstatement of paintwork. The values of hidden, as well as visible, structure are now much better appreciated and respected. The wholesale replacement of a medieval roof above vaulting, rather than its repair, for example, would now be unthinkable.

Archaeologists also proved vital in the conservation response to a series of disastrous fires in major buildings, notably Hampton Court Palace, Uppark House and Windsor Castle. The systematic salvage of material from post-fire debris contributed to working out in detail the form of what had been lost, and recovering elements for incorporation in the reconstruction. This provided not only a greater degree of authenticity, but also a quality control on the new work. There was a considerable debate over whether enough of Uppark had survived the 1988 fire to make its rebuilding justifiable as a conservation repair, or unjustifiable as a somewhat speculative reconstruction. The archaeological approach to salvage ensured that it was the former, and a large interdisciplinary team discovered the importance of looking very closely at a wide range of artefacts, re-learning how they were made, as well as what they originally looked like. One of the positive side-effects was that freehand modelling in lime stucco was revived as an artistic craft skill. But perhaps the most important was greater investment in effective fire detection and prevention systems in historic buildings, especially during building work.

1983	1984
National Heritage Act passed.	English Heritage (officially the Historic Buildings and Monuments Commission for England [HBMCE]) set up as the Government's advisor on England's historic environment.
	The UK ratifies the UNESCO World Heritage Convention.

The identification and protection of historic areas form a third strand of English conservation. While it has many different threads, ideas of local character and distinctiveness are key. Groups of historic buildings in their topographical setting often have a greater value in aggregate than the individual buildings of which the groups are comprised. These can be of significance in the local context, and sometimes in the national context, even if few or none of their buildings individually meet the criteria for inclusion in the (national) statutory list.

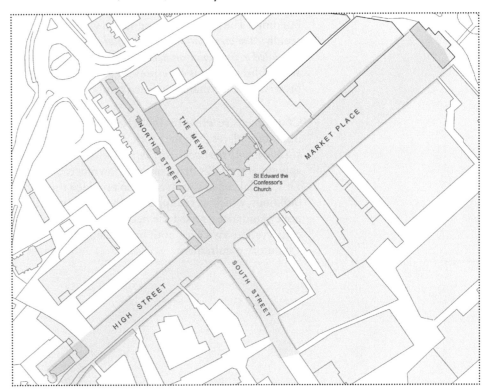

Early designation

The Romford Conservation Area was designated in 1968, shortly after the passing of the Civic Amenities Act, and it illustrates the narrow, early approach to designations. The main purpose of the conservation area was to protect the group of listed buildings at the west end of the Market Place, around and including St Edward's Church, and their setting, rather than the wider townscape.

1984

Establishment of *The Register of Parks and Gardens of Special Historic Interest in England*.

Updated conservation policy guidance issued as *DoE Circular 8/84*.

1986

The Greater London Council abolished, and responsibility for historic buildings in London transferred to English Heritage.

First seven UK sites added to UNESCO World Heritage list.

The Civic Trust, an umbrella body of local amenity societies formed under the leadership of Duncan Sandys, was fundamental in promoting the idea that the values of historic areas should be taken into account in land-use planning decisions. Formal recognition began when local planning authorities were empowered by the *Civic Amenities Act 1967* to designate conservation areas: 'areas of special architectural or historic interest, the character or appearance of which it is desirable to preserve or enhance'. Control over the demolition of unlisted buildings in conservation areas and the power to give repair grants was introduced in 1972.

Early official policy, set out in *Ministry of Housing and Local Government Circulars 53/67* and *61/68*, saw the role of conservation areas as safeguarding the context of listed buildings. The process of designation was expected to be finite, and completed very rapidly: *"the preservation of individual [listed] buildings is not enough…preservation must be related to their environment"*. In fact, conservation area designation remains an established means of responding to local perceptions of 'heritage' that is worth preserving and enhancing, listed or not, in line with its origins in the Civic Trust.

Comprehensive 'Historic Town Reports' were commissioned for four nationally important places (Chester, York, Bath and Chichester) *"to discover how to reconcile our old towns with the twentieth century without actually knocking them down"*. Chester boldly appointed the UK's first 'conservation officer', whose post was funded by a 'penny rate'. This special conservation rate also provided the city's share of repair grants for historic buildings, which were supported through national funds. From the outset, government policy guidance stressed the need for local authorities to have access to specialist conservation advice for both historic buildings and areas, although it has never been a statutory requirement.

1987

Policy changes make buildings more than 30 years old (and, exceptionally, 10 years old) eligible for listing.

Bracken House, London (1959), is the first post-war building in England to be listed.

UK ratifies the (Granada) *Convention for the Protection of the Architectural Heritage of Europe*.

Historic Towns Reports

In 1966, the Ministry of Housing and Local Government commissioned comprehensive reports on four nationally important places (Chester, York, Bath and Chichester), with the aim of discovering how the fabric of historic towns might be reconciled with 20th-century living. *Chester: A Study in Conservation*, was prepared by Donald Insall and Associates, and published in 1968.

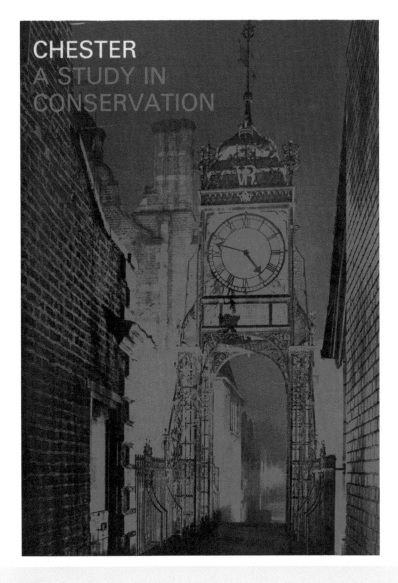

1988

First series of English Heritage's *Practical Building Conservation*, by John and Nicola Ashurst, published.

1989

Accelerated resurvey of listed buildings completed.

Association of [Building] Preservation Trusts founded.

Historic Royal Palaces established as a government agency to manage the five palaces no longer in regular use.

As specialist officers became more numerous in local government, the Association of Conservation Officers was founded in 1981. In 1997, it was transformed into the Institute of Historic Building Conservation [IHBC], with formal requirements for the competence and experience of its members.

Conservation areas gave local authorities the ability to designate at local level. This has contributed to what can be seen as a democratisation of heritage beyond the value judgement of the elite and expert. This trend includes things as well as places, and has been particularly championed by the National Lottery Heritage Fund [NLHF], formerly the Heritage Lottery Fund [HLF], which was established in 1994. For HLF, heritage is, in effect, anything that people value beyond mere usefulness. Professional opinion has increasingly embraced this approach, with the establishment of non-statutory registers of historic parks, gardens and battlefields, and, of course, the understanding of the landscape as an archaeological, as well as a visual and productive, resource.

These views are not universally shared; for some, they are symbolic of a lack of discrimination and represent cultural relativism, the consequence of a society now fearful of change and the future at every level from its own backyard to the planet. The association of change with 'harm' first began in the 1960s, with the public reacting against the wholesale redevelopment of historic towns and cities. Since then, it has tended to grow. Some of this local community rhetoric is certainly selfish 'nimbyism' hiding under the cloak of heritage protection. Nonetheless, the recognition of cultural heritage values in the wider historic environment, which gives context to designated heritage assets, and with which people identify, is likely to endure. It follows that those values should be taken into account in managing the historic environment, particularly through planning decisions.

1990

1990

The *Planning (Listed Buildings and Conservation Areas) Act* consolidates the law relating to historic buildings and conservation areas.

The *Care of Cathedrals Measure* brings works to cathedrals under the control of the Cathedrals Fabric Commission for England and individual fabric advisory committees.

1991

The 1975 Willis Faber Dumas building, Ipswich, becomes the first building under 30 years old to be listed.

Registering battlefields

Site of the Battle of Blore Heath, 23rd September 1459 (Wars of the Roses). For many people, these sites are places of memory and evocation, which fire the imagination; for others, they are simply empty landscapes, no different from any other. Historic England's non-statutory *Register of Historic Battlefields* identifies 43 important English battlefields which have not been transformed by later development, to offer them some protection and promote their better understanding.

1992

The Thirties Society becomes the Twentieth Century Society.

Begining of thematic listing of post-war buildings.

1993

The *National Lottery Act* enables grants for heritage projects from the lottery through the Heritage Lottery Fund.

1994

Government policy set out in *Planning Policy Guidance 15: Planning and the Historic Environment (PPG15)* and *Planning Policy Guidance 16: Archaeology and Planning (PPG16)*.

The *Ecclesiastical Exemption Act* limits exemption from Listed Building Consent to denominations operating their own, comparable, controls.

CONSERVATION BASICS
THE EVOLVING CONCEPT OF BUILDING CONSERVATION

THE EUROPEAN DIMENSION

This values-based approach accords with the ideas embodied in the *European Landscape Convention* (Florence 2000), which has been ratified by the UK Government. It defines a landscape as 'a place, as perceived by people', and seeks to sustain and enhance the cultural values of all European landscapes. More recently, the *Framework Convention on the Value of Cultural Heritage for Society* (Faro 2005, not yet ratified by the UK) deals with heritage as a human right, promoting a wider understanding of its relationships to community, society and nation. By relating the question of heritage values to the wider concerns of society about economic, social and cultural values, the Faro convention helps to inform the climate in which building conservation does (or does not) take place, rather than its technical practice. Like the *Convention for the Protection of the Architectural Heritage of Europe* (Granada 1985, also ratified by the UK), the Florence and Faro conventions were developed under the auspices of the Council of Europe.

The European Union does not have 'competence' (powers) in cultural policy, which remains with its member states, but Title XII, Article 151 of the *European Community Treaty* (2002) gives it powers to promote action in the field of cultural heritage. These relate mainly to encouraging or enhancing awareness in the European Community of its common cultural heritage, supporting and encouraging cooperation between member states regarding dissemination of cultural knowledge, the protection of significant European heritage, and artistic and literary creation. However, the *European Union Directives* on *Strategic Environmental Assessment* (*SEA Directive 2001/42/EC*) and *Environmental Impact Assessment* (*EIA Directive 85/337/EEC*, amended by *Directives 97/11/EC, 2003/35/EC* and *2009/31/EC*) are intended to ensure that cultural heritage impacts are taken into account in large or sensitive developments. This is supported by the United Nations Economic Commission for Europe *Aarhus Convention on Access to Information, Public Participation in Decision-making and Access to Justice in Environmental Matters* (1998).

1994	1995	1997
Heritage Lottery Fund established.	Non-statutory *Register of Historic Battlefields* is established.	The Association of Conservation Officers re-founded as the Institute of Historic Buildings Conservation [IHBC].
ICOMOS conference in Nara, Japan establishes the *Nara Document on Authenticity*.	Christopher Brereton's *The Repair of Historic Buildings: Advice on Principles and Methods* published by English Heritage.	

In the UK, the creation of the Department of the Environment in 1970 brought responsibility for the three strands of historic environment conservation – monuments, buildings and areas – within a single ministry. Two years later these functions were administratively merged into the Directorate of Ancient Monuments and Historic Buildings, but this was not to last. The incoming Conservative Government of 1979 began consultation on the remit of the Department of the Environment. From this arose the idea of devolving national responsibility for the historic environment to a semi-autonomous agency or 'quango' that would be able to operate with greater efficiency and enterprise, but under ministerial guidelines and to government policy.

English Heritage (officially the Historic Buildings and Monuments Commission for England), was set up under the terms of the *National Heritage Act 1983*. In Scotland, 'Historic Scotland' and in Wales, 'Cadw' (Welsh Historic Monuments) were given similar responsibilities, but remained part of the Government. The 1983 Act also dissolved the bodies that had hitherto provided independent advice – the Ancient Monuments Board for England and the Historic Buildings Council for England – and incorporated these functions in the new body. Another advisory body, the Royal Commission on the Historical Monuments of England [RCHME], which was responsible for documenting the records of English historic monuments, was not merged with English Heritage until 1st April 1999.

In 2015, management of the National Heritage Collection of state-owned historic sites and monuments transferred from English Heritage to a new charity, the English Heritage Trust. At the same time, the Historic Buildings and Monuments Commission for England was given a new common name – Historic England. Historic England is now the leading advisory body on all aspects of the historic environment in England. It advises the Government on policy and helps people care for, enjoy and celebrate England's historic environment.

1998

Publication of the *British Standard BS 7913:1998 Guide to the principles of the conservation of historic buildings*.

Historic Royal Palaces becomes an independent charity.

1999

RCHME administratively merged with English Heritage.

Commission for Architecture and the Built Environment [CABE] founded, replacing the Royal Fine Art Commission [RFAC].

HERITAGE-LED REGENERATION

Investment in the conservation of historic areas, particularly in urban centres, emerged as a social and economic, as well as cultural, benefit. Conservation, rather than comprehensive redevelopment, came to be seen as a source of quality and distinctiveness which could make areas more attractive places in which to live or work. As public appreciation of historic areas grew, so did demand for property within them, often reversing a long-term post-war decline. The approach soon extended beyond traditional 'historic centres' like those in the historic towns studies to former industrial areas, embracing a wide range of historic building groups. A realisation emerged that heritage could lead regeneration, and tended to add value in every sense.

Heritage-led regeneration can be defined as *"the improvement of disadvantaged people or places through the delivery of a heritage focused project"* (***Ela Palmer Heritage 2008***). It was soon seen as environmentally friendly, since the re-use of buildings makes the most of energy embodied in their fabric and reduces the amount of waste that becomes landfill. Heritage was promoted by English Heritage *"as a catalyst for better social and economic regeneration"* (***English Heritage 2005***), ensuring that some of the considerable public resources and private investment in this 'growth industry' were applied to the repair and re-use of historic buildings. The economic climate has changed dramatically since that time, however, and limited public investment is now focused directly on economic objectives, rather than indirectly on property-led regeneration. The potential of the built heritage to contribute to the reinvention of places and a positive identity for communities nevertheless seems firmly established.

2000

2000

English Heritage publishes *Power of Place: The Future of the Historic Environment,* a review of historic environment policy.

The UK ratifies the *European Convention on the Protection of the Archaeological Heritage (Revised).*

The *European Landscape Convention* adopted by the Council of Europe in Florence.

The Market Building, Covent Garden, London

Designed by Charles Fowler (1792–1867), the Market Building opened in 1830 to house London's principal wholesale fruit, vegetable and flower market. The market quickly outgrew its new premises and spilled into the surrounding streets, where further buildings were erected or adapted for market use. By the 1960s, traffic congestion in the area had reached a point where the market could no longer function effectively. It relocated to a new site in 1974, and the Greater London Council [GLC] purchased the Market Building, along with other market properties.

Public opinion strongly favoured preserving the historic character of the area by retaining and reusing the redundant buildings, and proposals for comprehensive redevelopment were rejected. The GLC's Historic Buildings Division prepared plans for the repair and conversion of the Market Building into shops, restaurants, bars, studios, and a venue for street entertainers. When it re-opened in 1980 it became the focal point of the regeneration of Covent Garden, proving that historic areas could make a positive contribution to the economic vitality of towns and cities.

2001

The Department for Culture, Media and Sport [DCMS] and the Department for Transport, Local Government and the Regions [DTLR] together publish *The Historic Environment: A Force for Our Future*.

2002

First annual *State of the Historic Environment Report* (subsequently *Heritage Counts*) is published by English Heritage.

2003

DCMS publishes the consultation document *Protecting our Historic Environment: Making the System Work Better*.

THE IDEA OF CONSERVATION PLANNING

The idea of 'conservation planning' was pioneered by James Semple Kerr in Australia, and underpinned (with the *Venice Charter*) the *Burra Charter,* first promulgated by ICOMOS Australia in 1979. While the *Venice Charter* and its precursors prescribed what was necessary to protect a narrow range of heritage values, the *Burra Charter* set out a process for managing change in ways that seek to retain *"all aspects of [its] cultural significance"*. The heritage values of places are seen as often both multiple and mutable. Heritage practitioners therefore need to be advocates and enablers as well as conservators, particularly in relation to values attached to places by communities that identify with them.

Both Kerr's work and the *Burra Charter* soon began to be referenced by practitioners in England, and the 'conservation planning' approach was taken up by the Heritage Lottery Fund. The idea is a simple one: understand the range of values that people attach to a place, and seek to manage the place so as to sustain as many of those values as reasonably possible. In the English statutory context, it was the idea of 'place' as a meaningful unit for heritage management decisions that was perhaps the most radical. The reason was the accumulated 'hierarchy' of existing legislation, which differentiates between scheduled monuments (national designation and control), listed buildings (national designation and local control), conservation areas (local designation and control) and registered landscapes (national designation and limited control). Within English Heritage, the amalgamation of its different statutory advisory teams into multi-disciplinary regional teams in 1990 created the organisational basis for an integrated approach to managing the historic environment, while conservation planning increasingly provided the intellectual underpinning for it.

As the scope of designation and recognition has widened to include historic buildings and areas significant for their design or associations, rather than simply their age, and which are sustained by remaining in use, there has been a *de facto* (but not universal) acceptance that 'minimum intervention' does not, of itself, provide an adequate response to the range of conservation issues faced by practitioners or regulatory authorities. This wider concept of heritage demands discrimination and a sense of proportion to inform attempts to identify and balance conflicting public interests (the essential concern of public policy) in a methodical and transparent way.

2004

DCMS publishes its *Review of Heritage Protection: The Way Forward*.

2005

Institute of Conservation [ICON] established by merger of UKIC and others as the primary UK organisation of conservators.

At a meeting in Faro, Portugal, the Council of Europe introduces the *Framework Convention on the Value of Cultural Heritage for Society*.

The *Nara Document on Authenticity* (ICOMOS Japan, 1994) examines concepts such as the tangible and intangible nature of culture, cultural diversity, and the legitimacy of all cultural values. It concludes that statements about cultural heritage should, in the first instance, be the responsibility of the community that has generated the culture, stating that *"all judgements about values attributed to cultural properties differ from culture to culture and therefore it is not possible to base judgements of values and authenticity within fixed criteria"*. In other words, authenticity can be seen as those characteristics that most truthfully reflect and embody the cultural heritage values of a place, rather than residing solely in its inherited fabric. The Document also addresses the question of how to balance the different – and potentially conflicting – cultural requirements *"provided achieving this balance does not undermine [any one community's] fundamental cultural values"*. This is to be achieved by judging the value of heritage properties *"within the cultural contexts to which they belong"*. Furthermore, management *"belongs, in the first place, to the cultural community that has generated it, and subsequently to that which cares for it"*.

These ideas, as well as established English conservation practice and public policy, provided the background to the publication of English Heritage's 2008 *Conservation Principles, Policies and Guidance for the Sustainable Management of the Historic Environment*. This attempted to domesticate the concepts of conservation planning and a values-based system of assessment, and respond to discussion papers about the scope and purpose of public conservation. The immediate context for the document was the need for an agreed framework for making decisions about managing the historic environment and any 'heritage asset' within it, taking into account the full range of cultural (and natural) heritage values attached to the asset by different interest groups. To do so, it was necessary to draw together and reconcile the objectives of the three established strands of English conservation embedded in legislation and practice.

2007

UK ratifies the *European Landscape Convention* [ELC].

The White Paper *Heritage Protection for the 21st Century* is published, proposing fundamental changes to the designation system.

To find common ground between the monuments, the buildings and the areas or landscapes, English Heritage attempted through internal and external workshops to define principles and a framework for their application to any and all elements of the historic environment, whether or not statutorily designated. The process proved to be a salutary reminder, not only of the different threads of thought rooted in established 'schools' of conservation, but also how the education, profession, thought and personal experience of practitioners shape their conservation philosophy. The potential of these differences to create adversarial positions between practitioners and organisations is obvious.

In a field of increasing diversity of subject and approach, the aim is to encourage intellectually coherent thinking acceptable to a majority of practitioners, particularly those with statutory roles. *Conservation Principles* could not, of course, possibly succeed in meeting such an ambitious goal. The immediate practical objectives were more limited: to achieve some consistency in English Heritage's response to casework, and provide a shared framework for arbitrating conflicts between sustaining particular cultural values.

As in all statements of conservation philosophy, *Conservation Principles* was the product of a specific place, time, and cultural and political context. It embraced the concept that a wide range of abstract cultural values beyond mere usefulness are attached by people to places. Heritage values are a subset of wider cultural values, attributed to things inherited from the past, or selected from current cultural artefacts valued sufficiently to warrant passing them on to future generations. *Conservation Principles* was based on the view that, rather than being intrinsic, cultural values are attached or attributed to places, even when the attribution stems solely from a fact that can be established more or less objectively, such as age.

2008

English Heritage publishes *Conservation Principles, Policies and Guidance for the Sustainable Management of the Historic Environment*.

2009

English Heritage designated by government as a provider of official statistics, including the *Heritage at Risk Register* [HAR].

HERITAGE PROTECTION REFORM & BEYOND

English Heritage's *Conservation Principles* anticipated, and was intended to inform, a more integrated approach to managing cultural heritage assets in the historic environment. The Government announced its intention to replace existing legislation with a single reforming *Heritage Act*, but, in the event, the draft bill (April 2008) was not taken forward. In March 2010, the Government published a new policy document, *Planning Policy Statement (PPS) 5, Planning for the Historic Environment*. This replaced the long-standing *Planning Policy Guidance Note 15: Planning and the Historic Environment* (1994) and *Planning Policy Guidance Note 16: Archaeology and Planning* (1990). *PPS5* stated, as the Government's overarching aim, *"that the historic environment and its heritage assets should be conserved and enjoyed for the quality of life they bring to this and future generations"* (*DCLG 2010*). Rather than providing prescriptive solutions, *PPS5* sought to establish principles, applicable to all recognised 'heritage assets', regardless of the legislation under which they are designated, recognised or managed.

PPS5 was supported by English Heritage's government-endorsed *Historic Environment Planning Practice Guide* (the *Practice Guide*) (2010), whose purpose was to *"assist local authorities, owners, applicants and other interested parties in implementing [PPS5] and to help in the interpretation of policies within the PPS...."* The *Practice Guide* *"supports the implementation of national policy, but does not constitute a statement of Government policy itself"*.

However, *PPS5* was short-lived. In March 2012 the successor government published a *National Planning Policy Framework* [NPPF] that replaced all existing planning policy statements, including *PPS5*. Nevertheless, this carried forward, at a more strategic level, the integrated, values-based approach established in *PPS5*.

2010

2010

PPS5: Planning for the Historic Environment issued as official conservation policy guidance within the planning system, replacing *PPG15* and *PPG16*.

CABE merged with the Design Council.

2012

National Planning Policy Framework replaces all former topic-based policy guidance, including *PPS5*.

2015

Common name of the Historic Buildings and Monuments Commission for England changed to Historic England.

Management of state-owned historic sites and monuments transferred from English Heritage to a new charity – English Heritage Trust.

Arguably, a wider concept of cultural heritage values in the historic environment leads to an operative, instead of a prescriptive, philosophy of conservation. Rather than applying a predetermined solution, it should be understood why a building or place has public values beyond utilitarian functionality, and manage it so as best to sustain those values. Only such an approach can accommodate the range of heritage assets now subject to conservation, and the diverse values, public as well as expert, which must be taken into consideration in doing so. This does not mean, however, that answers will necessarily be different from those that Carter or Morris, or the Historic Buildings Investigators of four decades ago, would have reached. If a place is valued for its significance and potential as a historic document, its illustration of a distant age and the skill of its craftsmen, and the weathering acquired in the centuries since, minimum intervention is still likely to be the appropriate response to sustaining those values. But a values-based approach to conservation also means looking for the potential to reinforce and reveal the significance of places. That in turn means making value judgements about design value and interventions: removing things, and even restoring documented elements, if the benefits decisively outweigh the losses.

Differing cultural responses

The *Nara Document on Authenticity* states: *"…all judgements about values attributed to cultural properties differ from culture to culture and therefore it is not possible to base judgements of values and authenticity within fixed criteria"*.

These examples – the new Coventry Cathedral (*facing page*), of 1956–62, by Sir Basil Spence (1907–1976), and the Frauenkirche, Dresden, of 1726–43 (*this page*), as reconstructed in 1994–2005 – show two very different cultural responses to war-time destruction.

The further one moves from the kind of place that prompted the responses of earlier generations, the more the practice of conservation may differ, and the more negotiation it is likely to involve. The point, as James Semple Kerr expressed so clearly and simply, is that the outcome needs to be reached by a logical and transparent process, not the application of a predetermined simplistic solution. It is too early to say whether this 'conservation planning' approach will find general acceptance, or if (particularly in the absence of unified legislation) existing approaches developed in relation to particular types of physical heritage or its statutory protection will continue to dominate day-to-day practice. The idea that doing the minimum is always the ideal response remains potent, even in the face of pervasive designation which necessitates discrimination.

The importance people attach to the cultural heritage values of places has grown exponentially over the past half-century, but that does not mean it will continue to do so. There is a new ethical imperative, environmental sustainability, that began (like heritage conservation) as a minority position, but which has become widely accepted and so adopted as public policy. That it may be in conflict with sustaining inherited cultural values in the historic environment is already becoming clear, and the drivers may not be so much technical as polemical.

Abbey Road zebra crossing, St John's Wood, London

The crossing at Abbey Road in north London is famous for its association with the English rock band The Beatles. It featured on the cover of their last recorded album *Abbey Road*, released in 1969 and made at the studios of the same name.

The crossing has become a popular destination for Beatles fans, and in 2010 was listed at Grade II for its cultural and historical importance.

Further Reading

Brereton, C. (1995); *The Repair of Historic Buildings: Advice on Principles and Methods* (2nd edition); London: English Heritage

British Standards Institution (1998); *BS 7913:1998 Guide to the principles of the conservation of historic buildings*; London: BSI

Brown, G. B. (1905); *The Care of Ancient Monuments*; Cambridge: University Press

Croad, S. (1992); 'The National Building Record: The early years'; in *Transactions of the Ancient Monuments Society*; Vol.36, pp.79–98

Department for Culture, Media and Sport (2001); *The Historic Environment: A Force for Our Future*; London: DCMS

Department for Culture, Media and Sport (2003); *Protecting our Historic Environment: Making the System Work Better*; London: DCMS

Department for Culture, Media and Sport (2004); *Review of Heritage Protection: The Way Forward*; London: DCMS

English Heritage (2000); *Power of Place: The Future of the Historic Environment*; London: English Heritage

English Heritage (2002); *State of the Historic Environment Report (Heritage Counts 2002)*; London: English Heritage

English Heritage (2005); *Regeneration and the Historic Environment: Heritage as a Catalyst for Better Social and Economic Regeneration*; London: English Heritage

English Heritage (2008); *Conservation Principles, Policies and Guidance for the Sustainable Management of the Historic Environment*; London: English Heritage

Evans, J. (1956); *A History of the Society of Antiquaries*; Oxford: University Press

Feilden, B. M. (1982); *Conservation of Historic Buildings*; London: Butterworth Scientific

Gillon, J. (1996); 'Conservation charters and standards'; in *Context*; Vol.51, p.20; available from *www.ihbconline.co.uk/context/51/#20*

Harvey, J. H. (1993); 'The origin of listed buildings'; in *Transactions of the Ancient Monuments Society*; Vol.37, pp.1–20

Harvey, J. H. (1994); 'Listing as I knew it in 1949'; in *Transactions of the Ancient Monuments Society*; Vol.38, pp.97–104

Jokilehto, J. (1999); *A History of Architectural Conservation*; Oxford: Butterworth-Heinemann

Miele, C. (ed.) (2005); *From William Morris: Building Conservation and the Arts and Crafts Cult of Authenticity 1877–1939 (Studies in British Art 14)*; New Haven and London: Yale University Press

Pendlebury, J. R. (2009); *Conservation in the Age of Consensus*; Abingdon: Routledge

Powys, A. R. (1929); *Repair of Ancient Buildings*; London: J. M. Dent

Robertson, M. (ed.) (1993); 'Listed Buildings: The National Resurvey of England'; in *Transactions of the Ancient Monuments Society*; Vol.37, pp.21–94

Stanley Price, N. *et al.* (eds.) (1996); *Historical and Philosophical Issues in the Conservation of Cultural Heritage*; Los Angeles: The Getty Conservation Institute

Stubbs, J. H. (2009); *Time Honored: A Global View of Architectural Conservation: Parameters, Theory, & Evolution of an Ethos*; Hoboken (New Jersey): John Wiley and Sons

Summerson, J. (1949); 'The past in the future'; in *Heavenly Mansions (and Other Essays on Architecture)* (Reprinted 1998); New York and London: W. W. Norton & Co; pp.21–42

Thompson, M. (2006); *Ruins Reused: Changing Attitudes to Ruins since the Late Eighteenth Century*; King's Lynn: Heritage Marketing & Publishing

Watkin, D. (1980); *The Rise of Architectural History*; London: Architectural Press

CURRENT LAW, POLICY & GUIDANCE

This chapter reviews the current legislation, policies and guidance that apply to heritage assets in England.

Works to protected buildings and sites are subject to two main areas of legal control, each with their associated policies and regulations. The first concerns heritage conservation, which is largely, but not wholly, embedded in the planning system, although all is now subject to the *National Planning Policy Framework*. The second concerns fitness for purpose and safety, which is largely set out in *Building Regulations*, but supplemented by other measures relating to health and safety in construction and the protection of other public interests, including natural heritage (see **Ecological Considerations**). Practice in both areas is underpinned by standards and professional guidance.

This section provides an overview at the time of writing (2012), but policy and regulation, particularly, are subject to frequent revision.

HERITAGE CONSERVATION

LEGISLATIVE & POLICY FRAMEWORK

The preceding chapter explained how primary legislation intended to protect specific historic buildings and structures developed incrementally in the UK. It is mostly applied in the context of a comprehensive town and country planning system, with the objective of regulating the use of land in the public interest (the right to develop land was, in effect, nationalised under the Town and Country Planning Act, 1947). It is the principal means by which cultural and natural heritage values as a public interest in land are protected and enhanced, and it is a specific objective of the current overriding *National Planning Policy Framework (DCLG March 2012)* which replaces all previous *Planning Policy Statements* and related guidance.

Over the last half a century the interpretation and implementation of legislation has inevitably grown in complexity and extent to the point where the standard textbook on the law relating to the historic environment in the UK, Charles Mynors' *Listed Buildings, Conservation Areas and Monuments*, 2006 (now in its fourth edition) runs to some 950 pages. The current framework of controls for different types of heritage assets is set out below, insofar as it relates to the repair of historic buildings and structures. It should be noted that statute and policy do not define precisely how these regimes for managing change within the historic environment should be applied to individual cases, which is, therefore, ultimately a matter for the courts.

Managing change in the historic environment

St Pancras Station, London (1864–68), designed by William H. Barlow (1812–1902), has been successfully adapted to meet the requirements of the 21st century in a redevelopment that includes the Eurostar rail terminal, retail premises, and leisure facilities.

For example, what constitutes 'demolition' as opposed to 'alteration' has been determined by the *Shimizu (UK) vs. Westminster City Council* case. The extent of 'curtilage' and 'setting' in individual cases, and the determination of works that do not require Listed Building Consent, rely on case-law and precedent (*Mynors 2006*). What constitutes 'substantial harm' as opposed to 'less than substantial harm' in terms of the *National Planning Policy Framework* has yet to be tested.

CONSERVATION BASICS
CURRENT LAW, POLICY & GUIDANCE

PLANNING POLICY FOR THE HISTORIC ENVIRONMENT

The *National Planning Policy Framework* (*NPPF*) sets out the Government's planning policies for England, including the conservation of all aspects of the historic environment, although the underlying (historically separate) legislative frameworks remain. The *NPPF* requires a proportionate approach based on understanding the significance of the heritage asset or assets concerned, in line with the general principles of conservation planning. This approach should underpin all interventions in heritage assets, even if they are not the subject of statutory control.

Section 12 of the *NPPF*, 'Conserving and enhancing the historic environment', is used as the policy basis when the relevant authorities are making plans or taking decisions. It is intended to allow a unified, strategic approach to managing change in the historic environment, regardless of the type of heritage asset under consideration or permission needed. At its heart is a presumption in favour of sustainable development, with a particular emphasis on the need for economic development. It recognises that the historic environment is an irreplaceable resource that brings social, cultural, economic and environmental benefits, and that heritage assets should be conserved in a manner appropriate to their significance, so they can be enjoyed for their contribution to the quality of life of present and future generations.

The *NPPF* applies to all elements of the historic environment, using the single term 'heritage asset' defined (in Annex 2) as *"a building, monument, site, place, area or landscape identified as having a degree of significance meriting consideration in planning decisions..."*. 'Significance' is defined as the *"...value of a heritage asset to this and future generations because of its heritage interest. That interest may be archaeological, architectural, artistic or historic. Significance derives not only from a heritage asset's physical presence, but also from its setting."* The *NPPF* defines 'designated heritage assets' as world heritage sites, scheduled monuments, listed buildings, protected wreck sites, registered parks and gardens, registered battlefields and conservation areas, but also recognises the significance and potential of undesignated heritage assets.

The *NPPF* does not contain an express presumption in favour of conserving designated heritage assets. It recognises there will be situations where substantial public benefits might outweigh the harm to and loss of significance caused by a proposed development. However, it states that 'great weight' should be given to the conservation of designated heritage assets, and that harm or loss should not be permitted without 'clear and convincing justification'. The importance of understanding significance sufficiently when assessing the potential impact of proposals is recognised, as is the value of gathering information to contribute to the knowledge and understanding of heritage assets, and making this publicly accessible.

Legislation governing listed buildings and conservation areas is largely consolidated in the *Planning (Listed Buildings and Conservation Areas) Act 1990* (the *1990 Act*). The Secretary of State has a duty to compile a statutory list of buildings of 'special architectural or historic interest' and, in doing so, is informed by the expert advice of Historic England. Detailed selection criteria have been developed around the concept of 'national importance' and are set out in *Principles for Selection of Listed Buildings* (DCMS, March 2010), supported by English Heritage's *Selection Guidelines* (March 2007) for listed buildings set out by building type. Virtually all buildings predating 1840 are listed, with increasing selectivity thereafter. Local authorities hold copies of the lists for their areas, and there is a national online database.

Listed buildings are designated in three (non-statutory) grades: Grade I, covering the most significant 2.5 % of buildings; Grade II* ('two star'), covering the next 5.5 %; and Grade II, covering the remaining 92 % of listed buildings. Listed Building Consent is required from the local planning authority for works *"for the demolition of a listed building, or for its alteration or extension in any manner which would affect its character as a building of special architectural or historic interest"* (1990 Act, s7). Regardless of the grade, when a building is listed, the entire building, both inside and out, is protected, as is *"any object or structure fixed to the building"* and *"any object or structure within the curtilage of the building which, although not fixed to the building, forms part of the land and has done so since before 1st July 1948"* (1990 Act, s5 (a), (b)). Exceptionally, a list entry may specifically exclude a part of a building (for example, a very large recent extension) as being *"not of special interest"*.

Curtilage structures are normally buildings or other structures, such as boundary walls or gates, historically associated with the principal listed building, but generally ancillary to it. If such a structure was in separate ownership when the principal building was listed, it will not usually be regarded as a curtilage structure. It is, however, always advisable to seek advice from the local planning authority in writing about whether works to a particular curtilage structure require Listed Building Consent.

Listed building control does not prohibit change to listed buildings, but is intended to ensure that, so far as possible, changes do not harm their significance. The *NPPF* sets out the basis on which applications for Listed Building Consent should be determined. It relies on two key principles: that the significance of a heritage asset must be adequately understood before a decision about proposals affecting it can be made (policy 128); and that when *"... considering the impact of a proposed development on the significance of a designated heritage asset, great weight should be given to the asset's conservation. The more important the asset the greater the weight should be."* (policy 132). Applications for Listed Building Consent must therefore clearly identify the significance of the listed building, both as a whole and particularly in relation to the elements affected (which will not normally be clear from the list description); and justify the proposed works against an objective of eliminating or minimising any harm to that significance. *NPPF* (policy 132) goes on to state that *"Substantial harm to or loss of a grade II listed building... ...should be exceptional"* and for Grade I and II* listed buildings *"...wholly exceptional"*.

The wide range of listed buildings

The term 'building' is defined broadly in the legislation and may include such things as lamp posts, telephone kiosks, lime kilns, bridges, cranes, railings, garden walls, statues, milestones, and man-made landscape features such as ha-has and grottos. Listing formalises a structure's special architectural and historic interest, and brings decisions about its future under the consideration of the planning system. All buildings built before 1700 that survive in anything like their original condition are listed, as are most of those built between 1700 and 1840. The criteria become tighter with time, so that post-1945 buildings have to be exceptionally important to be eligible for listing; normally, a building must be over 30 years old.

Since the 1990s, English Heritage has undertaken thematic assessments for listing that are focused on particular types and locations of 'heritage assets' (for example, railway buildings, military structures, post-Second World War buildings, cinemas, schools, agricultural buildings, letter boxes and many others). It has also published a series of selection guides setting out the assessment criteria for a wide range of buildings and structures (see **Further Reading** at the end of this chapter).

Clockwise from top left: Trellick Tower, Royal Borough of Kensington and Chelsea, London (Grade II*); Second World War tank obstacles, Isle of Grain, Kent (Grade II); hand-operated dockside crane, Bristol (Grade II); E. Pellicci café, East London (Grade II); K2 telephone kiosk (Grade II).

Many local authorities produce guidance tailored to local building types and traditions. Interpretation can vary between authorities, especially concerning the need for consent for repair which follows such good practice. Routine maintenance and minor repairs are unlikely to require Listed Building Consent if they do not affect the significance of the heritage asset. Proposals for substantial intervention and the introduction of new materials are much more likely to require consent than those which minimise replacement of historic fabric, and use materials and techniques which match the originals (*Mynors 2006*).

The advice of the local authority conservation officer should therefore always be sought before embarking on any significant repair or periodic renewal of the fabric of a listed building, since undertaking works requiring, but not having, consent is a criminal offence. If work is undertaken without consent, the local planning authority can issue a listed building enforcement notice, as well as prosecuting those responsible. Where a Listed Building Consent application is necessary, every effort should be made to respond to the local planning authority's pre-application advice and to provide the information required to support the application.

Listing for group value

New Bolsover Model Village (Grade II). Buildings may be listed for 'group value' where they are part of an important architectural or historic group, or are fine examples of planning, such as terraces, squares or model villages. Bolsover Castle, visible in the background, is a listed building (Grade I), a Scheduled Ancient Monument, and a Registered Park and Garden (Grade I).

The granting of Listed Building Consent does not obviate the need for planning permission if the proposals involve a material change to the external appearance of a listed building. Few conservation repair projects that follow the good practice necessary to achieve consent (or agreement that consent is not required) are likely to do so. Where it is required, however, Listed Building Consent and planning permission should normally be sought at the same time. Whilst applications for Listed Building Consent should be determined against national policy and guidance, applications for planning permission for the same works must also comply with local planning policies. Obtaining Listed Building Consent does not imply that planning permission will be granted for the same works, or vice versa.

While local planning authorities process and determine Listed Building Consent applications, those involving work to Grade I or Grade II* listed buildings, or the total or substantial demolition of any listed building, must be referred to Historic England. In London, Historic England has the power to direct the local authority to refuse Listed Building Consent in such cases; elsewhere, if Historic England or a National Amenity Society has unresolved objections to particular types of application, the local planning authority, if minded to grant consent, must first refer the application to the Secretary of State for Housing, Communities and Local Government to give him the opportunity, advised by Historic England, to call it in for his own determination. Historic England officers may therefore need to inspect the building for which consent is sought. Local authorities cannot grant themselves Listed Building Consent for their own buildings and must seek consent directly from the Secretary of State (who will take advice from Historic England).

There is currently no fee payable for a Listed Building Consent application. The target period for determination is eight weeks, but it can take much longer, particularly for work to Grade I and II* listed buildings, for the reasons explained above. Refusal of consent by local authorities is subject to a right of appeal to the Secretary of State for Housing, Communities and Local Government; most decisions are delegated to Planning Inspectors.

Listed Building Management Agreements can provide clarification as to what types of change to a listed building may or may not require Listed Building Consent. Such agreements do not have statutory weight, but are based on an informal consensus between all stakeholders, including the local planning authority, on the acceptability of (usually minor, repetitive) change within a listed building that does not affect its special character. Management agreements are particularly relevant to large, post-war listed buildings, where there is ongoing pressure for change. The first agreement was produced for the then Willis Faber Building in Ipswich. English Heritage published a study report, *Streamlining Listed Building Consent: Lessons from the Use of Management Agreements* (see **Further Reading**).

Curtilage structures

The Coach House, Audley End, Essex, is not listed in its own right, but since it is ancillary to Audley End House – a listed building – it is subject to listed building control.

Under section 60 of the 1990 Act, some places of worship can be exempted from listed building control. *The Ecclesiastical Exemption (Listed Buildings and Conservation Areas) (England) Order* (2010) limits this to places of worship actively in use by the Church of England, the Church in Wales, the Methodist Church, the Roman Catholic Church, the United Reformed Church and the Baptist Union. These bodies operate their own systems of control for work to their places of worship, and in return have 'ecclesiastical exemption' from the need to obtain Listed Building Consent (but not planning permission, nor approval under the *Building Regulations*). Where works are proposed to such a place of worship, an application must be made to the relevant denominational body. Applications will be referred to Historic England according to the same criteria as apply to other listed buildings. Ecclesiastical exemption generally applies only to buildings in active use for worship, not to associated, but separately listed, structures, such as churchyard monuments or walls.

For all but minor routine maintenance work to Church of England churches, a 'faculty' is required: this applies to all church buildings in active use for worship, not just listed buildings. After informal consultation, the parish must apply formally to the relevant Diocesan Advisory Committee [DAC], which then makes a recommendation (by way of a certificate) to the Diocesan Chancellor, to whom formal application (a 'petition') is then made for a faculty, on which Historic England and other interested parties may be consulted. DACs normally expect a high standard of conservation repair to historic churches. Anglican cathedrals have separate arrangements under the *Care of Cathedrals (Amendment) Measure* (2005).

The other exempted denominations operate similar systems of control; all other places of worship are subject to the need to obtain Listed Building Consent in the usual way. *The Operation of the Ecclesiastical Exemption: Guidance* (DCMS 2010) provides details about whom and what is exempt, and the relevant procedures.

WORKS TO SCHEDULED MONUMENTS

Legislation governing scheduled monuments was consolidated and updated in the *Ancient Monuments and Archaeological Areas Act 1979* (the *1979 Act*), which gives the Secretary of State for Digital, Culture, Media and Sport a power (rather than, as with listed buildings, a duty) to maintain a schedule of monuments of national interest: in doing so, the Secretary of State is informed by the expert advice of Historic England. Works to scheduled monuments are controlled by the Scheduled Monument Consent [SMC] regime, which is wholly separate from the planning process. *The Scheduled Monuments Policy Statement* (DCMS, March 2010) sets out the Government's policies in relation to the identification, protection, conservation and investigation of archaeological sites under the 1979 Act, and explains the basis on which monuments are designated and the requirements for SMC.

Scheduled monuments are not categorised by grade: they are all considered to be of national importance. However, not all nationally important archaeological remains are scheduled, because scheduling is primarily a management regime. Most scheduled monuments are archaeological sites, some of which include upstanding structural remains, but the schedule particularly includes defensive works of all periods, and medieval buildings and structures which are largely unused. Dwellings (except for caretakers' accommodation) and buildings in ecclesiastical use cannot be scheduled.

The most important practical difference between scheduling and listing in relation to repair is that virtually all works (including maintenance) require Scheduled Monument Consent. There are some minor exceptions for which consent is granted by the *Ancient Monuments (Class Consents Order)* (1994). For structures, these relate primarily to works by Canal & River Trust essential for the functioning of a canal. However, recurring maintenance works (which of course are essential to sustaining historic structures) can be permitted under a management agreement between the occupier and the Secretary of State or Historic England, which may provide financial assistance.

The schedule of ancient monuments is held by Historic England and in the regional *Historic Environment Records* [HERs], and will be online in the near future. HERs also hold archaeological data about scheduled monuments, and applicants for SMC should refer to the relevant HER. Applications for SMC must be made to the relevant regional office of Historic England, from which forms and guidance may be obtained. All applications for SMC are determined by the Secretary of State for Digital, Culture, Media and Sport, advised by Historic England. Consent will usually be conditional on appropriate impact assessment and recording by professional archaeologists. Irreversible changes should be avoided and intervention should generally be the minimum necessary to conserve the monument.

Any proposed repairs to a scheduled monument should be discussed with the regional office of Historic England at an early stage. As with listed buildings, undertaking unauthorised works to a monument is a criminal offence, but at least there is clarity as to the need for consent. Unless repairs to a scheduled monument constitute development, planning permission will not be required and the local planning authority will not be formally involved. Until the 2006 revisions, scheduled monuments were completely exempt from the *Building Regulations*. The revised *Building Regulation 21(3)* now states that they are exempt from the energy efficiency requirements of the *Building Regulations* where compliance would unacceptably alter their character or appearance (*Building Regulations, 2010*).

A small number of scheduled monuments are also listed: in such cases, scheduling takes precedence and only SMC is required. The anomaly of dual designation is gradually being removed in favour of the listing of buildings in use and the management of change through Listed Building Consent.

Dual designation

Some scheduled monuments are also listed buildings. In such cases, scheduling takes precedence and change is managed through Scheduled Monument Consent [SMC].

Clockwise from top: The Iron Bridge, Ironbridge in Shropshire; Hampton Court Palace, Surrey; Fairbairn Steam Crane, Bristol.

CONSERVATION BASICS
CURRENT LAW, POLICY & GUIDANCE

WORKS TO BUILDINGS IN CONSERVATION AREAS

Apart from specific controls over demolition, the management of the 'special architectural or historic interest' of conservation areas (which are designated under section 69 of the 1990 Act) is achievable principally through planning control of material alterations to the external appearance of buildings. Certain works that are normally exempt require planning permission. These include the demolition, in whole or part, of most buildings and structures, including: walls and outhouses; additions or alterations to the roof of a dwelling house (including dormers); cladding of external walls of a dwelling house; installation of satellite dishes; and the lopping or felling of trees.

Since, however, dwelling houses in single occupation have extensive 'permitted development rights' even in conservation areas, normal planning control is only effective in residential areas if those rights are curtailed by a direction made by the local authority under Article 4 of the *Town and Country Planning (General Permitted Development) Order* (1995) [GPDO] and subsequently enforced (see *Understanding Place: Conservation Area Designation, Appraisal and Management*; *English Heritage, March 2011*).

Conservation areas are 'designated heritage assets' under the terms of the *NPPF*, so any development or alterations to buildings within them needing planning permission should avoid or minimise harm to those aspects of the character or appearance of the area that warranted its designation. Applications are determined by local planning authorities (with the advice of Historic England, in cases of demolition and major development), and must have regard to the *NPPF* and adopted local policy. Planning applications are made to the local authority in the usual way, but if the work involves demolition of a building, conservation area consent is also required. Applications will normally need to include additional information proportionate to the impact of the works on the significance of the building or place and their setting.

Permitted development rights

Unsympathetic replacement window installed in an unlisted dwelling in a conservation area under 'permitted development rights'. This is an all-too-common sight in conservation areas where permitted development rights have not been curtailed by 'Article 4 direction'.

Unlisted buildings which are considered by the local planning authority to make a positive contribution to the character or appearance of a conservation area are 'non-designated heritage assets' for purposes of the *NPPF*. Such buildings can be identified, for example, in a published character appraisal of the conservation area, or by being 'locally listed' by the local authority. The relevant policies need to be applied to development proposals affecting them, or their setting (policy 135).

In most cases, however, conservation repair alone (like-for-like repair in original or matching materials), rather than any associated alteration or reinstatement, will not normally require planning permission, even in a conservation area. Statutorily listed buildings, of course, remain protected by the separate Listed Building Consent regime (see **Works to Listed Buildings**) whether or not they are in a conservation area.

GUIDANCE

BS 7913:2013 GUIDE TO THE CONSERVATION OF HISTORIC BUILDINGS

"This British Standard is intended for those who own, use, occupy and manage historic buildings, the professional team's contractors and others employed to work on their behalf, and can be used by decision makers and funders. It is intended to provide building owners, managers, archaeologists, architects, engineers, surveyors, contractors, conservators, planners, conservation officers and local authority building control officers with general background information on the principles of the conservation of historic buildings and sites, when setting conservation policy, management strategy and procedures."

The British Standard Guide represented an important step forward in formulating general principles and expounding good professional practice that were separate from, and more comprehensive than, national planning policy, but compatible with it. The Guide recognises the importance of the aesthetic values of historic buildings, whether designed or fortuitous, as well as the historical and archaeological values of their fabric, and emphasises the importance of understanding the building prior to works and recording further information discovered during them. Criteria are set out for judging the acceptability of proposals for repair, restoration, alteration, addition and new work in historic settings and areas. While seeking to minimise intervention in historic fabric, the Guide urges quality in new work.

One aim of the Guide was to prompt statutory authorities to adopt consistent approaches to the conservation of historic buildings. In the event, its influence was and is limited, primarily because the British Standards Institution [BSI] needs to cover its costs through a restrictive approach to copying and a high purchase price. An updated edition of the 1998 British Standard was published in 2013.

Working with British & European Standards

A standard is a published document containing precise criteria, designed to be used as a rule, a guideline or a definition, which presents an agreed and repeatable way of doing something. They are 'achievable' rather than 'best' practice; in many cases, best practice is significantly more rigorous.

In Britain, standards are produced by the British Standards Institution [BSI], and nearly all carry the prefix 'BS'. British Standards are drafted by committees comprising manufacturers, consumers, government departments and research organisations. All European Standards must be adopted by the BSI, and these are prefixed by *BS EN*. Adoption of other international standards is optional, but in practice most are incorporated into the BSI, and are published with the prefixes *BS ISO* [International Organisation for Standardisation] or *BS IEC* [International Electrotechnical Commission].

It is not obligatory to follow British Standards, but laws, regulations and contracts may refer to certain standards, and these may include clauses that make compliance compulsory.

There are six types of British Standard:

- Specifications list the detailed requirements that products, services or processes must satisfy.
- Methods give a complete account of the way in which something should be done, and a list of the necessary tools or equipment.
- Guides give background and general information about a subject (an example is *BS 7913:2013 Guide to the conservation of historic buildings*).
- Vocabularies give definitions of the terms used in a particular field.
- Codes of Practice give recommendations for acceptably good practice (for example, *11/30241242 DC BS 8221–1. Code of practice for cleaning and surface repair of buildings. Part 1. Cleaning of natural stones, brick, terracotta and concrete*).
- Classifications give descriptions of and designations for different grades of product or material.

Very few standards relate specifically to conservation, and many standards can prove inappropriate or impossible when applied to historic building materials or systems. In this case historical evidence of durability and performance usually provide a more convincing argument for choosing particular methods or materials, but British Standards may still be useful as a reference point for developing a suitable specification for particular projects.

HISTORIC ENGLAND: CONSERVATION PRINCIPLES, POLICIES & GUIDANCE FOR THE SUSTAINABLE MANAGEMENT OF THE HISTORIC ENVIRONMENT [ENGLISH HERITAGE 2008]

The background to this document (*Conservation Principles*) is explained in the preceding chapter. Its stated aim is to *"set out a logical approach to making decisions and offering guidance about all aspects of the historic environment"*, and it was intended to help Historic England (formerly English Heritage) *"ensure consistency of approach in carrying out [its] role as the Government's statutory advisor on the historic environment in England"*.

Conservation Principles sets out an intellectual framework for making decisions about change to significant places, which applies as much to decisions about the repair and maintenance of historic buildings as it does to those concerning major change affecting the wider historic environment. The process involves a step-by-step approach, enabling users to arrive at logical, defensible responses to complex issues. It combines the definition of six core principles with more detailed policies and guidance on their application when making decisions about changes to the historic environment. Under Principle 4, conservation is defined as *"the process of managing change to a significant place in its setting in ways that will best sustain its heritage values, while recognising opportunities to reveal or reinforce those values for present and future generations"*. It is explained that, as used in *Conservation Principles*, the term 'place' means *"Any part of the historic environment, of any scale, that has a distinctive identity perceived by people"*.

Historic England Conservation Principles

1. The historic environment is a shared resource

2. Everyone should be able to participate in sustaining the historic environment

3. Understanding the significance of places is vital

4. Significant places should be managed to sustain their values

5. Decisions about change must be reasonable, transparent and consistent

6. Documenting and learning from decisions is essential

Conservation Principles identifies and elaborates on a range of heritage values, grouped under the headings of evidential, historical, aesthetic and communal. These are considered in more detail in the following chapter. Under Principle 3, the 'significance' of a place is defined as embracing *"all the diverse cultural and natural heritage values that people associate with it, or which prompt them to respond to it. These values tend to grow in strength and complexity over time, as understanding deepens and people's perceptions of a place evolve."* The idea of significance and how it is evaluated is discussed in the following chapter.

The final section of the document sets out practical policies and guidance for evaluating and advising on proposals involving heritage places, ranging from routine management, maintenance, renewal and repair, to changes intended to increase knowledge, restorations, new works, and enabling development. While produced primarily to guide Historic England (formerly English Heritage) staff, *Conservation Principles* was subject to wide public consultation, and its adoption and application by the sector has been encouraged (not least by the document being freely available).

Although its language may still be unfamiliar, *Conservation Principles* nevertheless builds on earlier statements and experience in formalising an approach to assessing the impact of change that takes account of a wide range of heritage values. The overarching message of *Conservation Principles* is that *"balanced and justifiable decisions about change in the historic environment depend upon understanding who values a place and why they do so, leading to a clear statement of its significance and, with it, the ability to understand the impact of the proposed change on that significance"* (Conclusion, para.161). As previously noted, national policy on deciding planning permissions, listed building and conservation area consents is set out in the *National Planning Policy Framework* [*NPPF*] (March 2012). It defines 'sustainable development' as development that adheres to the principles and policies within the document, which as far as the heritage assets are concerned are materially the same as those set out in PPS5, its predecessor.

The *NPPF* defines 'heritage assets' as buildings, sites and areas that have heritage 'significance'. This is defined as: *"The value of a heritage asset to this and future generations because of its heritage interest. That interest may be archaeological, architectural, artistic or historic. Significance derives not only from a heritage asset's physical presence, but also from its setting."* This broad definition covers all the cultural significance of a place and is coextensive with the definition of 'significance' within *Conservation Principles*, which is the sum of the 'heritage values' of a place.

During the 1980s, historic environment policy began to be seen as a linked but distinctly separate strand from land-use planning policy. In the first decade of the 21st century, this gap started to close. This is likely to be an ongoing trend, at least for the next few years. Currently, the national planning policy framework clearly places the historic environment as just one of the many elements that need to be factored into planning decisions. Economic, social and environmental considerations are now all taken into account when forming planning policy for sustainable development. At present, economic considerations are being given greater emphasis, as the reality of the global financial crisis still continues to be felt. Too much emphasis on any of the three elements at the expense of the others will not result in long-term, successful and appropriate solutions, and heritage 'outrages' are likely to occur.

With a likely continued squeeze on funding for the public sector, the decline in the number of specialist historic staff in local authorities and heritage agencies seems set to continue. While this might limit the influence of the historic environment sector in the wider planning world, not-for-profit and voluntary groups are growing in size and influence, and they are likely to fill a part of the void, if not all of it. The so-called 'Third Sector' (the voluntary and not-for-profit sector) is already a powerful lobbying force that demonstrably influences elements of policy, helping to set and refine parameters for historic environment policy in England. This is likely to continue in the coming years. In fact, the first few years of the 21st century may well be seen with the benefit of hindsight to be the 'golden age' of public sector historic environment intervention. The private sector has always had by far the largest stake in the historic environment for a simple reason: it owns the vast majority of it. The Third Sector has become increasingly willing to work directly with the private sector without the intervention of the public sector. This may well permanently change the dynamic of the relationships between the key players in the historic environment.

Whether the current Government's 'Localism' drive for community-level planning becomes the default level for planning remains to be seen. Community-level planning might become an important, but limited issue, concerning mainly areas undergoing major change, or ones with residents with the time, resources and willingness to become fully engaged in planning their neighbourhood. The debate around who makes decisions about the future of heritage assets that are of more than local importance will never be wholly resolved. Owners, local communities, local, regional and national special interest groups, national politicians and technocrats and, on occasion, international interests all have a stake. It will be the mark of a highly robust policy if it manages to balance all these interests, while continuing to conserve and enhance our historic environment.

FITNESS FOR PURPOSE & SAFETY

In addition to legal requirements for heritage conservation, historic buildings are also subject to a range of building and other legislation that is concerned, mainly, with fitness for purpose and safety.

RECONCILING LEGISLATION WITH SIGNIFICANCE

Meeting the requirements of legislation concerned with functional performance and safety in historic buildings, in a way that avoids diminishing their significance, often involves ingenious and bespoke solutions developed in close consultation with controlling authorities. In some cases, it may be desirable to appoint a specialist consultant to assist in this process; for example, a fire engineer who will develop an alternative fire safety strategy for a building. In general terms, the aim should be to achieve an appropriate balance between meeting the functional requirements of the legislation to a reasonable level while sustaining heritage values. This process requires not only a thorough understanding of the purpose of legislation and the heritage values and significance of the building, but also knowledge of the building as a physical entity (structure and fabric) and how it behaves as an environmental system. In practical terms, deciding what is appropriate or reasonable involves considering a range of factors, including the likely effectiveness of measures, their impact on heritage values, the associated technical risks (for example, their adverse effects on the durability of the building fabric), whether the measures are reversible, and whether there are overriding environmental or community benefits to the proposals.

THE BUILDING REGULATIONS & RELATED PROVISIONS

The *Building Regulations* set standards for the construction of buildings to achieve a minimum level of acceptable performance. The regulations apply mainly to new buildings and there is no general requirement for existing buildings to be upgraded to meet these standards. However, certain changes, such as alteration, extension or changes of use, can trigger the need for existing buildings to be upgraded to comply, and related legislative requirements concerned with the use of buildings (also considered here) can prompt the need for physical change. Advice should always be sought from the building control provider about whether or not proposed works are subject to control. The regulations themselves are a statement of performance requirements and are not prescriptive. *Approved Documents* give practical guidance on how compliance with the functional requirements of the *Building Regulations* can be demonstrated. That historic buildings may require alternative approaches to meeting requirements is recognised in the *Approved Documents* for some parts of the building regulations.

Approved Document B (Fire Safety) allows for variation of the provisions in historic buildings where adherence to the guidance might be unduly restrictive. In such cases, it advises that it would be appropriate to take into account a range of fire safety features, and to set these against an assessment of the hazard and risk peculiar to the particular case.

Approved Document B also acknowledges that fire safety engineering can provide an alternative approach to providing reasonable levels of fire safety. This approach, based on site-specific risk assessment, can be beneficial in historic buildings (see the box on **Fire Safety in Historic Buildings**). It often allows greater emphasis to be placed on active fire protection measures, such as automatic fire detection, and suppression systems and fire safety management, to avoid, or at least minimise, disruptive upgrading of building fabric.

The *Regulatory Reform (Fire Safety) Order 2005* applies to the use of all premises except single private dwellings (previous legislation applied to designated uses of premises such as hotel, office, shop, factory and railway premises). Fire risk assessments, carried out by the 'Responsible Person', are required for all premises falling within the scope of the Order. These are used to highlight any steps that are necessary to ensure the safety of relevant persons (employees, contractors, visitors and other occupants). A series of 12 guides has been produced to help the 'responsible person' undertake their fire risk assessments in various uses of premises, such as offices and shops, places of assembly, factories and warehouses, and educational establishments. The guidance shows ways of complying with the legislation and is not intended to be prescriptive. All of the guides contain an appendix referring to the special needs of historic buildings.

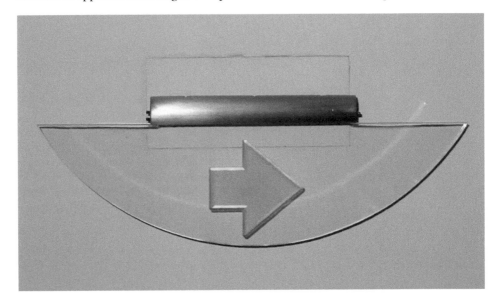

A site-specific, fire engineering approach can be helpful in devising strategies which meet fire safety requirements with minimal impact on heritage values. Here, directional signage and emergency lighting – to illuminate a means of escape in the event of power failure – have been designed to be as unobtrusive as possible. Although a non-standard exit sign, it nevertheless clearly indicates the way out.

Fire Safety in Historic Buildings

USE OF THE 'BUILDING FIRE PERFORMANCE METHOD'

Where adherence to guidance would adversely affect the significance of a historic building, a fire safety engineering approach can be helpful in identifying alternative, less disruptive, strategies for meeting requirements to provide reasonable levels of fire safety. The 'Building Fire Performance Method' is a way of organising the complex and interactive network of fire dynamics, potential active and passive fire safety systems, manual intervention and timescales into a logical framework. It enables the effect of fire in a specific building to be visualised, taking into account all the relevant factors involved in fire spread. There are five distinct components to the building fire performance evaluation:

- *'Prevent established burning' analysis*
  This is concerned with basic fire prevention: eliminating ignition sources, keeping combustibles to a minimum, and having fire and fire safety management procedures in place. An understanding of the way fire spreads is necessary to allow this analysis to be reasonably accurate.

- *'Flame and heat hovement' analysis*
  This assumes that a fire has reached the stage of established burning and evaluates the building's existing fire safety potential to prevent further spread beyond the room of origin. Established burning is the point at which the fire prevention activities have failed and the building's fire safety design becomes critical.

- *'Smoke and toxic gas' analysis*
  The smoke and toxic gas analysis also assumes that the fire has reached the established burning stage. It assesses the ability of the building to keep escape routes available and safe during the evacuation period and beyond.

- *'People movement' analysis*
  This balances the time required for occupants to escape against the time taken for the escape routes to become hazardous.

- *'Structural frame' analysis*
  This analysis looks at the ability of the building structure to remain stable during a fire and to resist deformation or collapse.

The 'Building Fire Performance Method' is ideal for historic buildings because it enables site-specific solutions to be identified, based on the assessment of risks, and provides the basis for informed discussions about alternative solutions. Solutions can be compared for their effect on the fire performance of the building, and their impact on its heritage values and significance. A qualified fire safety engineer, experienced in working with historic buildings, should be engaged to carry out the evaluation and to make recommendations on possible solutions.

Proposed building works to a historic building will be subject to control under *Part A (Approved Document A – Structure)* of schedule 1 of the *Building Regulations* if they involve 'building work' as defined in Regulation 3. This includes underpinning, new work and 'material alteration'. Compliance is also required where certain classes of change of use of a building are proposed. The point at which repair becomes new work or a material alteration is a matter of degree and is to some extent open to interpretation. For example, the renewal of a pair of damaged or defective rafters in a roof could reasonably be regarded as repair. However, if more than half the rafters in the roof require replacement, perhaps involving temporary propping and shoring, this would most likely be considered as 'building work' requiring approval under the *Building Regulations*. Similarly, where replacement roof coverings are proposed which are heavier than the existing, or where additional members are needed to strengthen the roof structure (for example, purlins, beams and struts), this will be deemed a 'material alteration'. Similar principles apply to other building elements.

RESISTANCE TO THE PASSAGE OF SOUND

Certain material changes of use of a building (which are defined in Regulation 5 of the *Building Regulations 2010*) are subject to new regulatory requirements (set out in Regulation 6) which include, among other things, control under *Part E: Resistance to the passage of sound*. However, the current *Approved Document* acknowledges that in historic buildings it may not be practical to improve sound insulation to the standards set out. In such cases, an approach is allowed in which the sound insulation is improved to the extent that is practically possible without harming the significance of the building, or increasing the risk of long-term deterioration of building fabric or fittings. Section 4 of *Approved Document E* gives guidance and details of treatments that give reasonable levels of sound insulation in dwelling houses and flats formed by material change of use.

Radon

Radon is a clear, odourless radioactive gas produced by the decay of uranium, which is found in low concentrations in most soils, particularly over granite and limestone bedrock. It produces radioactive dust in the air and increases the risk of lung cancer.

If levels of radon exceed the 'action level' in radon-affected areas, remedial measures (either basic or full) will be a requirement under the *Building Regulations Part C*, where existing buildings are subject to alteration, extension or change of use. ⊕ENVIRONMENT ⊕MORTARS

Further information can be obtained from the Health Protection Agency: *www.ukradon.org/* and the BRE: *www.bre.co.uk/radon/*

VENTILATION

Approved Document F (Ventilation) requires that people in a building are provided with adequate means of ventilation. In historic buildings it will most often be the replacement of windows that triggers the need for compliance with Part F (along with Part L, and in some cases Part N). *Approved Document F (Ventilation)* gives guidance on the provision of background (trickle) vents in windows. However, if the window being replaced does not contain background (trickle) ventilation, there is no requirement for the new one to include such provision.

The approved document also gives guidance on the extent of purge (formerly rapid) ventilation in replacement windows. In meeting requirements, consideration should be given to the 'whole house' ventilation concept. For example, if a room has opening windows on two opposing walls, these will be more effective than a single window of greater proportions in one wall only. Also, a room with a working flue is likely to have more than adequate ventilation, even if the area of the opening lights in the windows is less than the norm. However, if the window being replaced is one that is required as a fire-escape window under Part B (*Fire Safety*), then very careful consideration must be given to the openable area, especially if there is no alternative escape provision in the dwelling.

CONSERVATION OF FUEL & POWER

Approved Document L (Conservation of Fuel and Power) applies to existing buildings where substantial alterations to thermal elements or controlled fittings or services are to be carried out, or where extensions or changes of use are proposed. However, listed buildings, buildings in conservation areas and scheduled monuments are exempted from the need to comply with the energy efficiency requirements of the *Regulations* where, and to the extent that, compliance would unacceptably alter their character or appearance. Special considerations apply to buildings that are locally listed, or in national parks or other designated historic areas, and buildings that are of traditional construction. Special considerations also apply to buildings used primarily or solely as places of worship. In these classes of buildings, the requirement is to improve energy efficiency up to, but not beyond, the point that the character and appearance of the building is unacceptably altered, or the long-term durability of the building fabric or fittings is reduced. English Heritage (now Historic England) has published detailed 'second tier' guidance on the *Building Regulations Part L*, which the *Approved Document* states should be taken into account in determining appropriate energy performance standards for building work in historic buildings (see **Further Reading**).

GLAZING

Approved Document N: Glazing – Safety in relation to impact, opening and cleaning outlines safety requirements concerning glazing, and the operation and cleaning of windows. It notes that the *"installation of replacement windows or glazing (e.g. by way of repair) is not building work…"* and therefore not subject to Part N requirements. However, the need to meet the requirements of Part N may be brought about by a change of use of the building.

Approved Document M (Access to and Use of Buildings) applies to existing buildings when they are materially altered or extended. It also recognises the need to conserve the special characteristics (that is, significance) of historic buildings. It advises that, in such cases, the aim should be to improve accessibility to the extent that measures do not prejudice significance or increase the risk of long-term deterioration.

The *Approved Document* recommends striking a balance between conservation and accessibility, and promulgates an approach which takes into account the views of local access groups and heritage authorities.

The *Equality Act 2010* replaces major parts of the *Disability Discrimination Act*, which was intended, among other things, to facilitate dignified access for disabled people to services. To meet requirements, the service provider has to take reasonable steps to remove physical features that impede access, or, where this would involve unacceptable alterations to a historic building, to provide reasonable alternative methods of making the services accessible to disabled people.

English Heritage (now Historic England) has produced guidance on approaches to meeting these requirements (see **Further Reading**).

A lift provided for wheelchair users at St Alban's Cathedral.

Thermal Performance of Traditional Buildings

The basic purpose of a building is to moderate the extremes of climate to create comfortable spaces in which to live, work and play. Its external envelope provides shelter from rain, wind and snow, and shade from the sun. It also keeps in warmth, and regulates the entry of light and air. There is a continual and complex interaction between the building envelope, heat and moisture that affects both the internal environment and the behaviour of the building fabric. Traditional buildings incorporate elements and materials that influence this interaction in particular ways. For example, the high thermal mass of solid masonry walls and chimney breasts act as a heat store, moderating fluctuations in temperature, and helping buildings feel warmer in winter and cooler in summer. Also, most traditional buildings incorporate permeable materials that are able to absorb and release moisture freely, thereby buffering short-term fluctuations in humidity and reducing the risk of condensation. In addition, cellular room plans regulate the movement of air and heat within the building.

Maintaining the permeability of the fabric is fundamental to the thermal performance of a traditional building, its continuing well-being and the health of its occupants. Unfortunately, in a large number of traditional buildings, this ability to absorb and release moisture freely will have been compromised by adaptations and repairs carried out with the best of intentions, but using inappropriate materials such as cement-based mortars and renders, plastic paints and impermeable waterproofing treatments. Moisture trapped by ill-advised repairs increases the rate at which building fabric deteriorates, and adversely affects thermal performance. This gives a poor impression of the durability of traditional buildings and their ability to provide a comfortable internal environment. Energy efficiency is also diminished in buildings that are poorly maintained. For example, gaps around worn windows and doors, open joints in masonry or between infill panels and timber framing allow uncontrolled air infiltration that significantly increases heat loss. A building that is in a good state of repair can perform to its full potential, and recent research has indicated that the level of performance of traditional buildings is usually much better than assumed.

When considering ways of improving energy efficiency in heritage buildings, the aim should be to achieve an optimal balance between energy conservation and building conservation. To this end, a strategic 'whole building' approach, based on conservation planning principles, is recommended. This is a logical process that entails the following steps:

- Understanding the:
  - heritage value and significance of the building
  - condition of the building fabric and building services
  - current energy use related to building occupation and behavioural factors
  - energy performance of the building envelope and building services
  - hygrothermal behaviour of the building fabric (interaction with heat and moisture)
  - user requirements and aspirations

- Identifying opportunities, and devising measures to improve energy conservation and sustainability

- Evaluating the effectiveness and value for money of such measures

- Assessing the impact of measures on heritage values and significance

- Assessing the technical risks (for example, increased risks of condensation)

- Implementing the optimum strategy

- Monitoring the outcome (for example, post-occupancy evaluation).

This approach helps to identify the package of measures best suited to an individual building or household, taking human behaviour into consideration as well as the performance of the building envelope and services. Inexpensive benign improvements should be carried out before more invasive and potentially harmful measures are considered. In practical terms this incremental approach involves:

- ensuring the building fabric and services are in good repair and properly maintained to obtain the optimum performance from them
- introducing energy-efficient housekeeping, such as closing existing window shutters at night, closing doors, turning off lights and electrical equipment when not in use, avoiding overfilling kettles, taking shorter showers, and so on
- making benign enhancements, such as improving the efficiency of lighting and heating systems, using heavy curtains and thermal blinds to reduce heat losses, or providing 'intelligent' controls for building services
- controlling draughts – particularly through flues, doors and windows (unless secondary glazing is proposed) – to reduce excessive air infiltration throughout the building
- considering low-risk insulation opportunities, such as loft (ceiling-level) and floor insulation and secondary glazing
- considering the provision of energy from renewable sources
- once other improvements have been explored or implemented, considering more expensive and higher-risk strategies such as wall insulation.

Historic England has published detailed information on energy efficiency in historic buildings (see **Further Reading**), and guidance is also available from *www.historicengland.org.uk/advice/technical-advice/energy-efficiency-and-historic-buildings/*
The other volumes in this series also address energy issues, particularly the *Building Environment* volume. ⊖ENVIRONMENT ⊖ROOFING ⊖GLASS ⊖TIMBER ⊖MORTARS

Left: Evacuated-tube solar thermal panels mounted horizontally on the flat roof of a Victorian terraced house. In this configuration their visual impact is minimal but their output is reduced. Careful consideration should be given to the method of fixing the panels, and the condition and maintenance requirements of the roof coverings beneath.

Right: Secondary glazing fitted to a mid-19th century tripartite sash window. Well-designed secondary glazing need not adversely affect the heritage values of a historic building and is reversible.

Ladders may be used for simple tasks of short duration (such as inspections and cleaning debris from gutters). A risk assessment is required to justify the suitability of a ladder as opposed to other access equipment options.

HEALTH & SAFETY AT WORK

A range of health-and-safety legislation exists that may affect the conservation of historic buildings. This includes the *Workplace (Health, Safety and Welfare) Regulations 1992* and amendments, which cover a wide range of basic health, safety and welfare issues, and apply to most workplaces. The regulations include requirements for workplaces to be maintained in efficient working order (for health, safety and welfare), and that safe routes are provided for circulation in and around the building. In practical terms, there are seldom major conflicts between this legislation and building conservation. However, there may be instances where problems arise, for example, when there is the need to eliminate trip hazards on historic floors or paved areas. Other requirements concerning windows (including cleaning), doors and stairways may affect the conservation of these elements, particularly when a building is brought within the scope of this legislation due to a change of use to a workplace.

Other relevant health-and-safety legislation covers contact with hazardous materials, such as asbestos, and working at heights. The *Construction (Design and Management) Regulations 2007 (CDM Regulations)* makes health, safety and welfare requirements in respect of building operations. The implications of this legislation are discussed in the **Survey & Investigation Methods** and **Managing Maintenance & Repair** chapters of this book. The Health and Safety Executive [HSE] publishes a wide range of detailed guidance (see **Further Reading**).

EUROPEAN UNION DIRECTIVES & STANDARDS

As a member state of the European Union [EU], the UK is obliged to abide by its rules, conform to all *European Commission* [EC] *Regulations*, and incorporate the requirements of any EC Directives into its own legislation. Building conservation is affected by EU Directives, mainly in areas such as general building regulations, procurement, materials standards and restrictions on the availability or use of some traditional building materials, like paint containing white lead, and these do not always take account of heritage impacts. In 2003, a Working Group on EU Directives and Cultural Heritage was set up to research, document, monitor and communicate the consequences of Directives for cultural heritage, with the hope that any adverse effects could be mitigated either at the drafting stage, or when implemented in national law.

By 2007, some 25 directives had been identified and reviewed by the Working Group, in subject areas ranging from restrictions on biocidal and hazardous products, fire safety regulations, and energy efficiency and performance in buildings, including work relating to *Directive 99/13* on the limitation of emissions of volatile organic compounds in paints [VOCs], and *Directive 98/8/EC* on biocidal products. Member states, including the UK, have also been able to enact derogations within their own legislation that specifically address cultural heritage assets and permit use of restricted materials. The Working Group has pressed for a standard 'clause of special consideration' for cultural heritage, which would support similar inclusions by member states in developing national laws.

THE EUROPEAN COMMITTEE FOR STANDARDISATION

Acknowledging the lack of broad standards for European practice, in 2004 the European Committee for Standardisation (*Comité Européen de Normalisation*) [CEN] began a project to formulate pan-European standards for the conservation of cultural property (*CEN/TC 346 Conservation of cultural property* (11 parts; standards under development)). The aim is *"to acquire a common unified scientific approach to the problems relevant to the preservation/ conservation of the Cultural Heritage. Moreover, this common approach and the use of standardised methodologies and procedures would promote the exchange of information, would avoid the risk of duplication and foster synergy between the European experts and specialists involved in the preservation activity."*

The CEN project aims to draft standards in eight areas, including terminology, materials characterisation, and the evaluation of conservation product performance and environmental conditions. Within the UK, input is being coordinated through the BSI Technical Committee for Conservation. Unsurprisingly, given the vast array of subject areas, and the need for significant volunteer input from specialists in participating countries, development has been slow and sporadic. Current progress can be tracked at the BSI website.

Keeping Up-to-Date with Changes in Legislation & Regulations

Changes in legislation and regulations are inevitable, and it is important that conservation practitioners keep up-to-date with current statutory requirements. There are several ways of doing this, including subscribing to automatically updated reference publications or to online information-handling services. Information may also be obtained from the websites of the relevant government departments, and agencies such as The Health and Safety Executive. The 'Planning Portal' provides online access to information and guidance about planning legislation and building regulations, including *Building Regulation Approved Documents.* Official guidance publications are generally available as online downloads, free of charge. Other sources of information include professional and trade journals, and professional conferences or courses organised by professional institutions, universities and other training organisations.

A list of useful sources of information about legislation is given in **Further Reading** at the end of this chapter.

Further Reading

Department for Communities and Local Government (2006); *Regulatory Reform (Fire Safety) Order 2005: A Short Guide to Making your Premises Safe from Fire*; Wetherby: DCLG Publications; available from *www.gov.uk/government/publications/making-your-premises-safe-from-fire*

Department for Digital, Culture, Media and Sport (2013); *Scheduled Monuments and Nationally Important but Non-scheduled Monuments*; London: DCMS; available from *www.gov.uk/government/publications/scheduled-monuments-policy-statement*

Department for Digital, Culture, Media and Sport (2018); *Principles of Selection for Listing Buildings* (web document); available from *www.gov.uk/government/publications/principles-of-selection-for-listing-buildings*

Health and Safety Executive (2013); *Workplace Health, Safety and Welfare: Workplace (Health, Safety and Welfare) Regulations 1992: Approved Code of Practice and Guidance*; Sudbury: HSE Books; available from *www.hseni.gov.uk/publications/l24-workplace-health-safety-and-welfare-gb-acop-approved-use-ni*

Health and Safety Executive (2015); *Managing Health and Safety in Construction: Construction (Design and Management) Regulations 2015: Guidance on Regulations*; Sudbury: HSE Books; available from *www.hse.gov.uk/pUbns/priced/l153.pdf*

Historic England (2008); *Conservation Principles, Policies and Guidance for the Sustainable Management of the Historic Environment*; London: Historic England; available from *www.historicengland.org.uk/images-books/publications/conservation-principles-sustainable-management-historic-environment/*

Historic England (2015); *Easy Access to Historic Buildings*; Historic England; available from *www.historicengland.org.uk/images-books/publications/easy-access-to-historic-buildings/*

Historic England (2017); *Listing Selection Guides* series (web document); available from *www.historicengland.org.uk/listing/selection-criteria/listing-selection/*

Historic England (2017); *Energy Efficiency and Historic Buildings: Application of Part L of the Building Regulations to Historic and Traditionally Constructed Buildings*; Historic England; available from *www.historicengland.org.uk/images-books/publications/energy-efficiency-historic-buildings-ptl/*

Historic England (2019); *Conservation Area Appraisal, Designation and Management Second Edition, Historic England Advice Note 1*; Swindon: Historic England; available from *www.historicengland.org.uk/images-books/publications/conservation-area-appraisal-designation-management-advice-note-1/*

Historic England (n.d.); *Listed Buildings* (web page); available at *www.historicengland.org.uk/listing/what-is-designation/listed-buildings/*

Ministry of Housing, Communities and Local Government (2019); *National Planning Policy Framework*; available from *www.gov.uk/government/publications/national-planning-policy-framework--2*

Mynors, C. (2006); *Listed Buildings, Conservation Areas and Monuments* (4th edition); London: Sweet & Maxwell

Useful Web Addresses

Health and Safety Executive (n.d.); *Guidance* (web page); available from *www.hse.gov.uk/guidance/index.htm*

Planning Portal (2012); *Building Regulations* (web page); available from *www.gov.uk/building-regulations-approval*

CONSERVATION PLANNING FOR MAINTENANCE & REPAIR

This chapter explains the principles of conservation planning, and especially how understanding the significance of a historic building or place can help to guide decisions about maintenance and repair. Subsequent chapters address investigative techniques, and the structure and management of conservation projects in practical terms.

"Conservation is the process of managing change to a significant place in its setting in ways that will best sustain its heritage values, while recognising opportunities to reveal or reinforce those values for present and future generations."

Historic England, *Conservation Principles* (2008), Principle 4.2

Decisions about the conservation of a heritage asset – a historic building or place – should always be based on a thorough understanding of its cultural significance (the sum of its heritage values), its physical condition, and how its significance may be vulnerable to physical deterioration or other threats. Developing this understanding, then using it as the basis for devising conservation policies or strategies to sustain the significance of the heritage asset, is part of a process known as conservation planning. This process is not new, but rather formalises what good conservation practitioners have always done instinctively. Experienced conservation practitioners will often develop a 'feel' for where the values of a historic building or place lie, even if they are not required to define them formally. Some form of written statement is usually desirable, however, most commonly in the form of a 'conservation plan'. Conservation plans generally cover the following basic concepts:

- understanding (of the heritage asset and its context, if relevant)

- evaluation of significance (of the asset as a whole and of its elements)

- identification of conservation issues (including potential conflicts and how significance is vulnerable to harm)

- policies and strategies to sustain significance.

The benefit of this structured approach is that it provides a record of the way conclusions about significance, issues and strategies have been reached. Also, it can help in the process of decision-making, particularly where complex buildings or places are concerned, by providing a firm basis on which alternative solutions or approaches can be tested and particular proposals justified.

THE IDEA OF SIGNIFICANCE

Principle 3.2 of Historic England's 2008 *Conservation Principles* states: "*The significance of a place embraces all the diverse cultural and natural heritage values that people associate with it, or which prompt them to respond to it. These values tend to grow in strength and complexity over time, as understanding deepens and people's perceptions of a place evolve.*"

There are many reasons why people value historic buildings or places beyond practical usefulness or personal association. They help us to understand aspects of the past and add to the quality of our present environment. We also expect that they will be valued by future generations. All types of historic building, structure or place, from scheduled monuments to historic gardens, are now described generically as 'heritage assets'. The significance of a particular heritage asset is the sum of the heritage values that people ascribe to it. These values usually stem from its physical substance – its fabric – and its setting, but can be reinforced by artefacts and documents associated with it. They lie also in the meanings it has for people through use, association and commemoration.

The values that make heritage assets significant are wide-ranging and interrelated. Historic buildings and places embody the material remains of past cultures that illustrate many aspects of the way people lived. For example, they may provide insights into developments in construction technology, reflecting distribution of materials, skills, ideas, knowledge, money and power, in particular localities and at particular points in time. Buildings and places can connect us intellectually and emotionally with past events and people: they may also provide aesthetic pleasure. In addition, they create a shared sense of place and identity for people who relate to them, or for whom they are part of a collective experience or memory. These values are not inherent or intrinsic to the building fabric, but are attributed to the building or place by people. Significance – the sum of the values we attach to places – therefore is mutable and may change over time, as may the relative importance we attribute to those values.

Visitors to Stonehenge observing the summer solstice. Monuments mean different things to different people. Understanding the values that are ascribed to a place is fundamental to conservation planning.

UNDERSTANDING SIGNIFICANCE

Understanding the significance of a building or place and the values that contribute to it is fundamental to the conservation planning process, and is vital when considering approaches to repair or other forms of change. In addition to evident architectural and aesthetic values, there are the less tangible attributes to consider. Historic England's *Conservation Principles* identifies four primary categories of heritage value.

EVIDENTIAL VALUE

Evidential value derives from the potential of a place to yield significant evidence, usually from physical remains, about past human activity. Physical remains provide the sole source of information about undocumented aspects of the past, making age a strong indicator of relative evidential value. It is frequently associated with the research potential of archaeological deposits. All buildings encapsulate unique information about their own evolution; for example, of their evolving form or design, chronology, or decorative schemes, whether visible or having the potential to be revealed through research. The relative evidential value of a building tends to be related to its historical and aesthetic values.

Evidential value

The evidential value of archaeological remains is equally important, whether they are below the ground or within the building structure and fabric above the ground.

HISTORICAL VALUE

Historical value stems from the ways in which the present can be connected by a place to people, events and aspects of life in the past. This may be illustrative, by demonstrating important facets of past lives and helping understand the historic environment, or it may be associative, through being linked to a notable historical person or event.

Historical value

Top: Herne Windmill, Kent, illustrates aspects of agriculture, technology and rural life in the past.

Bottom: Mendips, the childhood home of John Lennon in Woolton, Liverpool, is of historical value through its association with the life and early work of the internationally renowned musician and band-member of The Beatles.

CONSERVATION BASICS
CONSERVATION PLANNING FOR MAINTENANCE & REPAIR

AESTHETIC VALUE

Aesthetic value derives from the ways in which people draw sensory and intellectual stimulation from a place. This includes fortuitous qualities which have evolved naturally in a place over time; including the effects of weathering and the patina of age, as well as design values attached to a deliberately created building, group of buildings or landscape.

Aesthetic value

Top: Formal aesthetic value. The Ranger's House, Blackheath, London, was built in 1690 and extended in the mid-18th and early 19th centuries. Each phase and the resulting whole are the outcome of deliberate architectural intentions.

Bottom: Fortuitous aesthetic value. Many streetscapes in historic towns owe their aesthetic value to centuries of piecemeal but harmonious development.

COMMUNAL VALUE

Communal value stems from the meanings of a place for the people who relate to it, or for whom it is part of a collective experience or memory: a shared cultural frame of reference. This can include commemorative and symbolic values important to collective memory, social values which contribute to people's identification with particular places, or the spiritual values people associate with special buildings and places, whether attached to organised religions or not.

Communal value

Top: Grandstand and turnstiles (1905), Fulham Football Club, London, designed by Archibald Leitch (1866–1939), the foremost football stadium designer of the early 20th century. Fulham FC was founded in 1879 and moved to its present site in 1896. Its popular name –'Craven Cottage' – alludes to an 18th-century *cottage orné* which once stood on the site.

Bottom left: War Memorial on Remembrance Sunday. A place may have commemorative, symbolic and spiritual values for people who draw part of their identity from it, or have emotional links to it.

Bottom right: The Sphere (1971) by Fritz Koenig (*b*.1924). The sculpture once stood in the plaza between the World Trade Center towers in New York City and was damaged on 11th September, 2001. Now on display in Battery Park, this public artwork has acquired new meanings and cultural values because it bears the scars of a momentous historical event.

When evaluating a heritage asset, it is important to understand the range of values attached to it and their relative weight, since it may not be practically possible to sustain all of the values equally. Although age is often seen as a value in itself, particularly as survival tends to be inversely proportionate to age, rarity (of creation or survival) can make a building or structure of any period significant. The 'younger' the building, the more likely it is that aesthetic or formal design value will tend to make a greater contribution to significance than evidential value.

The 'values-based' approach underpins the Historic England *Conservation Principles*. It was also fundamental to *Planning Policy Statement 5: Planning for the Historic Environment (PPS5)*, which put understanding the significance of a heritage asset, whether designated or undesignated, at the heart of statutory decision-making about proposals affecting the historic environment. Post-dating, but drawing on *Conservation Principles*, *PPS5* defined significance as: *"The value of a heritage asset to this and future generations because of its heritage interest. That interest may be archaeological, architectural, artistic or historic."* This principle has been included in the *National Policy Planning Framework*, which supersedes *PPS5*. These are the terms included in current legislation under which heritage assets are designated ('artistic', perhaps surprisingly, being derived from the *Protection of Wrecks Act 1973*).

Evaluating and defining significance provides the starting point for managing change in the historic environment. Current policy requires this assessment to be done to a level of detail *"...proportional to the* [heritage] *assets' importance and no more than is sufficient to understand the potential impact of the proposal on their significance"* [*NPPF 2012; para 128*]. Assessment must take into account not only the heritage values of the fabric itself, but also of the surroundings or context, any historic contents, and the way the building was used in the past and is being used now.

CONSERVATION PLANNING

Conservation planning uses an understanding of a heritage asset and its significance as the starting point for devising measures to sustain that significance. It is a logical process that has four initial stages:

1. *Understanding the origins, evolution and associations of a building or place in its contexts (both spatial and temporal), including current use and management*

   When did the building originate, why and for what purpose? What did it look like; what materials were used; how, when and why has it evolved since? The primary record is the building fabric itself, but background documentary research is usually helpful. Building archaeological investigation may be needed where the fabric is complicated or sensitive, and other specialist expertise may also be required to inform understanding. This is looked at in detail in the next chapter.

2. *Evaluating the significance of a building or place, as a whole and in its elements*

   What is important about it, why and to whom? This involves identifying the range of values that people attribute to that building or place, and how these are expressed through the physical fabric. The results of the evaluation are usually set out in a 'statement of significance'. The aim should be to have sufficient understanding of significance, particularly of the elements of the building, to be able to assess the impact of proposed repairs or alterations on that significance.

3. *Identifying how significance is vulnerable to threats; for example, to decay processes, or simply wear and tear*

   This may involve investigation of:
   - the types and condition of materials and building components
   - the nature of deterioration or failure and its rate of progress
   - the causes of deterioration, both immediate and inherent in the way the building was constructed, has evolved, or is being used
   - how deterioration is affecting the significance and functionality of the elements in question and of neighbouring elements, and the building as a whole
   - the risk level of threats.

   The effects of past and current uses (particularly increases in intensity of use, or new uses), location, financial constraints, management and maintenance arrangements (or lack of them), and access and security problems, should be evaluated. In parallel, the opportunities presented by a building or place may also be considered.

4. *Devising strategies to safeguard the significance of a building or place (and sometimes recover or reveal it), by minimising harm to its heritage values*

   This will normally include considering approaches to maintenance, repair and ongoing management. The potential impact on significance of a range of options will usually be assessed in order to identify an 'optimum strategy' that balances practical considerations with the objective of seeking to sustain all the identified heritage values of the fabric.

In principle, the conservation planning process is applicable whatever the scale of the project. However, the range and depth of investigation and documentation should be in proportion to the scale and complexity of the proposed works, and the sensitivity of the building: it should also be efficient in its use of resources. In some cases, the preparation of a formal comprehensive 'conservation plan' or 'conservation management plan' may be necessary or desirable. In other cases, a less detailed (but no less comprehensive) 'conservation statement' may suffice. If works are proposed to a listed building or other designated heritage asset, a written statement of justification will be required in any case, to support an application for consent (or to demonstrate that consent is not required). Similarly, written statements of justification will be needed in support of grant applications. For major applications to the Heritage Lottery Fund [HLF], a 'conservation statement' is required for 'round one' grant applications, as a preliminary to a full 'conservation management plan', required at 'round two'.

The two final stages in the conservation planning process are:

5. *Implementing the option(s) that causes least harm to significance, applying mitigation to the extent that this is possible (that is, recording any features that may be lost)*
   This involves developing the optimum strategy into a detailed design, producing contract drawings and specifications, obtaining the necessary consents and procuring the works.

6. *Recording and evaluation*
   The final stage is to record the works actually undertaken, monitor the outcome of the project and evaluate it against expectations. These data can then be used to inform future work to the building or place.

CONSERVATION STATEMENTS

In effect, outline conservation plans and conservation statements provide an overview of the heritage asset concerned, an appraisal of its heritage values, a summary of its significance and an initial assessment of the issues likely to affect its future management. Conservation statements are usually based on existing knowledge and readily available data, but should identify gaps in information. This may result in a brief for additional research to aid understanding, and perhaps the preparation of a full conservation plan. A conservation statement may often be useful as an initial step in a longer-term project, when alternative proposals are first being considered, and there is a need to understand the significance and potential for change (or absence of it) of a heritage asset. Conservation statements are therefore most useful for relatively straightforward historic buildings or sites, or during the early stages of more complex projects, since they can pull together the major issues in a simple way. They may not always be sufficient, however, even in straightforward cases, as the basis for making major decisions affecting the heritage asset.

CONSERVATION PLANS

A conservation plan will describe the historical evolution and current nature and condition of a building and place in detail, including a careful analysis of the physical fabric (which is the primary source of information about the building and place). It will also define and evaluate all the heritage values attached to the asset in order to provide a 'statement of significance'. The plan will then assess the vulnerability of that significance to change, including ongoing maintenance requirements and repair, especially major repair, as well as future use and alterations and/or development. Finally, the plan will provide a set of policies aimed at retaining, and, where possible, further revealing the significance of the building or place, and guiding decisions about its future management, use or alteration.

A full conservation plan is likely to be required for more complex projects, especially where major physical interventions to the heritage asset or assets are involved, or where management requirements are likely to conflict with one another. Since they combine all the vital, relevant guidance and policies for the future management and development of the asset in one place, conservation plans represent a long-term investment that can inform interventions of any scale, many of which alone would not warrant the development of such a detailed knowledge base.

CONSERVATION MANAGEMENT PLANS

These provide a greater level of detail in the proposed policies than a conservation plan. They are likely to include detailed maintenance and management guidance which translates the policies into detailed actions. Conservation management plans are normally required as part of the application process for large grants from funding bodies, particularly the Heritage Lottery Fund, which has produced its own detailed guidance (see **Further Reading**). As they facilitate an integrated approach to management, conservation management plans are particularly valuable for heritage assets that involve more than one type of heritage, such as buildings, archaeology, landscape and historic collections. They are also useful where public consultation is needed to help understand the values that a place may have for particular communities, and where complex, multi-period heritage assets need to be evaluated, especially when major intervention is likely and little is initially known about their history.

HOW DOES SIGNIFICANCE RELATE TO MAINTENANCE & REPAIR?

Understanding the significance of a historic building or place in its entirety and in its context is important when assessing whether it meets the threshold for statutory designation or is a priority for grant aid. Understanding significance is also vital when considering the balance between potentially competing public interests, as in the construction of a new rail line or motorway which might traverse the (possibly registered) grounds of a historic building, or even the site of the building itself. In the context of repair, however, it is often the values attributed to particular elements of the building, and the contribution that these make to the significance of the building as a whole, that are of most practical importance.

Repair can be defined as *"work beyond the scope of maintenance, to remedy defects caused by decay, damage or use, including minor adaptation to achieve a sustainable outcome, but not involving alteration or restoration"* (Historic England *Conservation Principles*, 2008). Most repairs are undertaken in order to counter the effects of the natural decay of materials, usually as a result of weathering or wear-and-tear through use. While the scale and frequency of problems of decay are often influenced by human action or inaction, both general (such as atmospheric pollution) and specific to the building (for example, resulting from inadequate maintenance), periodic repair tends to be inevitable if the integrity of a building is to be maintained indefinitely.

Heritage values may be attributed to particular elements and details of a building

Top left: An overpainted late 17th-century scheme of marbled paintwork partially revealed when a planted timber moulding was removed in the course of repair works. Painted architectural decoration and wall paintings were common internal treatments in both grand and vernacular buildings in the past. Important painted schemes may survive beneath subsequent layers of paints or wallpaper.

Top middle: It can sometimes be difficult to assess exactly the point when 'vandalism' becomes a significant part of the history of the building, and worth of preservation in its own right.

Right: An 18th-century carved enrichment forming part of a door surround.

Bottom: Toolmarks can provide information about how materials were worked. Close examination of the hewing marks on a 17th-century timber beam shows that a nick in the hewing tool has left a slightly curved mark with each stroke; this shows that an axe, rather than an adze, was used to shape the timber. The tool marks also contribute to the visual interest of the surface and connect us, in a very simple and direct way, with past human activity.

Left: Wrought-iron quadrant stay. Hinges, locks, latches and bolts are among some of the earliest mechanisms to have been developed. Often they are objects of considerable beauty and historic interest, but many have been lost over the years through replacement because their value was not appreciated.

Right: Classic Roman *opus signinum* mortar with crushed fired clay products (brick, tile, potsherds), along with other aggregates and lime binder. These mortars are found throughout the breadth of the Roman empire. At Pevensey Castle the *opus signinum* mortar was used only in the construction of the faces of the wall, no crushed fired clay being added to the mortar used for the wall core. Where the wall face has been lost, the pink-coloured *opus signinum* bedding mortar can be seen tailing back into the core mortar. This shows that the core of the wall was raised in courses corresponding to the face work: valuable historical information concerning Roman building methods.

Repair (including periodic renewal of elements such as roof coverings) is normally undertaken with conservation as a primary objective: to sustain the heritage values and significance of the building or place. Equally important in most cases is keeping the building in use, so there is an incentive both to maintain it and provide the resources to do so into the future. That does not, however, obviate the need to devise and test proposals within the general principles applicable to all interventions in significant elements of the historic environment. If the aim is to sustain significance, then it is necessary to understand, first, the values that contribute to that significance and, then, how the elements that will be affected by repair contribute to those values. This in-depth understanding informs priorities in the retention of existing fabric. It enables the heritage impact of different repair options to be compared, so the one that will best sustain those values can be selected and implemented, then subsequently documented and monitored. As Christopher Brereton stated: *"A thorough understanding of the historical development of a building or monument is a necessary preliminary to its repair. This may involve archaeological and architectural investigation, documentary research, recording and interpretation of the particular structure, and its assessment in a wider historic context... Such processes may, where appropriate, need to continue during the course of repairs."*

The heritage values most likely to be affected by repairs are evidential value (which is associated with the surviving building fabric) and aesthetic value, particularly where this is expressed in design value. Evidential value will be directly affected by the alteration, replacement or removal of existing fabric that contributes to the building's significance, particularly where the fabric belongs to the original or a significant period of its construction. In general terms, the older the fabric concerned, the more important its evidential value is likely to be. In planning repairs, therefore, consideration should be given to the effect the completed work is likely to have on evidence for the evolution of the building and its past forms that is contained within its fabric. That does not mean, however, that every scrap of existing historic fabric should be retained regardless of its physical condition.

Sustaining heritage values

Top: The Victoria & Albert Museum, London, was damaged during the blitz in the Second World War. It has been left unrepaired as *"a memorial to the enduring values of this great museum in a time of conflict."*

Bottom left: Anderton Boat Lift. The significance of the historic operation of the boat lift was considered (by the then AMAC) to outweigh the value of the much-altered historic fabric, and SMC was granted for the restoration of the functionality of the monument, achieved by using modern technology and materials.

Bottom right: Replacement of a decayed carved stone at Whitby Abbey, Yorkshire. The work was carried out in this way to avoid the conjectural restoration of original mouldings that had become illegible. The shape of the inserted stone represents the block from which the original moulded stone might have been carved. While this approach upholds historical values, it has had an adverse impact on aesthetic values.

Impact of repairs on heritage values

Left: A poorly designed and executed repair to Roman masonry has diminished both its historical and aesthetic values.

Right: Stone steps leading to the Painted Chamber at Cleeve Abbey, Somerset. The pattern of wear on the treads connects us directly to the generations of people who used this stairway in the past. That connection is lost when original fabric is replaced.

Aesthetic value will be affected if the appearance of an element of a heritage asset that forms part of a conscious design, or which contributes to, say, an aesthetically pleasing, but fortuitous, ensemble of building elements, is altered as a result of repair. The likely degree of alteration and its effect on appearance should be considered as part of the process of devising the repairs. In some cases, it will be a factor in deciding whether a degree of restoration (in order to restore the integrity of the design) is justified or desirable.

Historical value can also be affected by changes to the fabric, but perhaps more narrowly. The elements of the building that illustrate the historic asset's connection to past people, events and aspects of life are likely to be quite specific, and the potential impact of repair or alteration therefore more limited and predictable.

Assessing the potential impact of repairs on the heritage values of a historic building is vital when planning a repair proposal, whether it is a single local intervention or an extensive programme of repair, upgrading and/or alteration. Once the extent of repairs needed has been established, and potential approaches and techniques determined, assessment of likely impact should be influential when considering options and making decisions. When selecting the optimum approach, the benefits and disadvantages of various options (including their estimated costs) will need to be weighed, but sustaining significance must always be a priority.

AUTHENTICITY & INTEGRITY

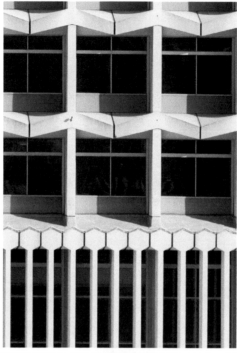

Aspects of authenticity

Top: Authenticity in design intention: pre-cast concrete components at the Centre Point building, London, 1961–66, designed by Richard Seifert and Partners.

Middle: Authenticity in the construction process: timber-to-timber repairs employ materials and craft processes that are similar to those which were used to bring timber-framed buildings into being. However, they involve a greater loss of original fabric than repairs using metal straps.

Bottom: Authenticity in the historic fabric: metal straps were commonly used to repair and stabilise timber-frame structures in the past. This type of repair is visually obtrusive and has an impact on aesthetic values, but it minimises the loss of historic fabric and is reversible.

Linked to the concept of significance are the ideas of authenticity and integrity. Historic England's *Conservation Principles* defines authenticity as *"Those characteristics that most truthfully reflect and embody the cultural heritage values of a place"*. If a building (or element of a building) has high evidential value, for example, or its aesthetic values derive from the effects of age, then sustaining authenticity requires the retention of as much inherited fabric as practicable, often through highly localised repair, thus adding another 'layer' to a long sequence. But if design value is paramount, as, for example, in the façade of the Centre Point building, London, then authenticity may lie in maintaining the building's intended appearance. This might mean replacing a damaged cladding panel in its entirety rather than repairing it in a way that might be visible, even if such a repair would be technically sound. Integrity, which is defined as 'wholeness, honesty', means considering buildings (or major elements of them) as entities. An elemental assessment of significance might rate the roof structure or the side elevation of a building as having less significance than the street front, but all are parts of the same entity. The integrity of the entity depends upon the wholeness of the three-dimensional structure. The idea of authenticity may also be applied to the materials and craft process by which a building is altered or repaired. In this context, the term 'authentic' refers to the use of materials and methods that are similar to those used to create the original building or element.

USING VALUES & SIGNIFICANCE IN DECISIONS ABOUT MAINTENANCE & REPAIR

The conventional wisdom in English building conservation practice is that treatments and repairs should generally aim to retain as much sound historic fabric as possible, and should not hinder the possibility of later access to evidence incorporated in the fabric. If a building is valued because it provides evidence of a distant age and hand, and the aesthetic qualities acquired through long weathering, then minimum or limited intervention is likely to emerge from the analysis as the appropriate response to sustaining those values. These are the types of building for which the concept of minimum intervention emerged in the 19th century as the preferred approach of informed people. Therefore, the conclusions of a values-based approach will not necessarily be different from those promulgated in the SPAB *Manifesto*.

Reconciling conservation of the natural and historic environments

Top: The remains of Wigmore Castle, Herefordshire; detail from *The South View of Wigmore Castle in the County of Hereford*, an 18th-century engraving by Nathaniel and Samuel Buck.

Bottom: Wigmore Castle today. When conserving the site in the 1990s, English Heritage set out to maintain its wildness, and to preserve the habitat of a number of rare and unusual species of fauna and flora. Prior to this, the conventional approach would have been to strip vegetation, clear away fallen masonry, consolidate the standing remains, and present them surrounded by neatly trimmed lawns. Instead, fallen masonry was retained, and the grasses, ferns and flowers that had colonised the walls were allowed to remain. Where necessary, to allow stonework repairs to be carried out, vegetation was carefully lifted, then replaced as 'soft-capping' to protect the walls from rain and more destructive deep-rooted plants.

The challenges of modern buildings

Top: Radar Training Station, Fleetwood, Lancashire, 1961–62 (Grade II), designed by Eric Morris Hart, and described by Sir Nikolaus Pevsner as 'a cute little piece'. The significance of a modern building is usually attributable more to its design value than any other heritage values. In such cases an approach to maintenance and repair that retains the building's intended appearance will be needed, although this may conflict with sustaining other values.

Bottom: Park Hill, Sheffield (Grade II*), built by Sheffield Corporation City Architect's Department and opened in 1961, was the first major scheme of slum clearance to be completed in England after the Second World War. It was also the first housing development to utilise external deck access ('streets in the sky'), which was conceived as a way of recreating the community spirit of traditional streets. Although Park Hill was initially successful and well-liked by residents, by the 1980s it had become dilapidated and unpopular.

The building was listed in 1998 amid controversy: many people were strongly in favour of its demolition. Park Hill is currently undergoing repair and rehabilitation to provide a mix of private and social housing. As part of this work the original brick façade panels are being replaced with storey-height glass and brightly-coloured aluminium cladding panels. While this alteration has an impact on heritage values, the justification is that it improves the environment of the dwellings and enhances the 'image' of the development, which is perceived to be an economic necessity.

However, the buildings and places that are now valued are far more diverse than in the past and no longer simply ancient. Many of these heritage assets require different approaches to conservation and more negotiation in order for their particular values to be sustained. It is as one moves from the more ancient to the more recent that a values-based approach becomes essential, particularly as design values make a greater contribution to the significance of a building listed as part of the 'national collection'. Retaining their design values (and, on occasions, recovering them from crude interventions) can conflict with sustaining other values.

Preservation and use

The Severn Bridge (which carries the M48 motorway over the River Severn) is listed at Grade I. It was opened in 1966, and was the first bridge in the world to use a streamlined, welded steel deck and inclined hangers.

It is interesting to consider the ways in which the approach to the maintenance and repair of the Severn Bridge must differ from that for another Grade-I listed bridge: the Iron Bridge in Shropshire, erected in 1779 and now preserved as a museum object.

Historic England Guidance on Maintenance & Repair

Historic England's *Conservation Principles (2008)* includes specific guidance on particular types of intervention and criteria against which proposals should be evaluated, both at the level of the historic environment as a whole and within the high-level *Conservation Principles*. The sections on *Maintenance*, *Periodic Renewal* and *Repair*, reproduced below, are particularly relevant. But there will be occasions where policies concerned with *Restoration* and *New Work and Alteration* will also be relevant, where reversion to earlier arrangements (for example, to an earlier type of roof covering) or new interventions (for example, the insertion of roof hatches to make valley gutters accessible for cleaning) arise as options for works primarily of repair.

Routine Maintenance & Management

The conservation of significant places is founded on an appropriate routine of maintenance and management.

Periodic Renewal

Periodic renewal of elements of a significant place, intended or inherent in the design, is normally desirable unless any harm caused to heritage values would not be recovered over time.

Repair

Repair necessary to sustain the heritage values of a significant place is normally desirable if:

- there is sufficient information to understand the impacts of the proposals on the significance of the place
- the long-term consequences of the proposals can, from experience, be demonstrated to be benign, or the proposals are designed not to prejudice alternative solutions in the future
- the proposals are designed to avoid or minimise harm if actions necessary to sustain particular heritage values tend to conflict.

DETERMINING CONSERVATION NEEDS

Sound justifiable decisions on appropriate types and levels of repair work will depend upon sufficient information being gathered during the assessment process. Inadequate or incomplete information may lead to incorrect diagnosis and poor choices being made about how best to secure the long-term care of the heritage asset. But while extensive investigations may be theoretically desirable, they can, if not clearly focused, lead to an excess of information and 'paralysis by analysis' as resources are diverted away from practical conservation. Usually, the depth of assessment is constrained by cost, difficulties in obtaining access, or restrictions on opening-up of building fabric or monitoring, and may be influenced by the requirements of funding bodies.

A two-stage approach can be helpful in deploying resources efficiently. The first 'reconaissance' stage involves establishing an overview of the significance of the building and its general condition, and helps to define the scope of a second, more detailed and often targeted stage of assessment, and the resources and facilities required to carry it out. In addition, the first stage assessment should consider, in broad terms, the strategic factors affecting the planning of any programme of maintenance and repairs which will be required, including access, timescales and resources.

Determining conservation needs tends to be an iterative, rather than a linear, process. As options emerge, a regular weighing-up of information is needed to see whether there is sufficient data to conclude that either no action is needed or a particular course of action should be adopted. If the information is inadequate, further investigation may be required before any decision is made. Apparent damage and defects can be stable, and may be attributable to past causes that have since been remedied. It is very easy for people who are not familiar with a building to find evidence of old problems and assume they are still active.

Different types and periods of structure present different problems, so solutions must always be site-specific. All of the practicable options should be considered, and the final choices should be those that either eliminate or more usually minimise harm to the significance of the building. The objective of repair is to reduce long-term decay or to preserve significant architectural detail. In some situations, therefore, the 'optimum solution' may be to take no action. Inappropriate or poorly-executed repairs may fail prematurely and accelerate the deterioration of original building fabric, increasing the extent and cost of future maintenance and repairs.

Once the building and its significance have been understood, the basic steps when devising any treatment or repair programme should be to:

- arrive at a comprehensive, soundly-based diagnosis of deterioration and its causes
- define the objectives of treatment or repair and any constraints
- determine how urgent the need for work is, taking into account threats, vulnerability and significance; assess the likely consequences if the defect is not treated, or the damage is not repaired
- establish the likely extent of the works needed to meet the conservation objectives, including mitigating the causes of deterioration (for example, moisture)
- assess the available resources (knowledge, skills, materials, finance)
- identify options that meet the objectives for treatment or repair
- assess the 'buildability', effectiveness, cost and maintenance implications of options (including site trials, if necessary)
- assess the impact of these options on the heritage values of the elements affected, and thus on the significance of the building as a whole
- select options that minimise harm to significance, while being 'buildable', effective and affordable
- determine priorities for implementation.

Reversibility & Retreatability

For many years, 'reversibility' (the ability to undo a treatment or repair without damaging or altering the original fabric) has been seen as a desirable objective. Reversibility is certainly relevant when considering alterations to a historic building (such as blocking up a window or building a partition wall). It is rarely very applicable to treatments and repairs, since these will usually involve permanent changes: surface treatments such as cleaning, consolidating or coating all involve changes that cannot be undone, and most repairs require that at least a small amount of the historic fabric will need to be removed (for example, a stone surface will need to be keyed before mortar repairs can be applied; timber will need to be cut to take a scarf repair).

'Retreatability' (the idea that interventions carried out today should not preclude future interventions) is often a more practical criterion, encouraging the use of like-for-like materials, and sacrificial repairs and coatings, for example, as well as facilitating the ongoing maintenance which should be a fundamental part of any conservation programme. It can still be useful, however, to consider relative reversibility when choosing between two different conservation options.

For repairs, the principal aim will be to remedy the cause of the defects. This might involve, for example, eliminating water penetration through re-roofing, renewal of flashings, gutters, or improvement of weathering details to help avoid water traps, in addition to correcting the damage caused by moisture.

With finite resources, it may not be possible to solve all the problems at once. It is often necessary to prioritise the required interventions, taking into account the severity of the problems, the significance of the building or feature concerned and what resources are available, and to prepare a phased schedule of repairs. Required interventions may be ranked as 'urgent', 'necessary' or 'desirable'. Evaluating priority will usually demand the skills and judgement of experienced historic building professionals, specialist conservators and conservation contractors. Prioritising works is discussed further in **Managing Maintenance & Repair.**

Temporary repair

Sometimes 'first aid' is needed to prevent further deterioration until more permanent repairs can be carried out. In this case, pre-stressed concrete lintels supported on a stack of concrete blocks are being used to provide needling and propping to a defective wall.

If resources are seriously constrained, it may be necessary to combine temporary measures to prevent further deterioration (for example, propping a failing beam or wall, or erection of a temporary roof) with permanent repair in other areas, but normally the aim should be to undertake sufficient work to achieve appropriate lasting repairs. The extent of a repair should normally be limited to what is reasonably necessary to make failing elements capable of continuing to fulfil their functions. However, the need to provide expensive access arrangements, such as a tall or complex scaffold, may influence the way in which works are prioritised, and may encourage a level of greater intervention if required to mitigate future maintenance needs in those areas.

It is important to look beyond the immediate requirement for action, to understand the causes or reasons for the need for repair, and to plan for the ongoing consequences of inevitable change and decay. It can be tempting only *"...to do small repairs when they seem to be needed"* (*Powys 1929*). But limiting repair to the minimum necessary at the time of interventions is generally not good practice in most situations, especially if it means that further repair will be needed in the short term. As Christopher Brereton put it, repairs should be undertaken *"... with the aim of achieving a sufficiently sound structural condition to ensure* [buildings' and monuments'] *long-term survival and to meet the needs of any appropriate use"*.

APPROACHES TO REPAIR

Repair techniques are addressed in the material-specific volumes of this series, but in general, and in line with the Historic England high-level principles:

- Only techniques and materials which have been demonstrated to be appropriate to the fabric should be considered. These will normally be the same as the original or parent material, or, where this is no longer available or appropriate, have compatible properties, both technically and aesthetically.

- Interventions should maximise the life expectancy of significant building fabric, consistent with sustaining its significance.

- Interventions should be reversible, if technically feasible and practicable, or, at least, 'retreatable', and should not prejudice future interventions when these become necessary.

- All works should be adequately recorded, and the records made available for others.

- Interventions should contribute to, or at least not compromise, the sustainability of future management and maintenance.

THE AMOUNT & DURABILITY OF REPAIR WORK

All buildings inevitably deteriorate with exposure to the environment and wear-and-tear from building users. It is important, therefore, to consider whether building fabric will benefit from repair, or whether repair itself will adversely affect or limit long-term durability. There are clearly instances where repair is essential; for example, refilling open mortar joints. The case for repair becomes less clear, however, when mortar joints are evenly, but not deeply, eroded. The decision whether or not to repoint must be based on a clear understanding of the durability of the existing mortar, the ongoing rate or erosion, and how erosion of the mortar joint or the physical properties of the mortar affect condition of surrounding building fabric.

For example, in brickwork where the original lime mortar joint has eroded back from the brick face by 5 or 6 mm over 230 years, the rate of erosion is approximately 2.6 mm every century. If the brick condition is good, with only minimal surface decay, and erosion limited to minor rounding of the brick arrises, it is reasonable to assume that the brickwork and mortar joints are still working together effectively. Assuming a reasonably consistent decay rate, repointing might not be required for at least 50 years, or when the joint erosion reaches 8–10 mm. At this point the increased exposure of the brick surface would be likely to reduce future durability of the brick. Periodic inspections enable the rate of decay to be monitored, and allow informed judgements to be made about the optimal time for repair to be carried out to maximise the life expectancy of the brickwork.

The following table compares the life-expectancy of various types of stonework repairs and illustrates the influence that workmanship has on durability.

DURABILITY OF REPAIR WORK

REPAIR TYPE	DURABILITY OF WELL-EXECUTED REPAIRS	DURABILITY COMMONLY SEEN IN SUB-STANDARD REPAIRS
Stone indent or piecing-in repair	60 to 120 years	10 to 20 years
Mortar repairs in sheltered locations	30 to 60 years	5 to 15 years
Mortar repairs in exposed locations	<20 years (not recommended)	<10 years
Mortar joint repointing	60 to 120 years	5 to 15 years
Shelter coating	<20 years	<10 years
Stone replacement	100+ years	<30 years
Brick replacement	100+ years	<30 years

TYPICAL LIFE HISTORY OF LIME MORTAR JOINTS

1: AS BUILT (1650)

Mortar: One part non-hydraulic lime with kiln slag: two parts sand. Joints finished flush with face of masonry.

2: AFTER 1 YEAR

First winter attacks weak, non-hydraulic lime before it has time to stiffen.

3: AFTER 2 YEARS

Early frost attack has weakened all joints, allowing rain penetration to core filling.

4: AFTER 2–10 YEARS

Masonry arrises, exposed by disintegration of mortar, become saturated during rain.

5: AFTER 10–50 YEARS

Frost causes further damage to exposed, saturated arrises, which are further weakened by acidic rain and stresses caused by crystallisation of soluble salts within the pores of the masonry.

6: AFTER 50–200 YEARS

Further weathering rounds off damaged arrises. By now the lime mortar has hardened sufficiently to resist further rain attack.

7: 1890

Repointing carried out in Portland cement mortar finished flush with the face of the weathered masonry. The apparent width of the joint is doubled, changing the visual character of the masonry. Voids are left due to inadequate joint preparation.

8: AFTER 20 YEARS

High strength, good adhesion and impermeability of cement mortar accelerate decay of vulnerable masonry. Rain penetrates hairline cracks between mortar and masonry.

9: AS FOUND

Further saturation and freezing loosens and then detaches most of the dense pointing, increasing vulnerability of masonry to further damage.

Diagram after John Ashurst

RESTORATION

An issue that often arises when devising a programme of repair is whether, or where, to include an element of restoration, and if this can be justified. Historic England's *Conservation Principles* defines restoration as returning a heritage asset *"to a known earlier state, on the basis of compelling evidence, without conjecture"*. The document sets out a number of criteria which, if met, would normally make restoration acceptable. Indeed, the distinction between restoration and repair may well become blurred when architectural details and/or decorative elements that are important to the character and appearance of a historic building become eroded or damaged. Understanding the values of these elements, particularly evidential and aesthetic values, and their relative contribution to overall significance, should guide decisions, so far as resources permit.

In relation to missing elements, Christopher Brereton's comment still holds good: *"Some elements of a building or monument which are important to its design, for example, balustrades, pinnacles, cornices, hood-moulds, window tracery and members of a timber frame or roof truss, may have been lost in the past. ...a programme of repair may also offer the opportunity for the reinstatement of missing non-structural elements, provided that sufficient evidence exists for an accurate replacement, no loss of historic fabric occurs, and the necessary statutory consents are obtained in advance. Speculative reconstruction is hardly ever justified."* There will, of course, be exceptions, for example, where the form in which the fabric has survived is due to, or provides evidence of, some historically important event.

Challenges of restoration

The façades of these 16th-century timber-framed town houses were modernised in the 18th century by inserting sash windows and rendering the walls to conceal the 'unfashionable' timber framing. The render was removed from the house on the left in the 20th century, as the appearance of timber framing had become desirable again. However, the 18th-century windows were retained; at no time in its history was the building intended to look as it does now.

Hill Hall, Essex

Hill Hall came into the care of the Ministry of Works after being largely gutted by fire in 1969 *(top)*. English Heritage, the successor to the Ministry of Works, decided to re-roof it and return it to use as the only long-term means of sustaining the fabric (the brick walls have cores bedded in loam rather than mortar), preserving the remains of 16th-century wall paintings that had survived the fire, and revealing the significance of Sir Thomas Smith's design (1569–77) *(bottom)*.

In some circumstances, restoration may provide conservation benefits that cannot be achieved through repair alone. For example, restoring the roof on a roofless building may be the most effective way of conserving valuable internal fabric, such as wall paintings or important early plasterwork. It may also help to make the building physically and economically sustainable in the long term.

ROLES & RESPONSIBILITIES IN CONSERVATION PLANNING

One of the first steps in the conservation planning process will be the selection of a project team (other than for very small or simple projects, which might be carried out by a single practitioner). The extent of the team, and the range of skills and expertise of team members needed, will be governed by the nature, size and complexity of the heritage asset, and of the anticipated works. The membership of the team may vary at different stages of the process according to the skills required. A successful project outcome will depend to a very large extent on the skills and experience, commitment, enthusiasm and open-mindedness of the people involved, as well as the availability of sufficient funding for the project. The roles and responsibilities of the individual members of the project team are discussed later in **Managing Maintenance & Repair.**

Options after disaster

Uppark, Sussex, one of the least-altered Georgian country houses in England, was severely damaged by fire in 1989. Fortunately, the collection of works of art, books ceramics, glass, metalwork and furniture, acquired specifically for the house by two generations of discerning collectors during the 18th century, was largely rescued.

After considering a range of options, including preserving the remains of the house as a ruin, The National Trust decided to restore the house to the condition it had been in 'the day before the fire'. The approach was justified because the relationship between the house and the collection that had been displayed within was of high significance. Furthermore, there was sufficient detailed evidence upon which to base an accurate reconstruction. This included photographic records of the main rooms as well as thousands of architectural fragments recovered by archaeologists from the salvaged debris that filled nearly 4000 dustbins. Much of the recovered material was subsequently incorporated in the repair work.

CASE STUDIES

The case studies on the following pages show how values-based conservation planning has been used to inform decisions about the repair and maintenance of historic buildings and places.

1. **Repair 'as found' or rebuild?**
   The conservation of a disused lime kiln on the Isle of Portland involves weighing the preservation of the structure in a flawed state 'as found' against rebuilding.

2. **Conservation management plans in action**
   The development of Forty Hall in Enfield, London, for new uses relied on a well-considered Conservation Management Plan to ensure its significance is sustained.

3. **Dealing with flaws in the original materials or design**
   The tiled roof of the Reform Club, on Pall Mall in London, was failing because of an inherently flawed design. Conservation options included improving upon this, or replacing it with a more reliable covering; historical research showed that the architect had originally intended to use a different type of tile, and this option was finally chosen.

4. **Adapting for modern use**
   Mansfield College in Oxford needed to cater for more students and for income-generating activities, such as conferences and wedding receptions. This meant finding ways to upgrade the kitchens, improve access and provide other facilities without compromising significance.

5. **Exposing early architectural features**
   The removal of a failing render at Prior's Hall in Widdington, Essex, revealed a complex early history. The decision was taken to permanently expose hitherto hidden elements such as a Saxon doorway and a 16th-century window, but at a cost to the significance of the building as a whole.

6. **Removal of early alterations**
   The 15th-century Guildhall in Finchingfield, Essex, had been rendered when the building was altered and enlarged in the 17th century. Proposals to remove the render were rejected for aesthetic reasons, as well as the requirement to preserve historic evidence of the building's history.

7. **Dealing with intractable decay**
   A fundamentally flawed and fragile but very important medieval stained-glass window in the Savile Chapel in Thornhill, West Yorkshire, is to be replaced with a replica, and the original glass displayed within the church.

8. **Converting to multiple uses**
   The nave of the church of St Peter in Peterchurch, Herefordshire, was refitted to serve as a community centre whilst still being usable for large services.

A DISUSED LIME KILN
ISLE OF PORTLAND

This disused lime kiln on the Isle of Portland – built and operated by convict labour during the latter half of the 19th century – had partially collapsed due to foundation damage caused by defective surface water drainage. What are the options for repair? Assuming that the cause of the collapse had been correctly diagnosed and corrected, it would be technically possible either to 'conserve as found' or to rebuild the damaged portion. Conserving the structure in its damaged state would entail works to stabilise the overhanging masonry, and consolidation of the exposed wall cores to enable them to remain exposed to the weather. On the other hand, rebuilding might be carried out using the fallen material as the accurate placement of the recovered ashlar components should, in theory, be possible. Alternative rebuilding options might also include reinstating the fallen masonry to the original design but using new stonework, or carrying out the rebuilding in a way that clearly differentiates the new work from the existing.

Each of these options has implications for different cultural heritage values: evidential, historical, aesthetic and communal. Deciding on the approach to repair depends on understanding these values and balancing them against other criteria, such as desired end uses and available resources. In this case, rebuilding the damaged portion might seem to be the obvious answer, particularly if the structure is to be put to some practical use. However, if the damage was the result of some significant historical event – an infamous prison riot, say – rather than the consequence of neglect, how might the approach to repair be different?

CONSERVATION MANAGEMENT PLANS IN ACTION

FORTY HALL
ENFIELD, NORTH LONDON

Forty Hall is a Grade I listed house. Built in 1629–32 for a wealthy London cloth merchant, Nicholas Rainton (1569–1646), its architect is unknown. The 17th-century house was a three-storey brick box, over a semi-basement. Externally, it was symmetrical and plain, showing a Classical influence notably advanced for its date. In contrast, its plan was old-fashioned, with rooms, including a hall and screens passage, arranged around a 'compressed courtyard'.

The present exterior appearance of Forty Hall dates largely from alterations of *c*.1708. The interior was partly reordered and redecorated with rococo plasterwork *c*.1760, and the house extended again in 1897, when a new main stair was inserted.

From 1951, Enfield Borough Council used the Hall as a museum and gallery, but, in 2005, it was decided to move the collections to a new building. The Council was committed to maintaining the house as a publicly-accessible historic and cultural attraction, but beyond this objective, plans were fluid.

The north front of the Hall.

A *Conservation Management Plan* was commissioned to bring together existing and emerging knowledge about the house and its historic estate, and inform a strategy for their conservation, development and long-term management. The plan was intended to inform and support a Heritage Lottery Fund grant application, but it is a free-standing document: a dynamic tool in the conservation and management of the site, as relevant to minor repairs as to a major development.

Plan showing the various phases of construction.

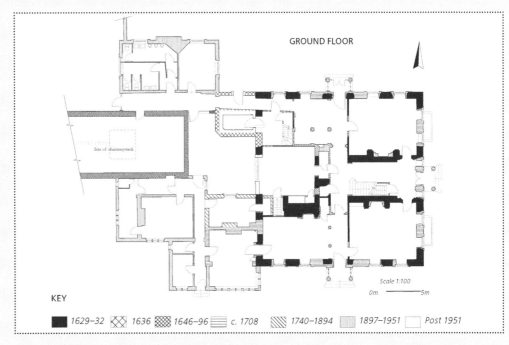

GROUND FLOOR

Site of chimney stack

Scale 1:100

0m — 5m

KEY

■ *1629–32* ⧂ *1636* ▨ *1646–96* ☰ *c. 1708* ⧄ *1740–1894* ⫲ *1897–1951* □ *Post 1951*

The precise nature of the Hall's original and potentially innovative villa form had long puzzled architectural historians. Although a fairly complete picture of the architectural development of the Hall emerged (especially from looking into voids), some aspects necessarily remained conjectural.

The significance of the Hall, its outbuildings and estate, in their architectural and historic contexts, was assessed on the basis of the 'Understanding' section of the *Conservation Management Plan*, using the values developed in Historic England's *Conservation Principles*. Understanding was based mainly on examination of the built fabric, since there was little detailed documentation. A gazetteer combined the understanding, assessment of significance and condition survey to create a room-by-room summary that could be used as a day-to-day management tool, updated as new challenges arose or discoveries were made.

Issues affecting the significance of the site and policies to address these were identified, alongside opportunities for beneficial change. Many were common to most historic houses: the need for regular maintenance, sensitive repair, specialist advice and integrated management. In addition, a lift was fundamental to ongoing public use of the Hall and Heritage Lottery Fund support. The plan identified a preferred option, involving replacing the 1897 main stair with one based on an earlier configuration, but it was clear that archaeological intervention would be needed to establish whether so radical a change could be justified.

The next stage of the conservation of Forty Hall was the appointment of a full professional team, who used the plan to help determine exactly how to repair and re-present the house without compromising its significance.

HERITAGE IMPACT ASSESSMENT OF THE MAIN STAIRCASE

The Issues and Policies section of the *Forty Hall Conservation Management Plan* identified opportunities for the building's development and future use. The preferred option included the installation of a lift in the centre of the house (where once there had been a back stair) to give easy access to all floor levels. To accomplish this, the existing 1897 stair would be replaced with a new one in the assumed form and location of the earlier primary staircase, thereby restoring the historic circulation pattern within the house, and providing space for the lift.

After public consultation, Enfield Borough Council adopted the plan. A full specialist professional team was appointed to develop a major grant application to the Heritage Lottery Fund that reflected the recommendations of the plan, and to obtain the statutory consents. The works required detailed justification, which was made with reference to Historic England's *Conservation Principles* (on which the current national planning guidance for the historic environment was subsequently based).

Justification for the proposals relied on two key conservation principles: understanding the significance of the building and the elements affected; and that change should be managed to sustain, reveal or reinforce the significance. To be acceptable, it needed to show that *"any resulting harm is decisively outweighed by the benefits"* (**PDP 2007**).

The 1897 staircase.

The conservation plan contained a comprehensive understanding of the site and the assessment of significance. It concluded that the 1897 stair was of much lower relative significance, and confounded and obscured both the historic plan and circulation pattern, and the 17th- and 18th-century decorative schemes of the house.

The practical benefits of inserting a lift supported the principle of sustaining heritage by making Forty Hall accessible and usable, rather than becoming 'at risk', or being put to less sympathetic uses. The new stair and lift would reveal and reinforce the historic plan and circulation of the house that had been obscured by the Victorian staircase, and thus, in principle, the scheme delivered heritage benefits that outweighed the loss of significance. The scheme also needed to conform to the policies and guidance of the *Conservation Principles*.

Conjectural reconstruction of the Hall as built, cut away to show the primary and secondary staircases.

Restoration of the 17th-century staircase demanded compelling evidence. Blocked doorways on the former half landings and the complete 17th-century plaster ceiling over part of the 1897 stair informed their 'reconstruction' in the conservation plan, but this remained, ultimately, conjectural. Therefore, a building archaeologist was commissioned to undertake a structural investigation and plaster was stripped from selected areas: this confirmed the initial hypotheses.

Justifying new work and alterations relied equally on the understanding of historic fabric in order to assess the impact of the proposals and on the quality of the new work. As there was no evidence for the detailing of the original stair, no attempt could be made to replicate it, and a new oak stair was designed, inspired by 17th-century precedents, but slightly adapted to comply with Part M of the *Building Regulations*. New joinery to doorways would exactly replicate that found in the adjoining rooms.

All new work in the stairwell will use traditional materials and techniques, which, from experience, carry a very low risk of unexpected consequences. The lift and its shaft use modern technology, but can readily be removed and replaced if unsatisfactory.

DEALING WITH FLAWS IN THE ORIGINAL MATERIALS OR DESIGN

REFORM CLUB
PALL MALL, LONDON

The Reform Club is a Grade I listed building, designed by Charles Barry and modelled on the Palazzo Farnese in Rome. It is thought by many to be his finest building. It was opened in 1841 and, remarkably, is still in its original use. By the 1990s it needed repair: in particular, the roof was leaking, which had caused damage to plaster and smoke detectors, and stained the magnificent stone cornice. Repair was also necessitated by other changes proposed to the third floor. Works began in 1993.

The roof was originally covered with single-lap patent slating. This was comprised of five courses of very large slate slabs (810 × 1220 × 19 mm), which abutted at the vertical joint and overlapped at the horizontal. By the 1990s, the system for weathering the vertical joints comprised flat timber strips, covered by lead which was bent around the timber and held with copper clips. Rain had penetrated under the lead, rotting the timber and allowing water to penetrate.

Replication of this inherently flawed design was rejected because of doubts about its fitness for purpose. A visually similar design in leadwork was suggested, but a full-scale mock-up revealed a number of technical shortcomings. A new, modified single-lap slate roof was designed. Although English Heritage (now Historic England) and the local planning authority were prepared to approve this proposal, it was not acceptable to the applicant because the design was not tried-and-tested, and would have demanded exceptionally high standards of workmanship that were unlikely to be attainable.

Italian tiles of the type originally intended for the roof covering of the club.

A detail from *Pall Mall: Club Houses,* an 1842 lithograph by Thomas Shotter Boys (1803–74), showing the Reform Club as it appeared shortly after construction.

Historical research showed that Barry's original intention was to use Italian tiles. Barry's 'specification for the contractor' of April 1838 states: *"TILING The front part of the roofs to be covered with tiles to the PC value of £5 per square….the whole to be left perfect at the completion of the works."* The 'penultimate design' in the *Survey of London* shows an Italian roof, and the remnants of one drawing in the RIBA drawings collection shows a section through the roof in what appears to be Italian tiles. The history of the Club by Louis Fagan (1887) states that the roof *"is covered with Italian tiles"*, but there is doubt as to whether this was the case. Minutes of the Building Committee in May 1840 record Barry's extras, including *"extra work not sanctioned as present by the Committee…slate instead of tile roof £300"* (*Reform Club Archives*). Although the tiles do not appear to have been used, they certainly were intended, at least until construction was very far advanced. Certainly, this would be consistent with Barry's aspiration to recreate the Farnese Palace, which he visited and drew in 1817.

Although Italian tiles were not common in England, today they are manufactured by Sandtoft, on the Humber. To get the detailing correct, the *tegulae* (unders), *imbreces* (overs) and specially-designed ridges and hips were produced on a full-scale mock-up at the works. Company representatives also visited and advised during installation on site.

This example illustrates the overriding responsibility to protect the building fabric effectively, rather than to follow flawed historical detailing. It prompted the selection of a design which was both technically sound and aesthetically consistent with the original Barry design.

ADAPTING FOR MODERN USE

MANSFIELD COLLEGE
OXFORD

A well-known firm of architects was commissioned by Mansfield College to look at options to upgrade the kitchens, improve access, and provide more dining facilities that would cater for an increased number of students and income-generating activities such as conferences and wedding receptions. The aim was to understand the significance of the different elements of the buildings, and to design solutions that would minimise harm to significance and address any existing damaging interventions.

The main buildings of Mansfield College, including the Chapel, date from 1886 and are listed Grade II*. They were designed by Basil Champneys, a renowned architect of his day, who won the RIBA Gold Medal in 1912. Champneys' late-Gothic design for the buildings survives largely intact, with very little alteration or addition. The most intrusive addition is a stair lift rising up the principal stair in the main entrance hall. The Main Range, Chapel and Principal's Lodging look onto the circular lawn, a major feature of the open quad, in contrast to the traditional closed quad arrangement found in many earlier colleges. This has been compromised by the construction of a residential block opposite the Main Range, which has truncated it. The setting is further damaged by car parking around the lawn. Even so, the buildings and their setting have designed aesthetic value.

A detail from one of the significance plans provided in the *Design and Access Statement*, prepared by Rick Mather Architects, to support applications for planning permission and Listed Building Consent. The plan shows the relative significance of different parts of the building fabric and spaces.

Facing page: The Tower and East Main Range.

KEY TO SIGNIFICANCE PLAN

Fabric Space

Considerable significance
Champneys' original design and in good condition or easily restored to the original condition.

Some significance
Champneys' original design but compromised by further additions and/or modifications to the fabric. Few features of special interest.

No significance
Recent modifications and additions generally post-World War II.

Mansfield was the first Nonconformist college to be founded in Oxford, after the *Act of 1871* removed the Anglican monopoly of the University, and is still the centre of the University's Nonconformist community. The Chapel was designed to be a dominant landmark on Mansfield Road, reflecting its importance, as religion was at the core of college life. It is therefore highly significant for its historic, illustrative and associative values, as well as its designed aesthetic values. It also holds communal value for past and present students who have worshipped here.

Internally, there is a clear hierarchy of spaces within the main block, with the dining hall and senior common room on a raised ground floor over the service rooms, which include kitchen, cellars, stores and cloakrooms. Interestingly, the latter, which are used by members of the college, and not staff, have subtle differences in joinery detailing from the rest of the basement, indicating the higher status of these rooms.

Some of the proposed solutions to the needs of the College are the demolition of some later service outbuildings, which harm the setting of the Main Range and Chapel, and their replacement with a new kitchen designed to be low-key visually, which will allow the principal buildings to dominate. Internally, the proposals involve the loss of original fabric in the basement kitchen and service areas that have illustrative historic value. The loss has been kept to a minimum to allow the original plan form of the building to remain legible. Where fabric will be lost, this is marked on the ground by a change in material.

The proposal also includes opening up the basement dining area to the outside to create a terrace for additional seating. This will disrupt the unity of the setting of the main buildings, however. Though the current proposal for the terrace is smaller than the one originally proposed, it will remain an intrusive feature in the quad, and this will harm the designed aesthetic value of the setting. In mitigation, the terrace will be screened to minimise its visual impact, and the remainder of the quad will be much improved by the removal of cars and the redefining of the formal circular layout.

On balance, English Heritage (now Historic England) considered that adequate justification had been put forward for the works, and that the benefits of improved access, removal of insensitive additions, repairs to fabric and the mitigation works outweighed the harm.

EXPOSING EARLY ARCHITECTURAL FEATURES

PRIOR'S HALL
WIDDINGTON, ESSEX

Prior's Hall, Widdington, was thought to be nothing other than a vernacular Essex house like many others. Then, about 30 years ago, the removal of failing render revealed the carcass of a flint rubble chapel dating from late Saxon times. It was later determined that the chapel had been converted into a house in the 14th century, floored in the 16th, and enlarged in the 17th century and later. When the building was re-rendered the characteristic 'long-and-short-work' (quoins formed of alternate vertical and horizontal stones) of the Saxon chapel was left exposed, a discreet indication of the building's antiquity. The recent failure of the modern replacement render, coupled with extensive structural problems, necessitated a further campaign of repairs and re-rendering. In the course of these works, more was done to expose the building's past: in the south elevation a Saxon doorway was re-opened and glazed, and a 16th-century hall window uncovered and reinstated.

The Saxon doorway and hall window of the Hall.

Has this work 'revealed' and 'reinforced' the significance of the building? In the light of the criteria for assessing the acceptability of restoration set out in *Conservation Principles*, the answer to this question is not clear. Values are by their nature debatable. In this case the act of 'restoration' was one of uncovering. Both the Saxon doorway and the 16th-century window are, in themselves, of greater value than the modern render that concealed them, and, indeed, than the 19th-century brickwork of a flue that had been built against the inside of the doorway.

The rendered elevation with Saxon quoins exposed.

The evidence for the work was compelling. The doorway has simply been revealed. The window had been partly destroyed by later alterations, but what survived provided enough evidence to justify its completion. The form of the building prior to this work was not the result of a historically significant event (unless this term is interpreted very broadly). However, it is doubtful that what has been done *"respects previous forms of the place"*. *Conservation Principles* states that *"the restoration of isolated parts of a place to an earlier form, except as legible elements of an otherwise new design, would produce an apparently historic entity that had never previously existed, which would lack integrity"*.

This criterion prompts a more searching consideration of the values of what has been exposed and what has been lost. The doorway and window are, no doubt, of greater evidential value than the plain render now removed; but their exposure has been achieved at the cost of compromising the historic character of the house as a building that has changed and developed through the centuries. Character itself is an amalgam of evidential, historical, aesthetic and communal values. To conclude, Prior's Hall has been reinstated in a condition of incompleteness that never existed previously. Some may prefer the interest added or exposed by the recent work, but if this exposure has revealed or reinforced significance, the act of revelation has itself detracted from the significance of the building as a whole. *Conservation Principles* defines significance as *"the sum of the cultural and natural heritage values of a place"*, and even if particular importance may be attached to the Saxon and medieval features revealed, the significance of the house lies in its totality.

REMOVAL OF EARLY ALTERATIONS

THE GUILDHALL
FINCHINGFIELD, ESSEX

The Guildhall, at Finchingfield in Essex, comprises a range of 15th-century almshouses, remodelled in the 17th century. Recently, as part of a major scheme of refurbishment, it was proposed to remove the rendering from the Guildhall's principal elevation to reveal its framing. The proponents of this course of action suggested that this would bring out the historic interest of the 15th-century frame. The implication was that it would both reveal and reinforce the building's significance.

In Historic England's *Conservation Principles*, the definition of 'conservation' acknowledges the existence of *"opportunities to reveal or reinforce (heritage) values"*; for example, by restoring a building to a previous form by removing later accretions: in the case of the Guildhall, its render. *Conservation Principles* sets out a number of criteria for assessing restoration proposals. The primary test is whether *"the heritage values of the elements that would be restored decisively outweigh the values of those that would be lost"*.

The Guildhall, Finchingfield.

Restoring a building to a previous form may be acceptable where *"the form in which a place exists is not the result of a historically significant event"*. In the case of the Guildhall this raises an important question: is not the change in local building tradition that resulted in the general rendering of East Anglian buildings (many of which had frames that were originally exposed) a historically significant event or process?

Conservation Principles provides a further test: whether *"the work proposed respects previous forms of the place"*. Analysis of the Guildhall suggests that it had been rendered in the 17th century when the original 15th-century building was altered and enlarged. Removal of the render would reveal an altered, multi-period framed structure, including later parts which had been rendered to begin with. Therefore, the result would not respect the *"previous forms of the place"* because at no time in its history did the Guildhall exist in the form proposed. It would also be incongruous in its village setting, where the timber-framed buildings remain for the most part rendered.

To return to the primary test – that of the balance of heritage values – it might be thought that this has been settled by what has been written so far. The Guildhall's present form embodies considerable evidential and historical value. The removal of the render would diminish both. The rendered finish is aesthetically satisfactory when compared with the incoherent appearance that would result from its removal. This conclusion may also be framed with reference to authenticity, a quality dependent on *"Those characteristics that most truthfully reflect and embody the cultural heritage values of a place"*. The authenticity of the Guildhall resides in its development, and that of the Essex vernacular building tradition, over time, and this would be gravely harmed were the render to be removed.

It must be acknowledged, however, that heritage values may be keenly debated. Past restoration of timber-framed buildings by the removal of later renders continues to shape many people's appreciation of this tradition. Few people would pause to question the exposed and limewashed framing of the very fine Guildhall at Thaxted, a few miles from Finchingfield, although its restoration in 1911 was itself controversial.

The Guildhall in Thaxted, Essex.

DEALING WITH INTRACTABLE DECAY

THE SAVILE CHAPEL
CHURCH OF ST MICHAEL & ALL ANGELS, THORNHILL, WEST YORKSHIRE

The east window in the Savile Chapel, at the Church of St Michael and All Angels, Thornhill, Dewsbury (West Yorkshire), dates from 1493, when the chapel was extended. Called the 'Doom' window, it depicts the Last Judgment. Ironically, the window was itself doomed because of the chemical composition of its glass. By 2008 it was in very poor condition. The surface of much of the clear painted glass was crazed, spalled and pitted. This reduced the legibility of the already very faded painted designs and left the glass in a fragile condition, vulnerable to ongoing and accelerated deterioration. In addition, the lead cames had become brittle and no longer provided adequate support to the glass; in many places, the glass had been cracked by mechanical damage.

Chemical analysis revealed that the composition of the clear glass was unusually high in potash and lime, and relatively low in silica, which made it particularly sensitive to atmospheric moisture. Thanks to a free supply of coal from nearby pits, the chapel had been well heated in the past, but it was poorly ventilated and damp. When combined with atmospheric pollution, these conditions led to the formation of a chemically-altered layer on the surface of the glass over the centuries. Dimensional changes in the surface layer, caused by fluctuations in temperature and humidity, caused crazing and cracking.

The significance of the window is recognised nationally. It is valued not only for its historical and artistic qualities, but as a resource for scientists and conservators to study the complex chemical deterioration processes affecting such ancient glass. Additionally, the Savile Chapel is home to an important collection of commemorative monuments to the Savile family that, with the glass, contribute greatly to the significance of the Chapel and the Church as a whole. The Church is a Grade-I listed building on account of its heritage interest, but it is also highly valued by its congregation and local people as a place of worship and focal point for the community. The Savile Chapel and its glass are especially cherished.

A range of options for conserving the window was explored. Following comprehensive and detailed investigations, including a period of diagnostic monitoring, the team of researchers and conservators concluded that the most effective and least risky method for protecting it from further deterioration would be to remove the glass panels to a controlled environment. Although this approach risked harming the significance of the Chapel, it was justified because it preserved the exceptional heritage values of the window; values that would otherwise be lost to future generations. There then followed a lengthy period of discussion between the many 'stakeholders' about how best to mitigate the impact on the Chapel of removing the glass from its original context.

Left: Detail of the original 15th-century stained glass in the east window of the Savile Chapel.

Top right: A detail from the cartoon of the stained glass produced by the company Burlison and Grylls before the reordering of the church in the 19th century. When compared with a recent photograph of the same section of glass (*middle right*), the extent of deterioration and loss of detail is clearly apparent. The cartoon is an important record and provides evidence on which to base an accurate reconstruction of the window.

Bottom right: Following extensive research by stained glass conservator Jonathan Cooke, a new cartoon has been produced in preparation for the reconstruction of the window. A detail of that cartoon is shown here.

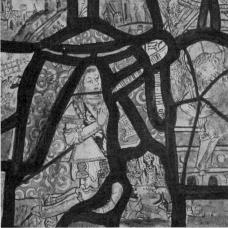

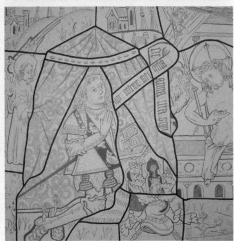

As it happened, accurate tracings of the Savile Chapel windows had been made during the 19th-century reordering of the Church. These were found in the local record office and showed how much detail had been lost from the design during the last century. Along with the surviving design, details in the window provided sufficient information on which to base an accurate reproduction. This was preferred by the congregation and local community to one of clear glass or a stained-glass window of modern design; they argued strongly that the iconography of the window was important to them in their acts of worship. It also played a significant part in the aesthetic experience of the Chapel, which they wished to retain.

The Chapel with the 15th-century stained glass in the east window removed for conservation. The temporary clear glazing seen here will be replaced with a replica of the original stained glass, and In due course the original glass may be displayed in an environmentally controlled showcase in the church.

Eventually, after much debate, it was decided to display the original glass in an environmentally-controlled display case, ideally in the Church, and to put in a replica window in the Chapel. The option to recreate the window was considered acceptable because the tracings done in the 19th century were sufficiently detailed to allow an accurate reproduction to be made. This option also met the criteria for restoration set out in *Conservation Principles*. However, if parishioners had favoured a replacement window of clear glass, or a contemporary design in stained glass, either one of these approaches would have deserved serious consideration in the debate. The most important outcome has been the preservation of the authentic glass.

CHURCH OF ST PETER
PETERCHURCH, HEREFORDSHIRE

Taking a Grade I-listed Romanesque church, stripping out and re-flooring the nave, installing a two-storey facilities block and matching furniture, all in minimalist style, may sound like a recipe for conservation conflict, but actually there was very little discord over this remarkable project.

The key to success here was a widely accepted assessment of the heritage significance of the Church, its characteristics, its elements and its furnishings. The first step was to understand the building and contents through observation, archaeology and research. The building itself has very high evidential, historical and aesthetic value, hence its high listing grade. It also has substantial communal value as the focal point of the village, its tall and slender spire a familiar landmark in the Golden Valley.

The interior fittings and finishes of the nave had, however, much more limited heritage significance in all values. Simple pews dating from around 1900 stood on a floor finished in plain pressed-clay tiles. Thus their evidential, historical and especially aesthetic values were very low, although – like almost all pews – they did in a few cases have communal value to long-established villagers for their personal associations. Parishioners took the initiative and proposed the complete refitting of the nave for use as a community centre, while it still remains available for larger services. The initial design of the scheme and its iterative passage through the Church of England's faculty jurisdiction process all happened unusually quickly, to secure major public funding before a deadline.

The nave of St Peter's Church has been adapted by Communion Design for multiple uses including a community centre and public library. The two-storey oak-clad 'facilities block', containing a kitchen, toilets and stores, can be seen at the west end of the nave.

The result is a widely welcomed new facility for the village and its environs, bringing social services provision, a branch library, concerts, films, and a wide and growing range of community activities and events. This fuller use of St Peter's has done much to keep the building cherished by local people, and through that to ensure its long-term conservation.

Further Reading

Brereton, C. (1995); *The Repair of Historic Buildings: Advice on Principles and Methods* (2nd edition); London: English Heritage

Clark, K. (ed.) (1999); *Conservation Plans in Action: Proceedings of the Oxford Conference*; London: English Heritage

English Heritage (2002); *Conservation Bulletin 43: The Value of Historic Places*; London: English Heritage; available from *www.historicengland.org.uk/images-books/publications/conservation-bulletin-43*

English Heritage (2007); *Conservation Bulletin 55: Heritage: Broadening Access*; available from *www.historicengland.org.uk/images-books/publications/conservation-bulletin-55*

English Heritage (2010); *Conservation Bulletin 63: People Engaging with Places*; available from *www.historicengland.org.uk/images-books/publications/conservation-bulletin-63*

English Heritage (2011); *Knowing Your Place: Heritage and Community-Led Planning in the Countryside* (web document); available from *www.stratford.gov.uk/doc/173665/name/English%20Heritage%20Knowing%20your%20Place.pdf/*

Feilden, B. M., Jokilehto, J. (1998); *Management Guidelines for World Cultural Heritage Sites* (2nd edition); Rome: ICCROM

Historic England (2008); *Conservation Principles, Policies and Guidance for the Sustainable Management of the Historic Environment*; London: Historic England; available from *www.historicengland.org.uk/images-books/publications/conservation-principles-sustainable-management-historic-environment/*

Historic England (2008); *Understanding Historic Buildings: Policy and Guidance for Local Planning Authorities* (web document); available from *www.historicengland.org.uk/images-books/publications/understanding-historic-buildings-policy-and-guidance/*

Historic Scotland (2000); *Conservation Plans: A Guide to the Preparation of Conservation Plans* (web document); available from *www.historicenvironment.scot/media/2786/conservation-plans.pdf*

Insall, D. W. (2008); *Living Buildings: Architectural Conservation, Philosophy, Principles and Practice*; Mulgrave (Australia): The Images Publishing Group

Kerr, J. S. (1996); *The Conservation Plan: A Guide to the Preparation of Conservation Plans for Places of European Cultural Significance* (4th edition); Sydney: The National Trust of Australia

The National Lottery Heritage Fund; *Conservation Planning Guidance*; (web page); available from *www.heritagefund.org.uk/publications/conservation-planning-guidance*

SURVEY & INVESTIGATION METHODS

This chapter provides an overview of survey and investigation techniques currently available, and discusses their applications and limitations. It considers measured surveys, as these form the basic template upon which a range of information can be recorded. The use of documentary sources and structural archaeology to provide data about the origins, evolution, and associations of heritage assets is discussed. The range of condition surveys, specialist investigations and diagnostic techniques, to assist in understanding the construction and condition of a historic building, is also reviewed.

Effective conservation planning depends on understanding both the significance of a historic building or place and how that significance is vulnerable to harm. Gathering and recording information about the origins, evolution and associations of a heritage asset, its heritage values, physical condition, and the processes of deterioration and other harmful agencies, is the starting point for devising appropriate strategies and measures to sustain significance.

The process of information gathering and assessment should always be well-focused and proportionate in scope to the significance and sensitivity of the heritage asset in question, and to the nature and complexity of the problems being addressed. As the assessment stage progresses, investigations may be scaled up or down to suit the circumstances of a particular case. For example, further and more detailed campaigns of investigation may be needed to answer specific questions that have arisen earlier in the assessment stage. On the other hand, the scope of investigations may be constrained by cost, difficulties in obtaining access, or restrictions on opening-up or monitoring; it may also be influenced by the requirements of funding bodies.

Systematic recording of data is an essential aspect of all conservation work, and is fundamental for the effective management and care of heritage assets. The existence of accessible records allows information about the history of the building, its construction and condition, and the treatments carried out, to be collated and used to inform future decisions. For example, the records of regular inspections can help to build up a picture of the way a building and its elements are working both as a whole and individually, and may be crucial to understanding the causes and consequences of deterioration.

A wide range of survey, investigation and analytical techniques is available to obtain an understanding of a heritage asset and the ways in which it is valued, and to provide the basis for its ongoing management, maintenance and repair. In recent years, developments in survey technology and digital imagery have made it possible to gather accurate data in quantities, and at speeds hitherto unprecedented. This makes it all the more important for the purpose and objectives of survey, investigation and recording to be clearly established at the outset, so that the optimum balance between precision, speed and cost can be obtained.

INTRODUCTION

Surveys, investigations and assessments of historic buildings and places entail a broad range of complementary procedures. The scope and objectives of these will be determined by the specific requirements of the project in hand. Surveys and investigations can be broadly categorised as follows:

- *Measured surveys*
  These provide the graphic base for recording, analysis and conveying information.

- *Surveys and investigations to understand history, evolution and significance*
  Including background documentary research, structural archaeology and other specialist investigations.

- *Surveys and investigations to understand construction, condition and behaviour*
  Including preliminary, periodic and detailed condition surveys; structural building services, environmental and ecological surveys; diagnostic investigations and analysis and characterisation of materials.

Many types of survey and investigation require specialised knowledge, skills and sometimes equipment. When commissioning specialists, it is important to ensure they are suitably qualified, and have proven practical experience in working with heritage assets (see **Managing Maintenance & Repair**).

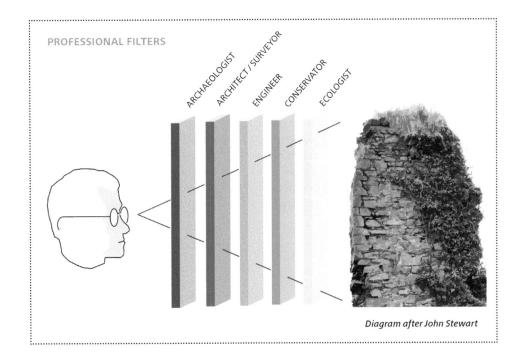

PROFESSIONAL FILTERS

ARCHAEOLOGIST

ARCHITECT / SURVEYOR

ENGINEER

CONSERVATOR

ECOLOGIST

Diagram after John Stewart

Practitioners from different professions utilise their own dedicated interrogative 'filters' to read evidence contained within historic buildings, such as chronological development, materials and structure and their condition, or flora and fauna.

MEASURED SURVEYS

Measured surveys (often referred to as metric surveys) serve a wide range of purposes in understanding, managing, maintaining, repairing and recording of buildings and places. For all but the simplest of buildings or works, they are usually commissioned at an early stage in a project, and provide the basic 'building block' that will be used and developed by the entire project team. In the past three decades, methods of measured survey and data capture have increased in accuracy, versatility and number. However, there is no simple formula for determining which survey techniques will be most appropriate in any given situation. Ultimately, the choice depends on the balance between the anticipated end use of the survey, the precision required, practical and technical considerations, the timescale, and cost. Therefore, before commissioning a survey it is important to be clear about its purpose, and to ascertain the differing requirements of the client and specialist members of the project team. It is helpful to write a short statement on why the survey is needed and what the anticipated end uses are; this should then form part of the survey brief.

The forms in which the survey data are produced and presented will be influenced by the end uses. For example, where alterations to a building are envisaged, accurate plans, sections, elevations and details will be required in the form of line drawings (although it may not be necessary to survey the whole building to the same level of detail if the alterations are confined to one area). The production of measured line drawings can also help in determining the sequence of building phases or the spatial relationship of building elements, which in a complex building may be difficult to discern, thereby aiding archaeological interpretation.

In other cases, such as when recording the colour of different types of stone in a façade, tool marks, or the intricate details of a wall painting, an image-based product might be more appropriate, because the textural detail of the subject would not be adequately portrayed by a line drawing. Some image-based surveys can be made without recourse to measurement. For example, annotated sketches or photographs with sketch location plans may well suffice if the purpose is solely to provide a template for mapping the extent of superficial deterioration, or to identify the locations where treatment or repairs are to be carried out, and to provide a pre- and post-intervention record. In practice, a number of different measured survey methods and products will often be required to suit the requirements of a particular project.

Facing page: The preparation of roughly-proportioned site survey sketches can be helpful in undertanding and recording the construction and condition of a building. Without supporting measurements they are of limited metric value.

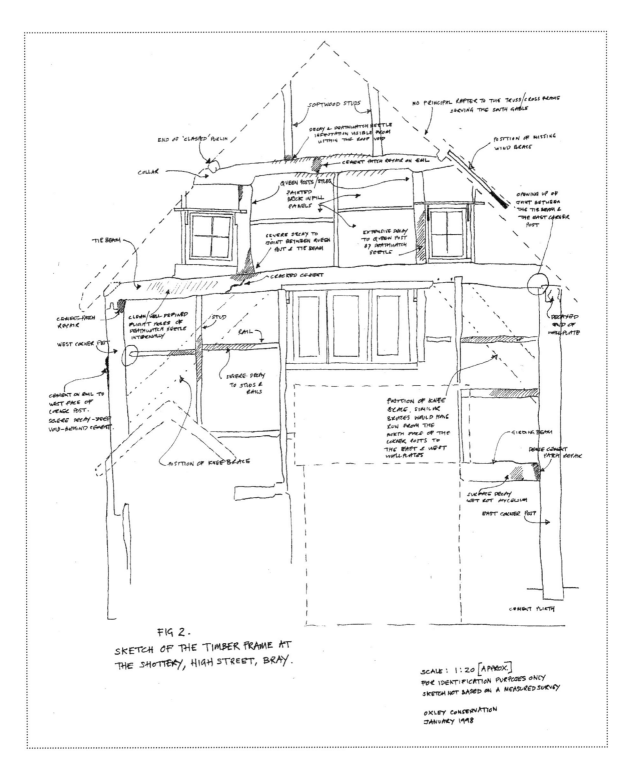

FIG 2.
SKETCH OF THE TIMBER FRAME AT
THE SHOTTERY, HIGH STREET, BRAY.

SCALE : 1:20 [APPROX.]
FOR IDENTIFICATION PURPOSES ONLY
SKETCH NOT BASED ON A MEASURED SURVEY

OXLEY CONSERVATION
JANUARY 1998

Photogrammetry

A composite internal elevation of the Battle Abbey courthouse showing raw photogrammetric data on the left, hand-drawn corrections in the centre (with the raw data shaded below), and the completed model on the right. The wall-head was completely obscured at the time of the photogrammetric survey and had to be recorded entirely by hand. It was also necessary to redraw stones that were either in shadow or otherwise indistinct in the photographs.

For most practical purposes, measured surveys of historic buildings should be capable of being used at a scale of 1:20 or 1:50. A larger scale of 1:5 may be needed for constructional details. For certain features, such as moulding profiles, a scale of 1:1 may be required.

It is important to understand the strengths and weaknesses of different metric survey methods. For example, the production of a detailed, hand-drawn survey entails the systematic observation of a building; a certain amount of interpretation is performed while the drawing is made. This can greatly assist in understanding, but is relatively time-consuming and expensive. On the other hand, surveys carried out using indirect techniques such as photogrammetry are fast, and can be dimensionally very accurate, but they may not identify features that would catch the eye of the experienced surveyor. In practice, a combination of direct and indirect methods is often used. For example, electronic or image-based techniques can be employed to provide an accurate dimensional base that is subsequently corrected, augmented and interpreted by direct survey techniques.

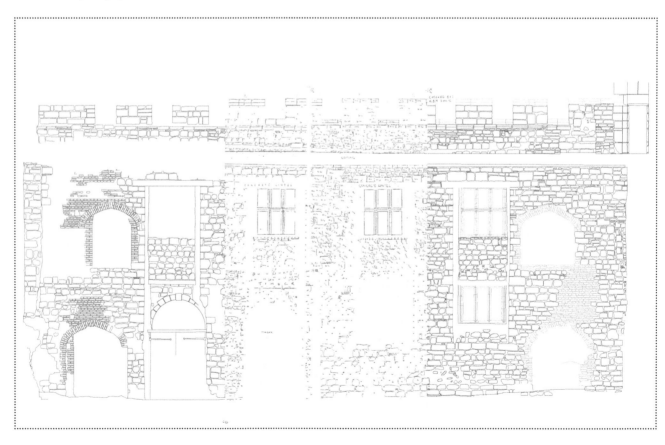

DIRECT TECHNIQUES

Direct survey techniques include hand surveys and electromagnetic distance measurement techniques using a total station theodolite [TST]. Direct techniques are dependent on the surveyor's observational skills and clarity about the purpose of the survey. As the data gathering process is selective, direct techniques can be very effective where the recording task has a thematic focus. However, this also makes them less flexible than indirect techniques.

HAND SURVEYS

Three types of drawing are used in heritage documentation: direct plotted drawings, measured drawings and sketch diagrams. Each type has its role according to the information requirements of the project in hand, and the methods used to produce them have their respective strengths and weaknesses. Direct plotted drawings are drawn to scale on site as the dimensional data are gathered. The production process is time-consuming, but direct plotted drawings are very reliable in terms of dimensional accuracy.

Making a plan from a dimensioned sketch, from site note to plot

1. The roughly-proportioned sketch plan is prepared by walking around the building, both inside and out.

2. Measurements are applied to the sketch systematically, working clockwise around the inside and then anti-clockwise around the outside.

3. Diagonal or brace measurements are added.

4. The plot is worked up in a computerised drawing system.

1. **2.**

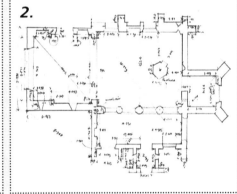

3. **4.**

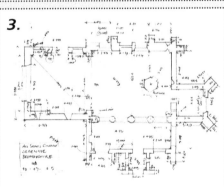
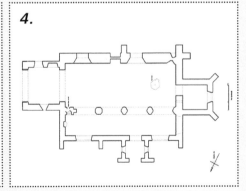

Top left: Diagnostic diagram of detail. The moulding is drawn up and measured as an indicative detail to assist in dating a building phase by joinery type.

Top right: Recording a wall using a rigid recording frame.

Bottom: Real-time CAD set-up being used. The tablet PC makes it possible to see and check the CAD work as the data are captured, a useful method for detail work using a total station theodolite.

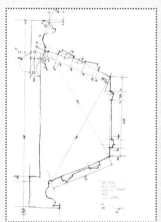

Measured drawings are produced off-site from dimensioned sketches and survey notes. This reduces the time spent on site, but missing or incorrect dimensions may mean additional visits. Sketch diagrams can be very helpful in understanding and recording building construction details, but without supporting measurements they are of limited metric value. Photographs can be used to support drawings and vice versa. Photography can be useful as an *aide-memoire* to reinforce the observations made in survey notes, and for recording colour, texture and undifferentiated form.

TOTAL STATION THEODOLITE [TST]

Survey with a TST is rapid and precise, but requires the surveyor to select the data to be recorded in the field. Instrument observations are made from fixed points or stations. Depending on the size and complexity of the job, further stations may be set out as required, or a traverse used to link sets of observations together. A TST combines precise electromagnetic distance measurement [EDM] and angular measurement with a data logging capacity to record three-dimensional points at ranges from approximately 0.25 m to 2000 m. A beam (usually infrared) is emitted, and distances are measured by recording the signal reflected from the selected target.

By combining the distance measurement with the horizontal and vertical angles between the instrument and the target, three-dimensional coordinates are calculated. TSTs are usually now available with a reflectorless function [REDM]. This is very useful for building recording work, as these TSTs will operate over a useful range of 0.5 m to 1000 m without needing a prism to return the signal. The TST is an invaluable tool for the production of plans, sections and elevations, deriving site-wide control by traverse, and providing control for rectified photography and photogrammetry.

The digital data from a TST can be used to produce CAD drawings directly on site. CAD may also be used to construct 'wire frames' to control hand-measured surveys, and the rectification and digitising of drawings and photographs, and to infill and supplement three-dimensional data from photogrammetric, laser scan or GPS data.

Indirect survey techniques include rectified photography, photogrammetry (measurement from stereo photography) and orthophotography, which are all image-based, and laser scanning. These techniques capture metric data in a non-selective and undifferentiated form. The captured data has to be subsequently processed to produce the required outputs. These techniques allow for very large amounts of data to be captured quickly. However, careful planning is required, and it is important that the purpose of the survey and its presentation remain clear at all stages of the process. Indirect techniques are also dependent on the interpretive skill of the photogrammetric operator or laser-scanning digitiser in accurately selecting the most appropriate point or sequence of points to represent the chosen feature.

RECTIFIED PHOTOGRAPHY

Rectified photography is a relatively quick and simple survey method, and is useful in circumstances where the subject is flat and contains a large amount of textural detail. The technique must, however, be used with caution. A standard photograph cannot usually be used to scale off accurate dimensions. To enable the printing of the rectified photograph to a specific scale, a method of dimensional control must be used. This can take the form of a simple scale bar or targets (usually four per image) that are attached to the façade. The distances between the targets can be found with a tape measure, or their positions can be coordinated using a TST. Rectification software is then used to produce an image in which the scaling variations caused by perspective are corrected.

Even without access to specialist rectification software, it is still possible to produce rectified photographs. Digital images can be manipulated, using photo-editing packages to remove distortions due to perspective. An image of control points can be exported from a CAD drawing and used as a layer in the photo-editing package, with the distorted image adjusted to fit the control. The image can then be imported into the CAD drawing of the control and scaled to fit, thus enabling printing to scale.

1.

2.

3.

4.

5.

Rectified photography of a sundial

The severe camera tilt has caused scale variation (1). The scale variation is removed by rectification (2,3). Once rectified (4), the image can be used for scale drawing (5), provided the tracing is only on the plane of rectification.

PHOTOGRAMMETRY

Photogrammetry is the technique of making precise measurements and drawings from two overlapping photographs ('stereo pairs') of the same subject, taken from slightly different positions. Photogrammetry is an economical and accurate survey method where a line drawing of a large area, such as a whole building façade, is required. It is most effective when applied to subjects where line detail is easily identified. The final product is a CAD drawing that can be easily edited by the end-user.

Digital cameras with calibrated lenses, where the precise focal length and distortion characteristics of the lens are already known, are generally used for photogrammetry. Ideally, the photographs should be taken as square-on to the subject as possible, but, if necessary, tilts of up to 30° can be accommodated. Digital and analytical photogrammetric systems now enable line drawings to be produced from very heavily tilted photographs. The further away the camera is from the subject, the greater will be the area covered in one stereo pair, but the image resolution will be lower. The image resolution required depends on the scale of the survey output and the amount of detail to be plotted. As a result, it will often be necessary to acquire several stereo pairs to cover one subject such as a building façade, with each pair overlapping its neighbours to ensure complete coverage.

A photogrammetric workstation is used to generate a stereo 'model' from the pairs of photographs. Measurements are obtained from the model in the form of three-dimensional coordinates, which are fed into a CAD system to produce a line drawing. The content of the drawing will depend on the detail visible in the photographs and the operator's interpretation of the detail.

To enable precise measurements to be taken from the stereo model, and to relate it to any other models, control points with known three-dimensional coordinate values are required. These may take the form of small plastic targets attached to the subject before the photographs are taken. Alternatively, existing points of detail may be used. The targets are dimensionally coordinated using a TST or other direct survey technique.

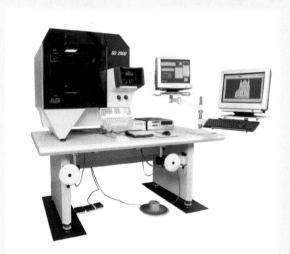

An analytical stereo plotter. The stereo pair is loaded, orientated and viewed before features are extracted by tracing using the hand wheels (x and y axes) and foot wheel (z axis). The linework is stored as CAD data.

The evolution of photogrammetric workstations, from optical-mechanical machines to PC-based digital systems, has significantly reduced hardware costs. However, fully-featured photogrammetric software is still comparatively expensive. In recent years, much less costly software for three-dimensional modelling, measurement and scanning, using normal cameras, has become increasingly available to the non-professional surveyor.

Photogrammetry at Whitby Abbey

Top: Eight stereo pairs were required to provide complete cover of the north face of the North Transept of Whitby Abbey church. The metric camera used for this pair was of 5 in × 7 in format, with a 300 mm lens. The red notation on the stereo pair marks the targets and detail points used for control. Note the fiducial marks on the border of the image; these are used to correct any film distortion that might have occurred. Here, they take the form of small circles, although the type of mark depends on the camera. Fiducial marks are not required in digital cameras, as the corners of the imaging sensor provide the same orientation functionality.

Bottom: Part of the photogrammetric drawing of the North Transept of Whitby Abbey church, plotted from a CAD file.

LASER SCANNING

A laser scanner is an automated range-finding device that rapidly captures a mass of three-dimensional data. It works by emitting a beam of laser light and, by detecting its reflection from the surface, can determine the precise distance between scanner and object. By simultaneously observing the angle of observation, a three-dimensional coordinate can then be calculated. Thousands, and in some cases, millions, of discrete points are generated per second to create a 'point cloud'. Each observation records the intensity of the laser reflection and, by using either an on-board imaging sensor or separate digital camera, colour information can also be collected. This 'point cloud' is then processed to produce a three-dimensional model of the scanned surface. Laser scanners used in building measurement are usually positioned on or near the ground (and are therefore referred to as 'terrestrial laser scanners'), but for other applications, such as wide-area topographic survey, the technology can also be airborne.

Laser scanners

Top: A phase comparison type, used for high-speed capture of large objects at long range (50–100 m³ at 0.3–180 m range).

Bottom: A time-of-flight type, used for large objects at long range (50–100 m³ at 0.1–300 m range).

The great value of a laser scanner is the comparative speed of mass three-dimensional capture. Although the technology is rapidly developing and becoming ever more appropriate to heritage recording applications, laser scanning is currently best suited to the three-dimensional recording of surfaces not suited to photogrammetric coverage due to the lack of defined edges; for example, caves and mine shafts, sculptural details, vault webs, dome and pendentive soffits. The technique is also useful in providing data for three-dimensional modelling, or to aid in the production of replicas.

Laser scanning is less suited to subjects where edge definition is important, as the edges and mouldings that characterise most architectural subjects can be indistinct unless the scan is of very high resolution. Subjects such as building façades are better dealt with by photogrammetry if a set of drawings is the required product. However, laser scanning used in conjunction with photogrammetry can be a useful, though comparatively costly, mapping technique for describing; for example, gross deformation or weathering of stone surfaces. Without recording textural information, laser scan data cannot record the condition of a site in the same way that image-based techniques can, so a co-incident programme of photography is a wise precaution when opting to use laser scanning.

Top: A triangulation scanner, used for high-density capture of small objects at close range (around 2-m³ object size at 2-m range).

Bottom: Phase-based scanner data with true colour RGB (red-green-blue) overlay. This image of a three-dimensional data set shows the coverage possible from a terrestrial scanner. The simultaneous capture of RGB data adds to the richness of the model. The patches of white in this image are areas where there is no data due to interference from obstructions (in this case the cornice and balcony railings).

Orthophotography

Top: A standard photograph of the south doorway of Kilpeck Church, Herefordshire, showing scale distortion.

Bottom: An orthophotograph of the south doorway of Kilpeck Church. The orthoprojection makes it possible to view the entire doorway at the same scale. Topographic detail can be combined with orthophotographs if required, with the line drawing appearing on top of the photographic image.

ORTHOPHOTOGRAPHY

An orthophotograph is a photograph that has been corrected for scaling variations arising out of the relief of the subject, as well as tilts of the camera relative to it. Orthophotographs are useful when an image-based output is required, but the subject is not sufficiently flat for rectified photography to be used. Using the digital photogrammetric process, it is possible to produce a digital terrain model [DTM] that is an accurate representation of the surface of the land or elevation of a building. The DTM can then be used to adjust the scale of an image, pixel by pixel, and thereby convert it from a photograph with perspective projection to one with orthographic projection: an orthophotograph.

The stereo photography for an orthophotograph of a building must be taken as square-on to the façade as possible, to minimise any gaps in the image that might be caused by, for example, a projecting feature that occludes the detail behind it. An orthophotograph can be printed out at the required scale, or imported into a CAD package. Within the CAD package, it can be combined with line data from conventional photogrammetry to produce a composite product. It should be remembered that an orthophotograph is two-dimensional, and so contains no depth information. However, an orthophotographic image can be 'draped' over a DTM using three-dimensional modelling or visualisation software.

INDIRECT METRIC SURVEY TECHNIQUES

TYPE	USES	SCALE	RANGE	REQUIREMENTS
RECTIFIED PHOTOGRAPHY				
2D SCALED IMAGES	Condition recording and assessment Works scheduling	1:20 to 1:50	2 m to 50 m	Metric or non-metric camera Precise control data or scaling information Rectification software
PHOTOGRAMMETRY				
3D STEREO PAIRS	Recording Condition monitoring	1:20 to 1:200	2 m to 100 m	Calibrated camera Scaling or precise 3D control data
LINE DRAWINGS, CAD	Architectural 'stone by stone' drawings Topographic surveys Landscape surveys Condition recording Works scheduling	1:20 to 1:200	2 m to 100 m	Photogrammetric plotting machine/software: analytical or digital
ORTHOPHOTOGRAPHS	Conservation plans, landscape survey, condition recording, condition assessments, works scheduling	1:20 to 1:200	2 m to 100 m	Operator experienced in stereo-viewing and image interpretation
DIGITAL ELEVATION MODELS [DEM]	Condition monitoring Surface and 3D modelling Reverse engineering Visualisations	1:5 to 1:50	2 m to 100 m	Image processing CAD and 3D-modelling software
LASER SCANNING				
3D TERRESTRIAL SCANNER Point clouds, meshed surface models	Surface and 3D modelling Record drawings Visualisations Reverse engineering	1:20 to 1:100	5 m to 500 m	Scanner, post-processing and 3D-modelling software
CLOSE-RANGE 'ARTEFACT' SCANNER Point clouds, meshed surface models	Condition monitoring	Actual size to 1:10	0 m to 5 m	Scanner Post-processing and 3D-modelling software

DIRECT METRIC SURVEY TECHNIQUES

TYPE		USES	SCALE	RANGE	REQUIREMENTS
DRAWING					
2D	SKETCHES	Diagnostics Support to 3D modelling	1:20 to 1:50	0 m to 30 m	Trained draughtsman CAD skills
	MEASURED DRAWINGS	Plans, sections and elevations	1:20 to 1:50	0 m to 30 m	Trained draughtsman CAD skills
TOTAL STATION/REDM					
3D	POINT DATA	Terrain models	1:50	5 m to 100 m	EDM set + field CAD unit, CAD skills
	WIRE-FRAME CAD DRAWINGS	Plans, sections, elevations	1:50	5 m to 100 m	EDM set Field CAD unit CAD skills
	CONTROL DATA	Monitoring and metric data integration	1:20 to 1:500	5 m to 100 m	EDM set Specialist survey skills
GPS					
	POINT DATA	Terrain models	1:100	20 m to 500 m	GPS set Specialist survey skills Open sky Height precision can be a problem
	WIRE-FRAME CAD DRAWINGS	Control data Site plans Landscape survey	1:100	20 m to 500 m	GPS set Specialist survey skills Open sky Height precision can be a problem

UNDERSTANDING HISTORY, EVOLUTION & SIGNIFICANCE

BACKGROUND DOCUMENTARY RESEARCH

Background documentary research provides the opportunity to learn more about the building and its history. It can be both challenging and time-consuming, as documentary evidence can be elusive. Even the best building accounts, where they exist, will have gaps and contradictions. However, decisions about conservation can be informed by even incomplete information, and missing parts of the jigsaw are often filled by evidence that comes to light later in the project.

Information gathered in this stage of investigation will help to determine the heritage values and significance of the building and its elements. In practice, historical information about a building will often be limited or anecdotal, even though its significance is acknowledged, for example, by virtue of it being listed or in a conservation area. List descriptions, particularly older ones, often give little or no historical information and do not explain why a building is significant, and thus further research will usually be needed. Documentary research can also provide useful information about past uses of a building, or events or changes that might have a bearing on its present condition. It can help in assessing the progress and rate of ongoing deterioration, and can pinpoint critical causes of decay that might occur only irregularly or infrequently and which might otherwise be missed, even during prolonged monitoring. An example of this would be periodic flooding. Although background documentary research is important for all these practical purposes, it is often under-resourced in many projects and sometimes skipped entirely. Ironically, background research is probably easier to do now than ever before, with many resources easily accessible online.

In planning for maintenance and repair, the aim should be to obtain a level of understanding sufficient to inform and justify the decisions to be made, rather than producing the definitive architectural history of the building, although this may be an objective in a larger or more complex project. This 'desk study' should be undertaken before any investigative work is commissioned. The first step in background research is to gather relevant existing data. The starting point will normally be the designation document (for example, list, schedule, conservation area designation statement, supported by a character appraisal, if one exists), which may provide references to more detailed sources. In addition, a national database, *Pastscape*, is available online, as are most local Historic Environment Records [HER]. These may lead to accounts of previous research, or to historic photographs, held in the National Monuments Record, or in the HER itself.

The National Archives is a further online resource that can help in sourcing historical information such as Land Tax and Hearth Tax (late 17th century) records. Easily accessible published sources, such as the **Buildings of England** ('Pevsner') series, the **RCHME Inventory** and **Victoria County History** volumes (the latter available via **British History Online**), may also include specific information about the heritage asset in question. Historic maps, including first and second edition **Ordnance Survey** maps, can be particularly useful. These are now available commercially, and some are accessible online. The local Record Office may also have older maps such as Tithe Maps (*c*.1840), and often older ones still.

The study of place name evidence – toponomy – can provide information about the geography of places and past land uses. For example, toponymic analysis of field names can provide historical evidence of agricultural practices and give insights into the relationships past generations have had with the land they farmed. This information can contribute to the depth of understanding of a building or place and its significance.

Photographs and drawings, especially old photographs and antiquarian images, can be extremely helpful in dating alterations, and, sometimes, in estimating the rate of deterioration. Old building construction manuals – the earliest of which date from the 18th century – can be helpful in understanding the anatomy of traditional buildings.

Maps

Left: Tithe maps (1836–1851) provide information about past land uses and, used in conjunction with other maps and documentary sources, can help in understanding the way a place has developed.

Right: First edition Ordnance Survey map (1875) plotted to a scale of 1:2500. Many historic maps are now available to be viewed online.

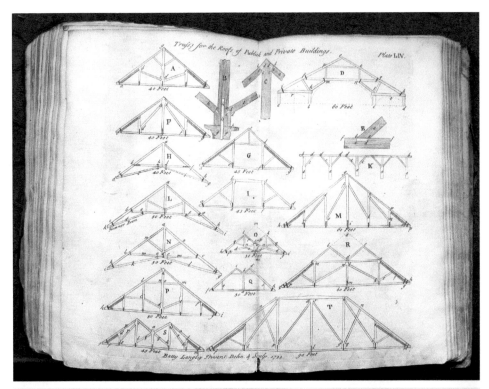

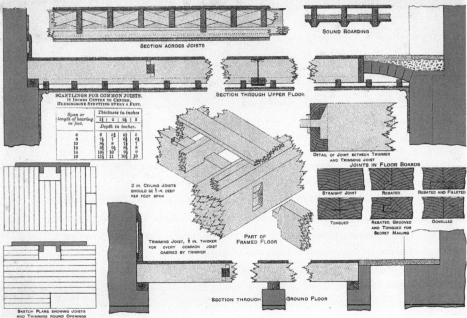

Construction manuals

Top: *"Truss's for the Roofs of Publick and Private Buildings"*, as recommended by Batty Langley in *The Builder's Compleat Assistant* (1738).

Bottom: Details of timber floor construction shown in *Building Construction* (1906) by Henry Adams.

CONSERVATION BASICS
SURVEY & INVESTIGATION METHODS

1.

2.

3.

4.

5.

6.

Building history in images

Old photographs and antiquarian images can be helpful in understanding the history of a building, and for dating alterations. This example shows the changes undergone by an early 18th-century town house, the Mansion House in Dodington, Whitchurch, Shropshire.

1. The town house in a coloured print of about 1840.

2. By the late 19th century, iron railings had been erected in front of the house, as this postcard shows.

3. When this photograph was taken of a procession in 1905, the house had become a school for girls (some of whom can be seen watching the parade from the windows).

4. By the 1930s, the building had become a petrol station and garage; note the changes in the ground-floor windows.

5. By the 1970s, it had changed from a garage to a supermarket.

6. In 1998 the house was converted into flats and the façade was restored (using the 1840s print as a guide).

Documentation of more recent interventions, such as the local planning authority's planning and development control records, can often be of great practical help in providing further information about the building. Information may also be obtained from the local planning authority about designation, and whether the building is within any form of designated area, either statutory or non-statutory (these include areas scheduled as ancient monuments, areas with archaeological significance, conservation areas, AONBs and national parks).

Other sources of information include building owners and custodians, and any records, drawings and/or specifications, estimates, accounts or contract documents they may have or know of. If a maintenance logbook, record or archive exists, this will provide an initial source of information. If a building record does not exist, however, this is the time to establish one. For churches, there should be a recent (within five years) quinquennial inspection report prepared by the Diocesan-approved quinquennial inspecting architect, and a church logbook (in which any works to the building should have been recorded).

The quality of information and sources must be taken into account. Although unsubstantiated reports and undated photographs may be helpful, firm conclusions can only be deduced from reliable data. Where warranted by the complexity or sensitivity of the building or place, a professional architectural historian may be engaged to carry out research and provide a report. This may be done in conjunction with a building archaeological investigation. Copies of documentary research and archaeological investigation reports should be lodged with the local authority Historic Environment Record [HER], or the local history section of the public library.

Old accounts

Where old building ledgers, accounts and estate records exist, they may provide information about works carried out to a building. Records of this kind may be used to assist in the interpretation of building archaeological evidence. The ledger books of John Eveleigh, an architect working in Bath (1787–1794), record building works he oversaw in the town.

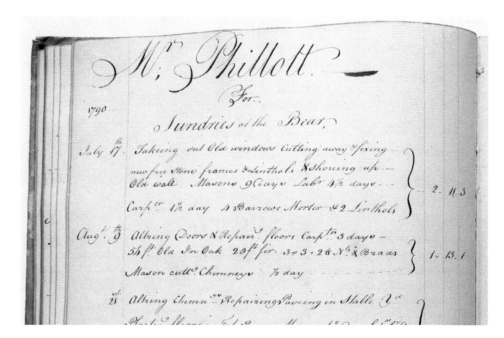

Architectural Reference Collections

Collections of historic artefacts, rescued from buildings and sites, can be an important source of information to help understand the history, evolution and significance of a building. Many national and local museums contain informative examples, but visits to other specialist collections can also be rewarding.

English Heritage's Architectural Study Collection is a substantial repository of architectural fragments, representing London's building history from the early 17th century to the mid-20th century. The Collection has been assembled over the past 100 years by English Heritage and its predecessors. It contains dated examples of decorative plasterwork; woodwork, including staircases, panelling and doors; metalwork, with examples of wrought- and cast-iron railings and balconies, as well as sections of structural ironwork. There is also a fine collection of historic wallpapers that range in date from c.1700 to 1939. Queries from researchers can be answered by appointment or by e-mail enquiry. The research data available provides information about provenance, photographs of the objects, historic and contemporary views of the original building location, and academic summaries of a range of building construction. From the summer of 2012 a selection of the Architectural Study Collection will be available online at *blog.english-heritage.org.uk/10-things-you-didnt-know-about-wrest-parks-collection-store/.*

The Brooking Collection of Architectural Details illustrates the evolution and construction of a wide range of architectural features spanning a period of nearly 500 years. Founded by Charles Brooking in 1966, the collection contains dated examples of window and doors, shutters and fanlights, staircases and period mouldings, along with ironmongery fire grates and rainwater goods. These have been retrieved from buildings across the British Isles, including the Channel Islands and Eire, ranging in status from large country houses to vernacular cottages. The Brooking Collection is available as a research and teaching resource by appointment. For details see *www.thebrooking.org.uk*.

Reference collections

The Brooking Collection includes dated examples of historic windows and doors from 1550–1960, as well as an extensive range of other architectural joinery, ironmongery, fire grates and rainwater goods.

While documentary research is important, the fabric of the building or structure itself provides the primary source of information about its history and evolution. Documentary records seldom provide information about how a building was put together: constructional details were usually left to the craftsman to resolve and, even where working drawings exist, the building fabric may tell a different story.

Therefore, a close examination of the fabric will normally be needed to reveal how a building was first built and its construction modified over time. The original construction of many historic buildings, such as late Georgian terraced houses, tends to be relatively standardised (albeit with regional variations) and easy to comprehend. Understanding subsequent interventions, their impact (both positive and negative) on heritage values, and their implications for repair proposals, is comparatively straightforward.

Building archaeology

A painted dado, revealed by the removal of later wallpapers, provides evidence for the location of a service staircase in an 18th-century house.

At the other extreme, some heritage assets are the highly complex outcome of numerous phases of development that have added layer upon layer, sometimes over a period of centuries. In such cases, a building archaeology-based investigation may be needed to gain a level of understanding about their history, evolution and significance sufficient to inform decisions about conservation. This will usually be carried out by a multi-disciplinary team with expertise in building archaeology, architectural history and historical research. Depending on the nature of the heritage asset, other specialist investigations may also be required, such as the dating of structural timbers by dendrochronology or paint analysis. Archaeological building analysis can be time-consuming and expensive. However, it is warranted where the level of complexity and sensitivity of a heritage asset is such that its significance, and the impact of proposed interventions, could not otherwise be fully or properly assessed.

Archaeological building analysis will often be carried out as part of the preparation of a Conservation Plan. The process usually begins (after background research) with visual analysis and interpretation of the fabric. This includes characterisation of buildings and setting; description of uses (past, if discernible, and present); plan form, layout and construction of buildings; and phasing of development and the main stages of alteration. It will normally involve looking into all accessible voids (the back of things can be as informative, in their way, as the front, since there will have been no attempt to disguise construction or change). For all but the smallest structures, a measured survey will aid interpretation, and provide a template on which to present and record findings.

The information gathered should be presented in written form, describing the main phases and sequence of development and alteration, and identifying features of particular interest or value. The report should be supported by annotated photographs and drawings showing the relative age of building elements (that is, a 'phased plan'), and, if relevant, drawings showing the form of the building in earlier periods. The report should also draw attention to the need for further investigations that may be required to answer specific questions, or reduce uncertainty about the history and evolution of the building. These might include documentary research, specialist investigation or opening-up building fabric to reveal concealed features (Listed Building or Scheduled Monument Consent, as applicable, may be required for this). In practice, further work of this type will often be commissioned as the assessment proceeds.

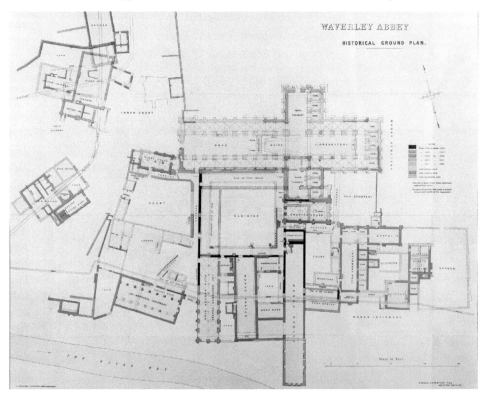

A periodised plan

The sequence of building construction, discerned through archaeological building analysis, is shown by indicating each phase in colour on a scale plan.

Phases of construction

Top: Photographs may be used as a recording template. Here, the archaeological interpretation of a timber screen has been combined with notes on its construction and condition.

Bottom: Derby House, Castle Rushen, Isle of Man. Rectified photographic image of the Outer Ward elevation of the lodgings of the Lords of Man. Period 3.5 is 15th-century work incorporated in the Period 4 building of *c*.1582–83, heightened *c*.1643–44 by James, 7th Earl of Derby. Late 17th- to early 19th-century interventions are shown in brown wash, those of *c*.1837 and later in grey wash.

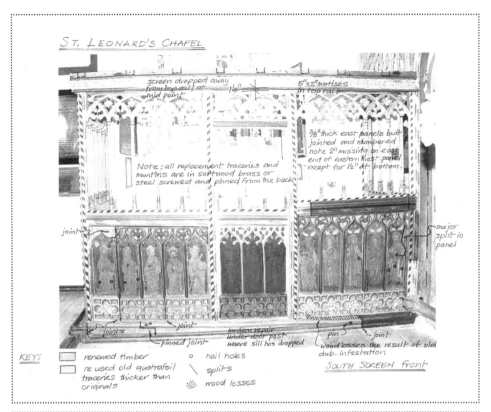

SUPPORTING INVESTIGATIONS

A range of specialised analytical techniques that can assist in understanding the history, evolution and significance of heritage assets is available. Some of these techniques are relatively expensive and time-consuming. Therefore, analysis should always be focused on providing answers to specific, clearly defined questions, in proportion to the complexity and sensitivity of the heritage asset under consideration.

Some techniques (dendrochronology, radiocarbon dating, thermoluminescence dating) provide absolute dating evidence. Others (mortar, paint and wallpaper stratigraphy) are used to determine chronological sequence.

DENDROCHRONOLOGICAL DATING

Dendrochronology – dating timbers or historic wooden objects by dating the trees from which they were made – uses the patterns of growth revealed by tree rings. Tree rings vary in width as a result of varying annual weather patterns. Wider rings result from more favourable growing conditions. These variations are shared to a greater or lesser extent by trees, locally, regionally and even nationally. Starting with living trees and working backwards in time through bog oaks and archaeological material, dendrochronologists have combined many overlapping sequences to develop 'master curves' of growth patterns covering thousands of years. To date a particular sample, the tree-ring widths are measured and the sequence is compared to a master curve, moving it along year by year until a good match is found between the two patterns. If the match is close enough, it pinpoints the period when the tree was growing. ⊖TIMBER

 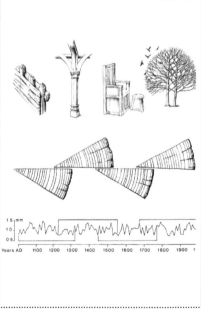

Dendrochronology

Left: A dendrochronologist recording tree-ring widths at the end of a board, using a microscope linked to an electronic measuring device. Where the end of the timber is not accessible, a core may be extracted.

Right: Diagram showing the principle of tree-ring dating.

Radiocarbon is an unstable isotope of carbon which decays at a known rate. It is created in the upper atmosphere by cosmic radiation, and is absorbed by plants through photosynthesis and thence into the biosphere. The ratio between radiocarbon and the two stable isotopes of carbon in living things is thus the same as that in the atmosphere: after death the proportion of radiocarbon in a sample declines and so, by measuring the proportion remaining, the time since death can be estimated. This involves a calibration process to convert the radiocarbon measurement to the calendar timescale. Radiocarbon dating can thus be applied to organic building materials: constructional timber (where dendrochronology has not provided absolute dating), wattle, thatch, hair or charred plant inclusions in plaster or mortar.

Single radiocarbon dates, when calibrated, will usually span more than half a century and will often span more than a century. Such analysis is expensive. It is suitable for short-lived materials, such as wattle or thatch, or the outer rings of longer-lived wood. For constructional timbers, or tree-ring sequences that cannot be dated by dendrochronology, a series of samples from known positions in the tree-ring sequence can be taken. A complex statistical process known as 'wiggle-matching' then allows the tree-ring sequence and the radiocarbon dates to be combined to give a date estimate which usually spans a few decades. Usually 5–6 radiocarbon dates are needed for this process.

It is also possible to date carbon in iron, which derives from the fuel used in smelting and smithing, and the carbon fixed from the atmosphere into lime mortar as it sets, although these methods are still experimental and should only be used with expert advice.

THERMOLUMINESCENCE DATING

Thermoluminescence dating can be used to estimate the time when a brick or tile was fired; again the method provides dates with an error range, and it is expensive. Quartz and other crystals found in clays store energy from ambient radiation, which is released by heating to a temperature over 500°C. When clay is fired, this stored energy is released; so measuring the energy present in minerals in a brick can date when the brick was fired (if the amount of radiation that the brick has received over the years since it was fired can be estimated). Generally, errors on luminescence dates are 5–10 % of the actual age (for example, AD 1900 ± 5, AD 1600 ± 20).

Sturminster Newton Town Bridge, in Dorset, was described by John Leland in 1542 as *"an excellent stone bridge of six arches newly built"*. By 1820, documentary records show the bridge in its current form, which is wider than expected for an early 16th-century bridge and does not conform entirely to Leland's description. The weathering and coursing of all the stonework is fairly consistent, which suggested that perhaps the bridge had been entirely rebuilt before 1820. However, closer investigation revealed subtle differences in surface finish of the stone – some were diagonally tooled with axe work, whilst others were vertically tooled with a bolster – suggesting that the original bridge had in fact been widened by erecting new arches on top of the cutwaters, probably in the 17th century.

During recent work to replace severely weathered stones, close examination of the bedding and core mortars confirmed the phasing of construction and alteration. The inner arches, *c.*1520, were bedded in a soft brown mortar, whereas the added arches were bedded in a hard white lime mortar with a harder brown core mortar. The 16th-century builders also appear to have coated the backs of their stones with clay before setting the core, a technique which is not seen in the later widening, but which was presumably intended to limit water penetration into the core. In places later repairs were apparent: also set in a hard white mortar, but with a less obtrusive aggregate.

MORTAR STRATIGRAPHY

The sequence of construction and alterations in a masonry building can often be discerned in differences in the materials used during specific phases or periods of building. Variations in the characteristics of mortars, plasters and renders (for example, aggregate size, particle shape, colour) can be helpful in differentiating building phases. Visual examination *in situ* with a ×10 hand lens is a quick means to establish very broad classifications. These can then be refined by the analysis of mortar samples extracted from the fabric. As the latter process is both destructive and expensive, it should be limited to resolving very specific questions. ⊕MORTARS

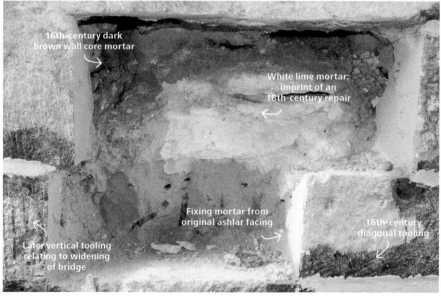

16th-century dark brown wall core mortar

White lime mortar: imprint of an 18th-century repair

Later vertical tooling relating to widening of bridge

Fixing mortar from original ashlar facing

16th-century diagonal tooling

History from paint layers

Top: Detail of an 18th-century baluster, one side of which has been stripped of paint. Although this reveals the original profiles produced by the wood turner, the information about the history of the building that the paint layers can provide is lost.

Bottom: A photomicrograph showing a cross-section through the layers of paint on the baluster. The earliest layers date from the 18th century and are all pale colours based on white lead. The darker colours were applied in the late 19th and early 20th centuries, and indicate changing fashions and the availability of a wider range of inexpensive pigments. Layers of dirt (the black lines) separate the successive schemes of redecoration. The comparison of paint stratigraphies from different architectural elements within a room can provide information about past alterations.

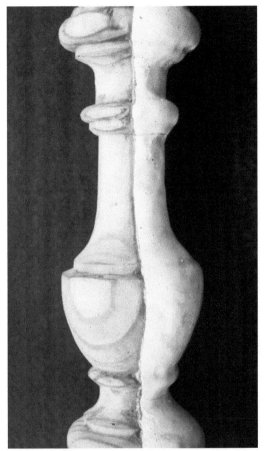

The practice of coating architectural surfaces with paints for both decorative purposes, and to protect substrates from the weather, has a long history. However, the importance of the architectural paintwork has really only been recognised in the last 30 years. Prior to that, it received little attention, and a great deal of painted decoration was lost. Much historic paintwork has been stripped or overpainted, but where early paintwork survives, it can yield valuable information. In fact, sometimes painted decoration may be of more significance than the substrate to which it has been applied.

Even where historic paintwork has been removed, residual traces may still survive. It is particularly important, therefore, that historic paintwork, or evidence of it, is recognised, and its significance assessed by an architectural historian or architectural paint analyst.

If a building element has been previously repaired, it is important to look out for and record what may only be microscopic remains of an earlier polychromatic scheme that was cleaned off in the course of the earlier work. Very small samples of the paint may be taken, set in resin, polished, and the sequence of layers examined in cross-section under the microscope. In addition to providing evidence of historic colour schemes, painted surfaces can be of great help in understanding the history of a building or object. For example, when paint stratigraphies on different elements in the same context are compared, if the earlier paint layers are missing in some samples, this suggests that the elements with fewer paint layers are of a later date and, therefore, represent a later intervention. Paint pigments may also be analysed to provide dating evidence.

⊖MORTARS ⊖TIMBER

WALLPAPER STRATIGRAPHY

Historic wallpapers may be uncovered in the course of exploratory or building works. Often there will be several layers of successive schemes of wallpapering; sometimes the evidence of wallpapering survives only in the form of strips or small fragments which have been covered over, and thus protected, by building elements added later. Early wallpapers often have datable designs or registration marks which can be helpful in understanding the history and development of the building, as well as providing insights to the tastes and status of the building's occupants. Where early wallpapers are found, the advice of a specialist conservator should be sought to assess their significance, and to make recommendations on conservation.

Decorative finishes

Top: Late 17th-century marbled paintwork revealed when a timber moulding applied in the 18th century was removed during repair works. Early paint finishes may often be found in hidden locations; for example, behind shutters.

Bottom: A fragment of a rare 18th-century crimson flock wallpaper, discovered in a town house in north Shropshire during repair works. Only small areas of the wallpaper had survived and were covered by many layers of later papers. Old newspapers were sometimes used as linings and may provide dating evidence.

Information gathered and analysed in background research and building archaeology enables the heritage values – evidential, historical, aesthetic and communal – of a building or place to be assessed. Communal values are often more difficult to evaluate because these can be wide-ranging and sometimes contradictory. In order to do so, it is first necessary to identify the various groups – people who share a common experience, interest or objective – for whom a heritage asset has value. These may be diverse and range, for example, from neighbours, local community groups, amenity societies and religious groups, to special interest groups at local, national and sometimes international levels.

The next step is to understand how and why these groups value it. This process should be proportionate to the scale and sensitivity of the heritage asset, and the expected impact of any works proposed. It is likely to begin with discussions with the building owners or custodians; then, where appropriate, more formal interviews, meetings or workshops involving the relevant groups or 'stakeholders' may be needed to gather a wider range of views. To be effective, workshops should be well focused and planned to stimulate comment and discussion. Often, it is in reacting to proposals for change, hypothetical or otherwise, that people reveal the ways in which they value a heritage asset.

Balancing conflicting responses may involve delicate value judgements, but this will be easier to do if the evaluation process has been inclusive and transparent.

Understanding communal values

Meetings or workshops give the various groups of people who share a common experience, interest or objective concerning a place the opportunity to say why they value it.

UNDERSTANDING CONSTRUCTION, CONDITION & BEHAVIOUR

The purpose of condition surveys is to gather the information needed to understand the construction and condition of a heritage asset, and to identify how its significance is vulnerable to decay. In the first instance, it will often be the appearance of a single defect that gives cause for concern, but before deciding on a course of action to remedy the apparent problem it is important to take a step back and consider the symptoms in their wider context. It is all too easy to be caught up in dealing with symptoms rather than the causes of problems, or to focus on specific issues rather than seeing the bigger picture. Condition surveys should seek first to understand the building as a whole – its construction, condition and behaviour – before assessing details and identifying the need for further specialised investigations.

Building condition surveys provide a snapshot of the construction and condition of a heritage asset at the time of the inspection, and are an invaluable baseline for later examinations. Changes in condition between successive surveys can reveal the progress of decay, allowing the rate of deterioration to be quantified. There are numerous types of condition survey. It is important to appreciate their relative merits and limitations when considering which type will give the information required, and be good value for money. Initial inspections and routine maintenance surveys may often be most efficiently and effectively undertaken by people who are familiar with the building, such as the owner or custodian, or by a retained architect or surveyor. On the other hand, formal condition surveys of heritage assets will normally be carried out by a suitably qualified and experienced conservation architect or surveyor. Additional specialist inspections, such as structural surveys, should be done by professionals with the relevant conservation experience. Surveys or inspections, carried out free of charge by companies with a product or treatment to sell (for example, damp-proofing or timber preservation), should be treated with caution.

A systematic and analytical approach to understanding condition, how significance is vulnerable to decay, and the assessment of risk, is important when devising strategies for conservation, including repair, alteration and re-use. This approach entails investigating and diagnosing the causes of decay, gauging the vulnerability to threats, estimating rates of deterioration, assessing the life expectancy of original and replacement materials, and weighing up the likely effectiveness of remedial work. Considered and informed decisions can then be made about appropriate conservation strategies to mitigate harm, and reduce the need for long-term intervention.

Photography

Photographs are an essential component of every programme of investigation, conservation and recording, both as an *aide memoire* and a final record. Since they capture the appearance of a building element at a particular time, photographs taken at different times can be compared to assess change. The rapid technological advances in digital photography and editing software have simplified the production of good quality, high-resolution images, and significantly reduced costs.

There is no single method for photographing buildings. A selection of approaches is described here, and a typical photographic record is likely to combine elements of some or all of these.

PICTORIAL PHOTOGRAPHY

The aim of pictorial photography is to present not solely the appearance of a building or place, but to convey something of its mood or atmosphere. This is highly subjective and is likely to be influenced – albeit sometimes subconsciously – by the conventions of fine art. Room interiors will aim typically to depict the overall appearance and character of the space. They may be taken in natural light or can be subtly enhanced by artificial means.

Where the intention is to evoke mood, deep shadows may be tolerated even where they obscure architectural detail. Sometimes a particular detail or ensemble will speak volumes about a building's history or current condition. Pictorial photography is well suited to presentational and promotional uses.

EVIDENTIAL PHOTOGRAPHY

Photography may be employed to document features and form, or to support observations recorded in words or drawings. Viewpoint, composition and lighting must all be carefully considered so that the desired information is captured. For clarity, views are typically square on to the subject (with the result that three-dimensional form is not prominent). Whenever possible, it is good practice to include photographic scale bars in the image, ideally one horizontal and one vertical, on the plane of the subject.

Lighting, whether natural or artificial, is selected to highlight the relevant detail. Images recording colour and shape are usually taken in bright light, free, as far as possible, from intrusive shadows. Oblique or raking light should be used to highlight texture or surface topography.

Close-up or macro photographs may be taken of small details. These should be taken in a sequence of images to allow the detail to be unambiguously located in the building. A photographic scale or an object of known dimensions, such as a coin, should be incorporated.

NARRATIVE PHOTOGRAPHY

Photographs may also be used to tell a story, record a process, phenomenon or event, or draw attention to associations that may not be obvious. For example, a sequence of views that relate rooms one to another through open doors might help to illustrate the way a building was used, or to appreciate the intentions of an architectural design. An image, or series of images, of a craftsman at work may provide insights about the form and character of the object being made.

ANALYTICAL PHOTOGRAPHY

Photographic techniques may be used to examine and monitor the behaviour of surfaces. Raking light, ultraviolet and infrared photography can reveal features that are indistinct or invisible to the naked eye. Polynomial texture mapping, a computer-based technique for imaging and displaying surfaces under varying lighting conditions, is used to reveal, monitor and record surface phenomena. Time-lapse photography allows changes in the surfaces of materials to be monitored over time. In addition, image enhancement software applications are capable of exposing features in old photographs that are otherwise indiscernible.

BUILDING CONDITION SURVEYS

There is no single 'standard' approach for each survey type; the level of detail and method of presentation will depend, to a large extent, on the experience of the building professionals and the brief they are given. Before commissioning a condition survey, it is advisable to ask a potential surveyor to provide examples of past survey work and a detailed summary of experience. It should always be remembered that surveys can be tailored to meet the specific needs of a particular case. It is essential to be clear about the purpose of a survey and what information is needed, as this will have a bearing on the level of detail and accuracy required. As with measured surveys, it can be helpful to include a short statement in the brief about the purpose of the survey, to make sure that the objectives are clear and that all the necessary information is obtained. Where site conditions or available resources (including access) limit the closeness of inspection, there will be a greater degree of uncertainty in the conclusions that can be drawn. Survey accuracy is discussed later in this chapter.

It is important that any past condition survey reports and maintenance records are made available wherever new condition surveys are commissioned. Past data can provide valuable information about decay rates, and the success and durability of past treatments and repairs. Commissioning surveys without reference to past work may result in inappropriate recommendations and unnecessary intervention. The main types of building condition survey are described here.

Quinquennial survey

Example of a survey schedule produced for the quinquennial inspection of a church. The location, material and condition of each element is recorded, along with a description of the nature and extent of any necessary remedial action. Each action is assigned a priority rating, in this case:

A = Within 12 months
B = Within 12–24 months
C = Within the quinqennium.

6.0 ROOF COVERINGS		
NOTES: See key plan		
PRESENT CONDITION	WORKS REQUIRED	PRIORITY
6.1 Porch - E Slope (1)		
6.1.1 Random slate in fair condition with some moss colonisation; crested clayware ridge in good condition.	Remove moss; minor repairs to replace/refix 15 no cracked/displaced slates.	A
6.1.2 Lead valley in fair condition.		A
6.1.3 Metal (not cast iron) OG eaves gutter in good condition but requires clearing.	Clear out gutter.	A
6.1.4 Plastic RWP in poor condition.	Replace defective RWP.	A
6.2 Porch - W Slope (2)		
6.2.1 Random slate in fair condition; some displaced slates next to valley gutter and 7 no on lead tingles.	Minor repairs to refix displaced slates.	A
6.2.2 Lead valley in fair condition.	Renew lead lining to parapet gutter in accordance with current Lead Sheet Association (LSA) recommendations.	C
6.2.3 Lead parapet gutter (E side of Tower); no evidence of present leakage but lead lining in poor condition; length of bay exceeds current recommendations and is ridged; numerous "Flashband" patches.		
Gutter discharges to concealed internal (probably lead) RWP. Plastic grating fitted to outlet. NB: Outlet from internal gutter in Tower discharges into this gutter	Investigate condition of internal rainwater pipe, and replace if required.	C

TYPES OF BUILDING CONDITION SURVEY

DESCRIPTION	ADVANTAGES	DISADVANTAGES
PRELIMINARY, BASIC OR RECONNAISSANCE SURVEY		
At the project inception stage, carried out as part of a feasibility study to establish and review options for repair, conservation, adaptation and alteration. A basic site survey is required to find out, in broad terms, key information such as condition, construction, site access, size, and so on	Relatively low cost Useful fact-finding exercise Can avoid potential expense by eliminating unsuitable options at an early stage	Only intended to provide broad understanding; further surveys are required to provide greater detail
PERIODIC SURVEY		
Also known as property management surveys (for example, quadrennial and quinquennial surveys) The primary survey type for maintenance planning and management	Information is prioritised into years when repair work is required and includes budget costs An excellent way of planning for and implementing cyclical maintenance Can provide a valuable data base of information on condition, and past maintenance and repair	Computer-based systems which require minimal data input, particularly for defect analysis, frequently have limited value Not suitable for improvement work or 'project work', which involves detailed or extensive conservation and repair
BUILDING FABRIC SURVEY		
Also known as a 'full', 'comprehensive' or 'detailed condition survey' The principal survey type when a detailed condition analysis is required. Surveys can be tailored to establish the amount and scope of future conservation and repair project work. Variations on this survey can include: • Prioritised recommendations for maintenance and repair over specific time periods. For example, urgent, within two years, within five years, within 10 years, within 20 years • Prioritised cost budgets • Tabulated 'defect schedules' which list and describe all visible defects • Recommendations for conservation and repair	A comprehensive understanding of condition, and the requirement for conservation and repair Can be tailored to specific needs	Relatively high initial cost, but when correctly used are very cost-effective
STRUCTURAL SURVEY		
Structural surveys investigate the structural performance of a structure and are provided in addition to building fabric surveys	Specifically investigate and report on structural performance of buildings, structures or structural elements	Structural surveys generally do not assess condition unless it directly affects structural performance Recommendations can be limited to structural repair and may not address cosmetic issues Structural surveys are often confused with building fabric surveys, but do not look at the condition of the building fabric as a whole and each of its individual elements

PRELIMINARY, RECONNAISSANCE OR BASIC SURVEY (GROUND-LEVEL SURVEY)

A basic survey can be a useful tool for providing an overview of the construction and condition of a building, structure or site. It will usually be a preliminary activity in a wider programme of survey and analysis, and is the starting point for understanding how significance is vulnerable. A basic survey is normally carried out from ground level. The construction and condition of the building are assessed in broad terms. The survey report may simply provide a list of key defects, and a summary description of building elements, construction methods, materials and condition. A basic survey is relatively inexpensive to undertake and can limit expenditure in the early stages of a project, should it not proceed further. It will also establish the need for further detailed survey and investigation, as well as helping to focus requirements for them.

A ground-level survey may also be used for relatively simple projects where the condition of a building structure or element is already known through periodic property management surveys. The survey can identify the scope of repair required, based on recommendations from a quadrennial or quinquennial survey. Success of this type of survey is dependent on defects being easily identified from ground level. Binoculars and fieldscopes can be used to assist visual assessment, but they should not be relied on entirely, since they are tiring to use and tend to flatten objects. The quality of the binoculars will greatly influence visual accuracy.

PERIODIC SURVEY OR PROPERTY MANAGEMENT SURVEY

Periodic surveys are undertaken on a four- or five-year cycle, and are used to monitor condition and manage maintenance and repair work on a planned preventative basis. Although the need for a major programme of work may be identified by a periodic survey, the level of information provided about construction and defects will not normally be sufficient for preparing detailed repair specifications or schemes for alteration or re-use.

The content and format of the property management survey is prescribed by the property owner, which may be a charity, non-government organisation, or company that has a large portfolio of buildings and sites in its care. Historic England employs a standard format and survey method for its property management surveys (see **Asset Management Planning**). Periodic surveys identify key defects, and can also be used to assess rates of decay, deterioration and soiling based on comparative analysis of the data set out in previous periodic surveys. The survey report will be composed of a concise descriptive account of the defect types and causes. These are usually grouped into building elements such as roofs, walls, floors, windows and so on.

Property management surveys are very useful tools for the long-term management of a building or place, since they allow property managers to assess accurately and monitor the decay and deterioration of a site, including the wear-and-tear caused by building occupants or users. It is an effective way of planning expenditure, and leads to more cost-efficient management by encouraging regular and timely maintenance to prevent both premature failure of materials and minor defects developing into major ones.

BUILDING FABRIC SURVEY

A full building fabric survey, also referred to as a detailed building or condition survey, involves the comprehensive visual examination of a building or structure to provide information about its original construction, subsequent alterations and present condition. Its purpose is to provide a detailed factual statement about the physical condition and physical history of all aspects of the building, including its structure, components and services, that is both descriptive and analytical. The objective is to gather the information needed to diagnose defects correctly, assist in the development of the conservation strategy, and, ultimately, to inform the preparation of specifications for treatment and repair. In principle, a condition survey should consider the building as a whole and in detail. Areas of concern should be examined more closely. Problem areas highlighted in previous survey reports, where these are available, should also be revisited. Studying the patterns of deterioration is one of the best ways of diagnosing underlying problems, and a systematic approach to inspection is necessary to avoid missing any symptoms.

Establishing the rate, or rates, of deterioration and decay is paramount in understanding building fabric condition, and future life expectancy of both specific elements and the building as a whole. For example, the condition survey may identify typical losses to the exposure face of a limestone cornice of, say, 20 mm over 170 years, which on average equates to 11 mm every 100 years. Surface erosion of this kind is unlikely to adversely affect the structural integrity. However, it could be a cosmetic issue, or a functional one, as in the case of a water-shedding feature such as a cornice. This knowledge will then inform whether replacement, repair, conservation or no action is required.

Assessing rates of deterioration

Comparing the current condition of carved stonework (*right*) with a photograph taken 30 years ago (*left*) shows the extent of deterioration that has occurred during that period.

Inspection at height
Safe and economical inspections
at high level can be made with
a camera attached to a remotely
operated aerial vehicle [ROAV].

It is also necessary to understand the condition of the building as a whole. This overall view is needed both to help formulate the conservation strategy, which may include options for adaptation and re-use, and to identify priorities. A detailed understanding of the extent, type and cause of each individual defect is required to provide this holistic view.

To carry out a detailed and accurate survey, it is necessary to have close access to the fabric of the building. This may mean erecting scaffolding or using a mobile-elevated access platform. Consideration must be given to security and safety, not only of the people carrying out the inspection, but also the occupants of the building and passers-by. Also, the likelihood of encountering hazardous materials – for example, asbestos and bird guano – should be taken into account and appropriate measures adopted to mitigate risks (see **Managing Maintenance & Repair**)

Survey accuracy is dependent on accessibility. Ideally, surveyors will use all of their natural senses, including sight, hearing, touch, smell and, in some cases, even taste, to analyse defects. Close access is therefore essential if all the senses are to be used. Sight is the obvious mechanism for detecting and understanding condition. Voids can be detected in masonry by tapping or scraping surfaces with metal objects, and listening to the sound generated. Hollow or 'dead' sounds indicate voids, whilst higher pitched ringing sounds indicate solid well-bonded fabric. Touch is also frequently used to help understand the density of surfaces. A sharp object may be pressed into timber to detect softening and decay, or touch may be used on friable surfaces in order to make a subjective assessment about the amount of deterioration. Smell can assist in detecting the existence of damp or decay in a building, or identifying the presence of animals on a site. Experienced surveyors sometimes use taste to identify the presence of salts on a masonry surface. Although this simple technique can provide immediate evidence to support other observations, it is a method that cannot be generally recommended because of the risk of toxicity from associated materials.

The following table illustrates various access options, and how, typically, they influence accuracy and the corresponding level of contingency funding that should be allocated to cover unknown or insufficiently understood defects.

EFFECT OF ACCESS ON ACCURACY		
SURVEY TYPE	**ACCURACY**	**CONTINGENCY FUNDING REQUIRED**
Ground-level inspection	50–70 %	30–50 %
Close inspection	70–90 %	10–30 %
Close inspection with site-based investigation	80–90 %	10–20 %

RECORDING SURVEY DATA

No matter how general or detailed a survey may be, it should always be fully recorded. Survey records may be in the form of written descriptions, photographs, and annotated drawings or sketches. Mapping of decay and other notable features is done on a base template which may be a sketch, photograph, measured survey or photogrammetric plot. The bulk of the data collection takes place on site, and observations are usually marked directly onto the template. Therefore, the template medium should be robust enough to endure the conditions on site. For exterior surveys, it is advisable to use polyester drawing film which can be worked on in pencil in the rain.

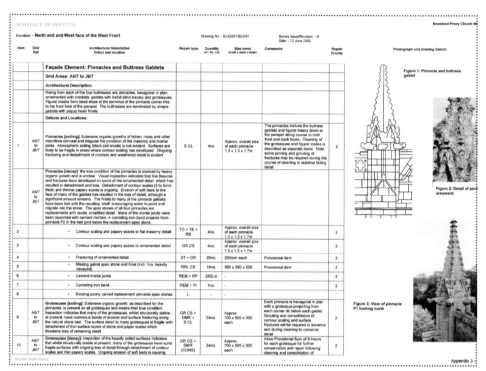

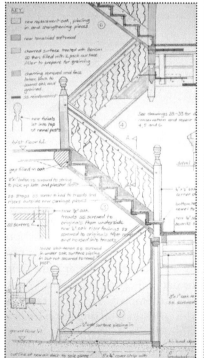

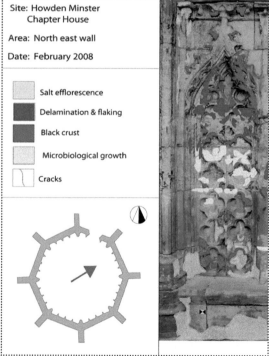

Condition surveys

Top: Example of a condition survey schedule. Each defect is described, along with its location and the cause of decay. The type and size of repair recommended is also recorded. Illustrations in the form of drawings, sketches or photographs may be used for clarity.

Bottom left: Mapping condition on a drawing base. This example shows a 17th-century staircase damaged by fire. It includes some archaeological interpretation, such as the identification of past alterations and repairs. It is also starting to consider likely treatments; for example, which components may be preserved and which will have to be replaced. Although, in theory, the assessment phase involves a sequence of discrete areas of research and information gathering, in practice, the experienced conservator, conservation architect or surveyor will often combine them.

Bottom right: Mapping condition on a photographic base. The use of colour codes to represent each type of decay shows overall patterns and can assist diagnosis. In this case, a condition map of a section of walling suggests that stone at ground level is constantly wet, and that there is an evaporation zone running across the face and rising slightly to the right. Sulphation ('black crust') in the more protected areas is also indicated.

Subsequent readers of the survey record may not be familiar with the site, so a clear key and consistent nomenclature should be established and adhered to. Measurements or scaled plotting of the extent of damage are useful, and can help to determine the relationship between the causes of deterioration and the actual damage. As a minimum, the survey should record the following data (as applicable):

- name and location of the building
- date of the survey, and the names, qualifications and contact details of the people involved
- method of inspection used during the survey (including methods of access)
- general environmental information, including local weather, neighbouring structures and land use, and nearby industries that might affect the building or other threats
- description of weather conditions at the time of the survey
- general description of the building, including the basic materials, features of interest and construction methods used
- specific elements (as required):
  - condition of the structure: identifying cracks and displacements; locations of movement monitoring devices
  - condition of materials/components: identifying material types, surface decay and defects; locations of sampling
  - the environment: identifying sources of moisture; locations of measurement and monitoring
  - materials and locations of previous treatments and repairs
  - information about building use and occupancy (including heating)
  - specific condition reports on particular elements or areas of concern
  - results of specialist investigations.

The survey report should also include a discussion about the recorded data and observations, and draw conclusions, including the diagnosis of faults as far as possible. Finally, it should make recommendations about the next steps towards achieving repairs, including further specialist investigation(s), if needed.

STRUCTURAL SURVEYS

Structural surveys specifically investigate and report on the structural performance of buildings, structures or structural elements. They should be carried out by a qualified structural engineer with experience in working with historic buildings. Such surveys may be required for a number of reasons. For example, investigations will be needed if there is evidence of continuing movement or excessive deflection in a structure or component.

Structural assessment involves a systematic internal and external inspection to understand how the loads are being transferred through the building to the foundations, and how the original structural design may have been subverted by decay or alterations. This is followed, if necessary, by investigations to gather more specific information, such as opening-up to reveal concealed parts of the structure. If problems appear to be associated with foundations, excavation may be necessary to determine the ground conditions (if the building or site is designated, listed building or Scheduled Monument Consent may be needed before invasive investigations are carried out). In some cases, there may be a requirement for short- or long-term diagnostic monitoring to help understand the behaviour of the structure. For example, the direction and magnitude of movements may be monitored using a demountable strain gauge, and reference studs fixed either side of a crack or joint. Where movement monitoring is carried out, it should take into account seasonal fluctuations, and the final diagnosis should be clear and accurate, as errors in interpretation may lead to expensive and unnecessary works.

Understanding how the building has performed historically, particularly in relation to loads imposed in the past by various uses, will be an important consideration. For example, a granary building is likely to have supported reasonably heavy loading, which is likely to exceed modern requirements for, say, domestic loading. Thus, knowledge of its previous performance makes it reasonable to assume that the building will perform satisfactorily in its new use.

Calculations should be used as a guide to help the experienced engineer make an informed judgement about the performance of the building. It should not be expected that elements of a historic building will necessarily comply with modern standards, but they may well have proved themselves through their past performance. Where informed judgment based on calculations cannot determine the adequacy of a beam, a load test may demonstrate that, for example, deflection is within acceptable limits. Load tests should not be undertaken 'lightly' and must be carried out with due care, but they are often a very useful and cost-effective test of a structure or an element within it.

In some instances it will be necessary to undertake further specialised surveys to fully understand the condition of a heritage asset and how its significance is vulnerable to harm. These include the following types of survey.

BUILDING SERVICES SURVEYS & INSPECTIONS

Building services are an integral part of most historic buildings. Typically, they include gas, water and electricity supplies; power, lighting, heating and ventilation installations; drainage systems. They may also contain lifts, fire alarm detection and installations, intruder alarms, and public address, telecommunications and IT systems. In addition, some buildings may have lightning protection systems. Building services that are poorly maintained or defective can put at risk the health and safety of occupants, and can seriously harm the heritage interest and significance of a building. For example, faulty electrical circuit wiring and components can start fires, and decay and damage may be caused by water leaking from defective pipe work. To minimise risks and ensure that legal requirements are observed, it is important that building services installations are regularly inspected and tested at the appropriate intervals by suitably-qualified and experienced people (see **Appendix 1: Typical Maintenance Checklist**).

Too often, building services are viewed as the least culturally significant part of a historic building or place. However, in some instances, building services installations may themselves be of significant heritage value. The lack of recognition of this has led to the loss of many early examples of heating, ventilation and lighting installations, and of equipment such as lifts.

While it is essential to ensure that building services meet the requirements of the present day, it is all-too-easy to overlook their historical significance, and to discard services and equipment that are an important part of our industrial heritage. Although their condition and modern standards of performance and safety often militate against the conservation of significant building services and equipment in their original working order, it may still be possible to incorporate components as part of a new or overhauled system. Original radiators and pipe work may be reused, historic lifts re-engineered to allow doors and cars to be retained, and historic light fittings adapted to allow for more energy-efficient lamps.

Before any work is undertaken, a thorough and systematic investigation should be carried out to establish:

• whether any of the existing building services are of historic significance

• if any part of the installation is sufficiently sound so that it can be retained and used within a modernised system

• where components of historical interest cannot be integrated into a new design, whether these should be retained in a redundant state, or at least photographed and recorded before being removed.

Historic building services

Although often overlooked, building services installations may sometimes be of significant heritage interest in their own right.

Top left: Temporary prefabricated bungalows, built between 1945–49 to ease post-war housing shortages, included fitted kitchens – a luxurious innovation for the period. This example in Stroud, Gloucestershire survived largely unaltered until 1996 when the building was demolished. Although a small number of prefabs in England have been listed, most of those surviving are under threat of demolition making such fixtures an increasing rarity.

Top right: Early 20th-century 'cage' lift re-engineered to comply with modern standards.

Middle left: Early 20th-century light switches, timber pattress plates and wiring casings preserved at the factory of J. W. Evans and Sons Ltd in the Jewellery Quarter, Birmingham.

Bottom left: Cast-iron radiator, part of a heating system installed in a house in the late 19th century to replace open fires. It has been overhauled and retained as part of a modernised system.

Bottom right: 'Box-end' heating coils *c.*1880. Although this 19th-century heating installation is now redundant, the component parts have been preserved.

Investigations should start with a desk study of documentation such as plans, record drawings, system test certificates, periodic surveys, where these exist. This should be followed by site surveys to provide a clear and coordinated picture of what services and service routes survive. This information will also be invaluable when designing new installations.

Unfortunately, many people involved in building services engineering pay little attention to the history of the industry in which they work. Moreover, concerns about professional liability for non-standard approaches, or the re-use of existing components, can be a disincentive for conserving historic services. However, the Heritage Special Interest Group of the Chartered Institute of Building Services Engineers [CIBSE] is working to raise awareness in this sector.

ENVIRONMENTAL SURVEYS

Environmental factors – the climate both outside and inside the building – play an important part in the behaviour and deterioration of all building materials, especially timber. Therefore, an environmental survey may form part of the condition survey or subsequent investigations. Particular attention should be paid to rainwater disposal arrangements and underground drainage, as these are often implicated in building defects and decay. Environmental monitoring may be needed to determine whether a problem is a continuing one, or the deterioration is the remnant of past problems that have been resolved. The *Building Environment* volume in this series examines environmental surveying in detail.

ECOLOGICAL SURVEYS

An ecological survey may also be required to establish whether the building or site is inhabited or visited by any protected wildlife species such as bats, or roosting or nesting birds. The presence of protected wildlife may have implications for the timing of works, and may be a significant project planning issue. Further detailed information about surveys, recording and the selection of appropriate techniques is given in the **Ecological Considerations** chapter.

DIAGNOSTIC INVESTIGATIONS

Information from building surveys may not provide sufficient information to fully understand the construction of a heritage asset, accurately diagnose faults, or plan its repair and care with confidence. In such cases, it will be necessary to weigh the risk of making a wrong decision based on incomplete or inadequate information against the costs (in time and money) of more detailed investigations requiring specialist expertise and equipment. It is important to understand the potential and limitations of the techniques available. Specialist investigations should always be precisely focused to provide answers to specific questions that have arisen in earlier stages of the assessment process. The most commonly used investigative techniques are described in this section.

Non-destructive or minimally invasive testing and assessment techniques offer ways of 'seeing' beneath the skin of the building, and can assist in the identification and diagnosis of faults. They avoid disturbance of the building fabric, but they can be expensive and the results can sometimes be very difficult to interpret. It may not be possible to draw firm conclusions without some degree of opening up or destructive assessment. Accurate interpretation demands a high degree of skill and experience on the part of the operatives, and must take into account the building context, materials and structure.

INFRARED THERMOGRAPHY

Infrared thermography, or thermal imaging, uses a camera to produce a 'thermograph', a visual image of the infrared radiation emitted by the subject. The amount of infrared radiation emitted increases with temperature so variations in temperature can be displayed. The most sensitive thermographic cameras have a wide range of applications in archaeological condition and structural assessments. Under ideal conditions, infrared [IR] thermography can provide an astonishing amount of detailed information, but it is entirely dependent on there being sufficient thermal variations. Sometimes surveying at night provides the best results. Externally, thermography is sensitive to weather conditions, and judging the optimum time to carry out a thermographic survey can be tricky. It is difficult to predict which combination of factors will produce the most useful results on any given day for any given building. A wet wall will yield little information while it remains saturated, and high winds tend to reduce thermal differences on the surface. Suspended ceilings, dry lining and insulation will all obscure the structure behind.

DIAGNOSTIC INVESTIGATIONS & THEIR USES

OBJECTIVE	TECHNIQUES	NDI/DI	FURTHER INFORMATION
STRUCTURE			
Locate and map structural discontinuities in masonry	Infrared thermography	NDI	STONE EB & T ENVIRONMENT
Locate concealed structural defects in masonry	Impulse radar and ground-penetrating radar	NDI	STONE EB & T ENVIRONMENT
Identify concealed construction; identify and locate hidden defects	Endoscopy and micro-camera systems	NDI	MORTARS ENVIRONMENT EB & T
Locate and map bond failures between renders, and pargetting and masonry substrate	Infrared thermography	NDI	MORTARS
Locate and map iron/steel corrosion in reinforced concrete	Half-cell potential corrosion survey	NDI	CONCRETE ENVIRONMENT
Locate discontinuities (such as voids and cracks) in homogeneous materials such as stone and metals	Dynamic impedance or impact-echo techniques, and ultrasound pulse velocimetry	NDI	STONE EB & T METALS ENVIRONMENT
ELEMENTS			
Locate concealed metal fastenings, cramps and reinforcement in stone, render or timber	Metal detection	NDI	STONE MORTARS
Locate and map concealed internal elements of masonry and concrete; locate and map buried structures	Impulse radar and ground-penetrating radar	NDI	STONE CONCRETE EB & T
Assess thermal performance of building elements; locate cold bridges, insulation failures, service runs, etc	Infrared thermography	NDI	ENVIRONMENT
Locate and map concealed timbers or timber frames	Infrared thermography	NDI	TIMBER
	Impulse radar and ground-penetrating radar	NDI	
Locate internal insect damage and decay voids in timber	Micro-drilling and ultrasound pulse velocimetry	DI NDI	TIMBER
Locate and determine condition of iron nails/armatures in timber or plaster	Radiography	NDI	MORTARS TIMBER

NDI = Non-Destructive Investigations
DI = Destructive Investigations

Thermal imaging can sometimes reveal hidden features. In this example, a timber frame concealed by a rendered façade is clearly discernible in the thermal image.

Thermography can also be difficult when there are thermal breaks in the wall construction, or where the wall materials do not conduct heat well. For example, if a brick skin has been built outside and over a timber frame, the frame will not always show up on the infrared [IR] camera. An IR image of a timber element is certainly clear evidence of its existence, but the converse is not true: if an element has not been detected, it does not mean that it is not there.

If a building has a working heating system, this can be used to enhance the temperature gradient or other heat sources can be introduced. Generally, no special access is required, and most work can be carried out from ground or floor level.

Thermography can be used to:

- locate timber frames and other structural elements behind render, plaster and other finishes, both internally and externally
- identify structural discontinuities, such as joints or blocked openings in rendered masonry
- locate bond failure and moisture ingress in renders and pargetting
- map the distribution of moisture in walls, floors and so forth
- identify areas of heat loss, cold bridging, insulation failures and service runs, as well as fire checks and heating systems.

Ultimately, the quality and quantity of the information obtained from IR thermography depend on the quality of the camera and the skill of the operator. The greater the sensitivity and resolution, the wider the range of conditions under which the technique can be used. The most sensitive cameras can capture temperature variations as small as 0.01°C. The cameras are hand-held and self-contained, and provide real-time images that can be recorded as a still or moving image. However, it has to be remembered that surface temperatures can be greatly affected by humidity, and can be distorted by external heat sources such as heaters, halogen lamps or sunlight, and this must be taken into account when interpreting the results.

Endoscopes allow areas inaccessible to the naked eye to be visually examined by penetration of the physical fabric using existing apertures or holes. Equipment varies from relatively simple and inexpensive scopes consisting of a light source, a small diameter rigid tube with built-in optics and eyepiece, to complex steerable systems capable of reaching up to 6 m and equipped with miniaturised video cameras set in articulated heads. The camera head and cable can be as small as 5 mm in diameter.

While endoscopes are a useful tool, they do have flaws. For instance, in the case of the inspection of floor and wall voids, it should be theoretically quite easy to identify concealed problems such as fungal growths, insect attack and structural weaknesses. However, while this is often possible, in practice many voids are filled with debris. Moreover, endoscopes are designed to focus on surfaces and objects only a few centimetres away, so their effective working distances are very short. It can be difficult to retain a sense of orientation and scale, and to understand what you are looking at. The wider the range of movements available, and the greater the length of probe that can be inserted, the easier it is to become disorientated.

Using an endoscope to examine the condition of laths and plaster nibs, prior to salvaging important decorative plasterwork following a fire. Due to collapse of the roof there was no access from above.

The amount of light that can be directed through the instrument into dark voids is restricted, and inevitably the smaller the diameter of the endoscope, the less light can be supplied. It is sometimes better to introduce a second light source through an adjacent hole: this has the added benefit of creating shadows, which can make identification of features much easier.

It is relatively straightforward to record images to a stills camera or video for later analysis, but interpretation of the images can be surprisingly difficult, and it is not unknown for fibreglass insulation and spiders' webs to be misinterpreted as dry-rot mycelium.

Micro-drilling in progress to examine the condition of the base of a corner post in a timber-frame building. The 'decay detecting drill' produces a graphic or digital plot showing the resistance of the timber to the probe. Correct operation of the drill and accurate interpretation of results require skill and experience.

MICRO-DRILLING OF TIMBER

Micro-drilling was developed specifically for assessing the condition of timber, and is perhaps the most accurate practical method currently available. It can provide measurements of the severity and extent of decay, the ratio of sound to decayed wood, and the position of decayed areas within the cross-section of the timber. This is extremely valuable for engineers assessing whether a damaged timber can still continue to fulfil its structural role.

The 'decay detecting drill' bores a tiny hole in a timber component, using uniform pressure, and measures the speed of penetration of the bit. The data is recorded graphically or digitally, and gives an indication of variations in the density of the timber. As with many testing methods, results can be easily misinterpreted, and accurate assessment depends on the skill and experience of the surveyor. ⊖TIMBER

ULTRASOUND PULSE VELOCIMETRY

Ultrasound is one of the most widespread non-destructive techniques for materials testing, routinely used in many industrial, scientific and medical applications. A transmitter emits ultrasound pulses on one side of the material, which are picked up by a receiver on the other side. The time the ultrasonic pulse takes to travel through the material between the two heads (the transit time), or to echo back from cracks, voids and variations in density, provides information on condition. In timber assessment, interpretation requires experience, especially when trying to differentiate between changes caused by decay and changes due to acceptable anomalies (such as shrinkage splits). Ultrasound is therefore particularly informative when used in conjunction with micro-drilling.

Tuned hammers are used to pass acoustic pulses into solid homogenous materials such as stone or steel. The lower-frequency sound waves are used to reveal discontinuities such as cracks and voids, and to assess mechanical properties such as integrity and strength.

METAL DETECTION

The location, approximate size and general shape of metal cramps, fixings and tie rods, embedded in materials such as masonry or timber, can be detected with a portable metal detector. Reinforcement in concrete can be determined as well. Metal detectors and 'cover meters' measure the currents that they induce into ferromagnetic objects. The location of iron and steel can be ascertained up to a depth of 250 mm, and can give an accurate depth measurement up to 100 mm. The ranges are considerably reduced for non-ferrous metals and austenitic stainless steels. ⊖METALS

HALF-CELL POTENTIAL CORROSION SURVEY

Corrosion of steel reinforcement in concrete causes electrical currents to flow in the concrete that covers them. These can be measured and plotted on the surface of the concrete, using a 'half-cell' probe and a high-impedance digital voltmeter to map the corrosion of the reinforcement bars. ⊖CONCRETE

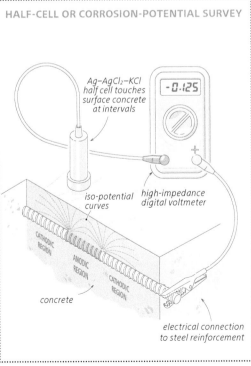

HALF-CELL OR CORROSION-POTENTIAL SURVEY

Ag–AgCl₂–KCl half cell touches surface concrete at intervals

-0.125

iso-potential curves

high-impedance digital voltmeter

CATHODIC REGION

ANODIC REGION

CATHODIC REGION

concrete

electrical connection to steel reinforcement

Left: Ferroscan being used to scan a 600 × 600 mm area on a soffit, and the resulting output (*inset*).

Right: A half-cell potential corrosion survey may be carried out to map the corrosion of steel reinforcement bars in reinforced concrete.

Radio waves can penetrate as much as 1.5 m into masonry walls. Different materials absorb and reflect radio waves differently, so impulse radar can be very useful for locating embedded timber and metals, or mapping the internal elements of masonry and concrete structures. Large areas can be surveyed quickly and the equipment can be operated in a confined space. However, in practice, the results obtained can sometimes be ambiguous and difficult to interpret, particularly in composite or complex elements which tend to produce spurious echoes. One of the most frequent and successful applications of radar is in mapping the position of steel in concrete. This can be done more rapidly than with metal detection, and can resolve multiple layers of steel through the full depth of a slab or beam. GPR may also be used to map buried structures, and has been used to locate the positions of burial vaults in a graveyard. Radar techniques may be used in conjunction with other non-destructive techniques.

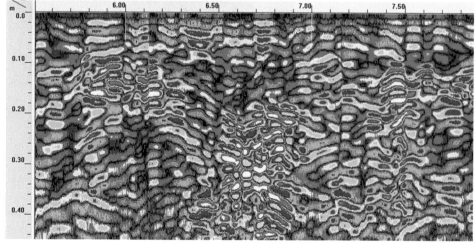

Undertaking a radar survey of a church tower (*top*) and the resulting 'radargram' (*bottom*), in which voids appear in purple. The complexity of the image shows clearly why expert interpretation is essential.

X-ray radiography used to assess the condition of iron nails in a medieval gate. Portable equipment (*top*) was used to produce the X-ray radiograph image (*bottom*).

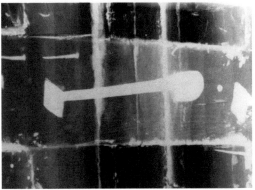

X-radiography and gamma radiography can be used to 'see' inside objects. This is of particular value when the precise form or construction of an object is obscured or concealed by, for example, corrosion products or coatings. Radiography is commonly used for archaeological artefacts and in fine art conservation, but its applications in building conservation are limited. However, it has been used on timber components with polychrome decoration of high significance to ascertain the method of construction, and to assess the extent to which iron nails have corroded. It has also been used for the non-destructive assessment of reinforced concrete floors to provide information about reinforcement size and placement.

INNOVATIVE TECHNIQUES

Methods of non-invasive and non-destructive investigation are currently advancing rapidly, with methods for timber assessment being developed from military, medical and engineering applications. For example, computer tomography can map a cross-section through materials using electrical conductivity, X-ray and gamma-ray attenuation, or nuclear magnetic resonance. As yet, most of these techniques are too slow, too unwieldy or too expensive for widespread use, but the rapid advance of technology means that new and useful tools may soon appear.

MOISTURE MEASUREMENT

Water is responsible for many forms of deterioration in a wide range of materials, so analysis of moisture content and identification of the source of moisture is often needed for accurate diagnosis, and to help in determining the remedial measures required. It may also be necessary after flooding and before decoration is undertaken.

The following table summarises the commonly available methods used for measuring moisture content.

MEASURING MOISTURE CONTENT	
TYPE	COMMENTS
MEASURING ABSOLUTE MOISTURE CONTENT	
GRAVIMETRIC ANALYSIS	The test requires a sample, so is destructive. Timber samples may conveniently be about 25 mm cubes; masonry or mortar samples 1 or 2 grams. Smaller quantities are sometimes all that is available, but require greater weighing accuracy and may be less representative.
GAS-CARBIDE METERS	These are used for masonry materials that may be powdered, usually by removal with a masonry bit. The sample must weigh 6 grams or more and each test takes a few minutes. The method is useful for a few samples where a rapid result is necessary, but gravimetric analysis is more efficient for a large number of samples. Used carbide must be correctly disposed of.
INDIRECT MEASUREMENTS OF MOISTURE CONTENT	
RESISTANCE METERS	These can accurately measure electrical resistance/conductivity, but calibration to moisture content is added by the meter manufacturer. They are usually reasonably accurate when applied to wood, but caution is needed in interpreting results in any other materials. If a reading suggests that the material is dry, then this is probably correct, but elevated readings might be caused by salt contamination, surface coating, or any other factor that would influence electrical properties.
CAPACITANCE METERS	These are probably less influenced by salt contamination than resistance meters, but they are affected by material density and surface irregularities. Best used as a comparative instrument to scan surfaces for change.
INFRARED THERMOGRAPHY	This method provides a non-destructive assessment of large areas. Disadvantages are that there must be sufficient thermal variation, and external thermography is heavily dependent on weather conditions. The greater the sensitivity and resolution of the camera, the wider the range of conditions under which the equipment may be used. Interpretation to moisture contents is comparative and dependent upon the skill of the operator. Confirmatory sampling may be necessary.

Devices used for assessing moisture in building materials

Top: A resistance meter probe in timber.

Bottom left: A gas carbide meter. The powdered sample, produced by drilling with a masonry bit, is weighed, then placed in a sealed flask with a measured amount of calcium carbide powder. When the flask is shaken vigorously, the moisture in the sample reacts with the calcium carbide to produce acetylene gas, which exerts a pressure on the flask that can be read on an external gauge calibrated to indicate moisture content. The residues from the test must be disposed of correctly.

Bottom right: Capacitance moisture meters generate a radio-frequency field through a sensor that is pressed against a surface. They respond to the dielectric properties of materials to determine their approximate moisture content.

CONSERVATION BASICS
SURVEY & INVESTIGATION METHODS

DIAGNOSTIC MONITORING

The results of the condition survey will often generate questions on how or why a particular condition has arisen. Information recorded at one moment in time is unlikely to explain the complex relationship between the building and its environment. Monitoring of the environment over an extended period may be needed to help provide some of the answers.

Monitoring can help to build a picture of how the condition of materials or components reacts to differing environmental factors such as fluctuations in air temperature and humidity, as well as ventilation rates. Whatever parameters are monitored, it will usually be necessary to collect data over a period of at least one year, to cover seasonal variations. An initial period of monitoring is usually needed to determine the best types of sensor and the optimum locations, and to calibrate instrumentation. The system for diagnostic monitoring should be designed to give answers to specific questions. It is all too easy to collect data that do not, in the end, contribute to understanding.

MONITORING MOISTURE CONTENT

To relate sources of moisture to deterioration, it is often necessary to record how moisture levels in porous materials change over time. Most moisture monitoring is carried out by making repeated measurements using the techniques described above. Wooden dowels inserted in a wall provide a practical method for monitoring changes in moisture content. Softwood rods (ideally 6–9 mm wide) are inserted into drilled holes, which are a few millimetres larger in diameter than the rods to allow for the wood to swell if moisture is present. They are left for about six weeks (but not longer than eight) to equilibrate, then their moisture content is assessed by cutting them into sections and using gravimetry or a resistance moisture meter. They are particularly useful for providing large numbers of measurements (which may be repeated using a new set of rods) on occupied sites where inevitably some sensors will be lost.

Many attempts have been made to develop equipment to monitor absolute moisture content, but as yet there is no reliable *in-situ* method commercially available that can give more than a broad idea of changes. Researchers are currently trying to develop *in-situ* versions of hi-tech moisture-monitoring equipment, such as nuclear magnetic resonance, time domain reflectometry and microwave attenuation.

Moisture measurement and monitoring are discussed in more detail in the *Building Environment* volume of this series. ENVIRONMENT

Diagnostic monitoring at Cleeve Abbey, Somerset

Top: The wall paintings of the Painted Chamber (15th century) are of national significance. Active deterioration of the painted surface and plaster is evidenced by flaking and powdering within a diagonal band in the lower part of the wall. A sheet of Melinex, on the floor abutting the wall, allows for the collection of debris which is used for quantifying the rate of loss.

Bottom left: A condition survey of the painting and wall provides a qualitative and quantitative description of relevant deterioration phenomena. It highlights clearly the area of active loss.

Bottom right: Following the survey, a programme of environmental monitoring was put in place, measuring ambient temperature, wall surface temperature, radiant temperature and relative humidity, in real time, as well as photographic imaging of solar radiation on the wall. Similar data was collected outside for comparative purposes. Small sensors are connected by wire to radio transmitters, which transmit data to a central datalogger. Environmental patterns were reviewed on a regular basis. Timed imaging of solar episodes on the wall allowed a direct correlation with environmental data. This showed that sunlight fell on the lower part of the painting for a few months in winter, heating the surface (the green peaks in the lower graph) and causing the plaster to spall around tiny dark flint inclusions in the aggregate. This deterioration can now be prevented by blocking the window over the winter period.

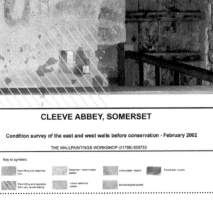

CLEEVE ABBEY, SOMERSET

Condition survey of the east and west walls before conservation - February 2002

THE WALLPAINTINGS WORKSHOP (01795) 538750

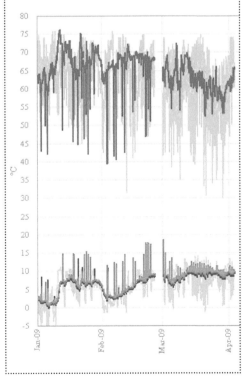

MATERIALS & ANALYSIS

Condition surveys and other diagnostic investigations in historic buildings often identify issues of material failure which require scientific analysis. Typically, investigations of this type are intended to help:

- accurately identify materials and their provenance, including original materials, added materials, residues of previous treatments and contaminants (such as salts), by looking at their mineral composition, chemical composition, pH, and in the case of timber, grain or cell structure

- understand the behaviour and performance of materials, including the causes and mechanisms of deterioration or failure, by evaluating their physical and chemical properties (such as mass, particle size, porosity, hardness, compressive/tensile strength, modulus of elasticity, thermal properties, melting temperature, conductivity, hygroscopicity, permeability and solubility; in addition, assessing adhesion, cohesion, corrosion potential, weathering, and biodeteriogens)

- determine the presence of any hazardous building materials, including their provenance, concentration and toxicity

- assess the suitability of new materials for replacement of failed materials.

Analysis of materials for the purpose of dating is described elsewhere (see **Structural Archaeology**).

A large number of techniques are available for the examination, identification and analysis of materials. These range from simple procedures such as basic chemical disaggregation of mortar and its examination with a hand lens, which can be done on site, to highly-sophisticated analytical techniques that are mainly carried out in a laboratory. Some portable analytical equipment is now available, but this has its limitations. The most common laboratory techniques include polarised light microscopy [PLM], scanning electron microscopy-energy dispersive X-ray [SEM-EDX], Fourier Transform infrared spectroscopy [FTIR] and gas-chromatography mass-spectroscopy [GCMS]. The techniques used to examine and analyse specific materials are described in detail in the other volumes of the *Practical Building Conservation* series. ↪CONCRETE ↪EB&T ↪GLASS ↪METALS ↪MORTARS ↪ROOFING ↪TIMBER ↪STONE

As with other forms of specialist research, material analysis is undertaken in response to very specific questions, which determine the scope, depth and methods of investigation. An ill-defined or imprecise analytical programme is a waste of time and resources, and can produce meaningless results (or at the very least, it does not provide optimal data). For this reason, it is good practice to engage the relevant materials analysts at an early stage in the process to establish an analytical strategy which is fit-for-purpose. A good analytical strategy will have clearly articulated research questions, which will ensure focus, precision, optimal use of samples where taken and value-for-money.

Some simple or more sophisticated analytical techniques using portable equipment can be carried out *in situ* by non-invasive and non-destructive methods. However, more commonly, the in-depth evaluation of materials will be invasive and require the extraction of samples for analysis in the laboratory. The size of sample required is usually very small, though it may still sometimes entail damage to highly significant surfaces. Therefore, sampling is often constrained in such contexts. Furthermore, the removal of samples from a Scheduled Ancient Monument and many listed buildings requires statutory consent; the application should be supported by the analytical strategy.

The analytical strategy should identify the location, size and number of samples to be taken, and include a description of the area to be sampled and its condition. The type of analysis has to be decided on in advance, as this determines both the size and characteristics of the sample required (for example, one in powder form, as opposed to one with a layered stratigraphy). Some analytical techniques do not destroy or affect the sample, allowing it to be analysed further using a different technique. Samples of both sound and deteriorated material should be extracted, and these should be representative of the area and its manifest conditions.

The area to be sampled should be accurately recorded. This also provides a template for logging the location of samples, which must be clearly labelled after removal. Where portable analytical tools are used, the exact locations where readings are taken should be noted, the areas described, and their condition characterised.

Proper and meaningful interpretation of the results of analysis relies on precise documentation, and rigorous integration and management of information.

ASSESSING VULNERABILITY & RISK

Information gathered through background research, surveys and investigations provides the basis for understanding the significance of a heritage asset, and how significance is vulnerable to harm. Gauging vulnerability, and identifying and assessing risk, are essential processes in the development of a conservation strategy for a historic building or place.

ASSESSING VULNERABILITY

Vulnerability is a function of the magnitude and frequency of threats, and the level of susceptibility to harm. For example, the vulnerability of building fabric to deterioration by weathering depends on its fragility, determined by its characteristics and condition, and the degree of exposure to which it is subject.

ASSESSING VULNERABILITY			
	NATURE & CONDITION OF FABRIC OR ELEMENTS		
EXPOSURE	FRAGILITY		
	ROBUST (1)	NORMAL (2)	FRAGILE (3)
MILD (1)	1	2	3
MODERATE (2)	2	4	6
SEVERE (3)	3	6	9
VERY SEVERE (4)	4	8	12

THREAT

Risk assessment is a method in which the seriousness of the harm that might be caused by a particular threat is estimated, then weighed against the likelihood of its occurrence.

There are many different forms of risk assessment, encompassing all parts of modern life. In one example, the probability of a particular threat, and the seriousness of its consequences, can be assigned numerical values which, when multiplied, give an indication of the level of risk. The relative significance of a heritage asset (or the particular elements of it in question) also needs to be factored into the equation.

ASSESSING RISK				
		SERIOUSNESS OF CONSEQUENCES →		
		INSIGNIFICANT	MODERATE	SEVERE
PROBABILITY ↓	NOT POSSIBLE	0.00	0.00	0.00
	POSSIBLE	0.00	0.25	0.50
	CERTAIN	0.00	0.50	1.00

This can then be used as the basis for prioritising actions and allocating resources when planning for maintenance and repair. For example, the harm to significance that might be caused by ignoring a building defect can be evaluated and compared with the likely impact and outcomes of alternative remedial measures. Similarly, determining the probability of extreme weather conditions can assist in planning appropriate and proportionate mitigation strategies.

A similar method of risk assessment is also used in emergency planning. For instance, the consequences of fire in a particular building might be judged to be manageable in terms of the safety of its occupants, due to the fire precautions provided. However, in the event of fire, damage to the heritage significance of the building and its contents could be very serious indeed. Risk assessment can be used to appraise the value of providing additional fire safety systems (which may never have to be used), to further reduce the risk of fire and mitigate its consequences.

Risk assessments are also fundamental in planning for the provision of required levels of health and safety. This is discussed in **Managing Maintenance & Repair**.

Survey Sheet

BUNAVONEADER WHALING STATION, ISLE OF HARRIS
Condition Survey Sheet

Feature ....Engineering shop / Blacksmith...............

RCHMS No....../.........................

Date ..27./.02./07.

Materials	Element	Condition	Comments
stone *rubble*	Plinth / platform	P	Only visible on N side
brick	base in SW corner	F/P	? for forge
mortar	}		
render			
concrete	Floor slab	? F	Covered with turf.
Iron/steel	holding down bolts ; pipe,		
timber	valve & supports	P	} slow corrosion
others			

Overall condition	good (4)	fair (3)	poor (2) ✓	very bad (1)

Rate of deterioration   FAST ☐ ☑ ☐   SLOW

Management issues

Drainage ; vegetation ;

Notes

Iron objects, coal & slag beneath turf.

Facing page: Bunavoneader Whaling Station, Isle of Harris, Scotland, is a Scheduled Ancient Monument. The site comprises the remains of the buildings and structures, mostly dating from the late 19th and early 20th centuries, used in the processing of whale carcasses to produce oil and other products. These include the sea wall, pier and slipway, 'flensing' platform (where whales were skinned and cut up), boiler houses and chimney, laboratory, oil and fuel stores, narrow gauge railway, manager's and workers' accommodations, and numerous ancillary buildings and structures. With a few exceptions, only foundations and building platforms have survived. A preliminary condition survey was carried out as part of the preparation of a *Conservation Management Plan*.

This page, top: A simple survey template was devised to quickly gather basic information about the construction, condition and vulnerability to threats of each of the structures. Documentary research and a building archaeological investigation were carried out concurrently. A gazetteer was produced that included an assessment of the significance of the individual structures.

Inset: The results of the preliminary condition survey were used in conjunction with the gazetteer to produce a 'triage' plan to indicate priorities for further detailed investigations and actions to restrain deterioration.

Risk category

HIGH

BUNAVONEADER WHALING STATION
West Loch Tarbart, Isle of Harris

LOW   Relative vulnerability to threats

Assessment of Buildings at Risk

Heritage assets that have not been adequately maintained or have been abandoned are at risk of total loss, or vulnerable to becoming so. Identification of assets either at risk or vulnerable, and developing a strategy to deal with them, is the key to securing their future.

Managing and resolving assets at risk requires working in close partnership with owners, local planning authorities, and other relevant organisations and partners. At a national level, Historic England provides practical advice and guidance. To assess and monitor the scale of the problem, and prioritise resources and action, the creation of a local heritage-at-risk register is the first step in tackling neglected buildings.

There are some general approaches that are relevant to all assets at risk. The Historic England template for a Buildings-at-Risk survey includes categories for designation, occupancy and use, condition, and priority.

CATEGORIES FOR DESIGNATION, OCCUPANCY & USE

DESIGNATION CATEGORIES		
PRINCIPAL DESIGNATION(S)	**RELATED DESIGNATIONS**	**LOCAL AUTHORITY REGISTER**
Listed Building: Grade I, II* or II	Registered Park and Garden: Grade I, II* or II	Locally listed buildings
Scheduled Monument	Registered Battlefield	
	Conservation Area	

OCCUPANCY / USE CATEGORIES	
CATEGORY	**DESCRIPTION**
VACANT	Not in use
PART OCCUPIED	Partially in use
OCCUPIED	In use
UNKNOWN	
NOT APPLICABLE	

CONDITION CATEGORIES

Buildings are assessed on the basis of condition and, where applicable, occupancy (or use). Buildings capable of beneficial use are at risk if they are in very bad or poor condition, and vulnerable if they are in fair (and occasionally good) condition and vacant, partially occupied or about to be vacated as a result of functional redundancy.

Buildings and monuments incapable of beneficial use are at risk if they are in very bad or poor condition, or vulnerable if they are in fair condition but lacking management arrangements to ensure their maintenance.

CONDITION CATEGORIES

CATEGORY	DESCRIPTION
VERY BAD	Building or structure where: • there has been structural failure • there are clear signs of structural instability • (where applicable) there has been loss of significant areas of the roof covering, leading to major deterioration of the interior • where there has been a major fire or other disaster affecting most of the building
POOR	Building or structure with: • deteriorating masonry • a leaking roof • defective rainwater goods, usually accompanied by rot outbreaks within and general deterioration of most elements of the building fabric, including external joinery • where there has been a fire or other disaster which has affected part of the building
FAIR	Structurally sound, but in need of minor repair or showing signs of a lack of general maintenance
GOOD	Structurally sound, weather-tight and with no significant repairs needed

PRIORITY CATEGORIES

Priority categories are used as a means of prioritising action.

PRIORITY CATEGORIES		
CATEGORY	DESCRIPTION	
A	Immediate risk of further rapid deterioration or loss of fabric	No solution agreed
B	Immediate risk of further rapid deterioration or loss of fabric	Solution agreed but not yet implemented
C	Slow decay	No solution agreed
D	Slow decay	Solution agreed but not yet implemented
E	Under repair or in fair to good repair	No user identified
	Under threat of vacancy (applicable only to buildings capable of beneficial use)	No obvious new user
F	Repair scheme in progress	End use or user identified (where applicable)
	Functionally redundant buildings	New use agreed but not yet implemented

For more details, see *www.historicengland.org.uk/advice/heritage-at-risk/*

Further Reading

Clarke, K. (2001); *Informed Conservation: Understanding Historic Buildings and their Landscapes for Conservation*; London: English Heritage

Curteis, T. (2008); 'The Survey and Identification of Environmental Deterioration'; in *The Building Conservation Directory 2008*; available from *www.buildingconservation.com/articles/envdet/environment.html*

D'Ayala, D., Smars, P. (2003); *Minimum Requirements for Metric Use of Non-metric Photographic Documentation*; available from *smars.yuntech.edu.tw/papers/eh_report.pdf*

D'Ayala, D., Fodde, E. (2008); *Structural Analysis of Historic Construction: Preserving Safety and Significance: Proceedings of the VI International Conference on Structural Analysis of Historic Construction, SAHC08, 2–4 July 2008, Bath, United Kingdom* [Volumes 1&2]; London: CRC Press

Eppich, R., Chabbi, A. (eds.) (2011); *Recording, Documentation, and Information Management for the Conservation of Heritage Places: Volume 2: Illustrated Examples*; Shaftesbury: Donhead

GBG (GB Geotechnics Ltd.) (2001); *Non-destructive Investigation of Standing Structures: Technical Advice Note 23*; Edinburgh: Historic Scotland

Historic England (2016); *Understanding Historic Buildings: A Guide to Good Recording Practice*; Historic England; available from *www.historicengland.org.uk/images-books/publications/understanding-historic-buildings/*

Hughes, H. (ed.) (2002); *Layers of Understanding: Setting Standards for Architectural Paint Research* [Papers taken from the proceedings of English Heritage's national seminar held in London on 28th April 2000]; Shaftesbury: Donhead

Letellier, R., Schmid, W., LeBlanc, F. (2011); *Recording, Documentation, and Information Management for the Conservation of Heritage Places: Volume 1: Guiding Principles*; Shaftesbury: Donhead

Pinchin, S. (2008); 'Techniques for monitoring moisture in walls'; in *Reviews in Conservation*; London: IIC; pp.33–45

Robson, P. (2005); *Structural Appraisal of Traditional Buildings* (2nd revised edition); Shaftesbury: Donhead

Swallow, P., Dallas, R., Jackson, S., Watt, D. (2004); *Measurement and Recording of Historic Buildings*; Shaftesbury: Donhead

Watt, D. (2007); *Building Pathology: Principles and Practice*; Oxford: Blackwell

Watt, D. (2011); *Surveying Historic Buildings* (2nd edition); Shaftesbury: Donhead

Useful Web Addresses

Doubleday, H. A. [and subsequently others] (ed.) (1900–); *Victoria History of the Counties of England* [series]; London: Archibald Constable & Co; also web page at *www.british-history.ac.uk/catalogue.aspx*

Historic England; *www.historicengland.org.uk/images-books/*

Historic England (2019); *https://archive.historicengland.org.uk*

Historic England (2019); *PastScape* (web page, searchable database of records held in the national historic environment record); available from *www.pastscape.org.uk*

Historic England, IHBC, ALGAO-England (2012); *Heritage Gateway* (web page, national and local records of England's historic sites and buildings, including images of listed buildings); available from *www.heritagegateway.org.uk/gateway/*

ECOLOGICAL CONSIDERATIONS

In this chapter, the current framework of legislation that exists to protect wildlife species and habitats is outlined. The flora and fauna that may inhabit or colonise historic buildings and places are described, and the practical implications they have for managing maintenance and repair projects discussed.

Why consider ecology in connection with the maintenance and repair of historic buildings and structures?

The relationship between the 'natural environment' and the 'historic environment' is not always an easy one, and apparent conflicts between nature and building conservation are not uncommon. For example, the dense ivy growth that is potentially damaging to a wall is also a diverse wildlife habitat. However, if it is possible to look after a heritage asset in a way that maintains, and possibly increases, its ecological interest, there is every reason to manage it in this way. Public interest in 'green' issues is considerable, and a place that combines both historic and wildlife interests is doubly attractive.

In addition, wildlife is subject to various legal requirements. Historic sites can provide a habitat for a wide range of flora and fauna; some of these may be damaging and will need to be removed, but many of them are harmless or even beneficial. It is, therefore, important to understand the ecology of a site, and to consider the implications for flora and fauna of any proposed works.

RECONCILING NATURAL & HISTORIC ENVIRONMENTS

LEGAL CONSIDERATIONS

There are at present five principal pieces of legislation that deal with wildlife in the UK of which site owners and managers should be aware:

1. Wildlife & Countryside Act 1981
2. Conservation of Habitats & Species Regulations 2010
3. The Countryside & Rights of Way Act 2000
4. Natural Environment & Rural Communities Act 2006
5. Protection of Badgers Act 1992

A site may be covered by more than one designation. Hurst Castle, a scheduled monument in Hampshire, sits within the 1077-hectare 'Hurst Castle and Lymington River Estuary' SSSI, the 5401-hectare 'Solent and Southampton Water' SPA, the 11,243-hectare Solent Maritime SAC and the 5306-hectare 'Solent and Southampton Water' Ramsar site. These designations cover different aspects of the wide range of coastal habitats surrounding the site, which in addition is home to four protected species of bats.

WILDLIFE LEGISLATION

1. Wildlife & Countryside Act 1981 (as amended)

This Act deals with a wide range of issues, but the most important is the protection of wildlife (plants and animals). The most relevant parts deal with:

- the designation and protection of Sites of Special Scientific Interest [SSSIs]
- the protection of wild animals (including birds and mammals: especially bats, reptiles and amphibians)
- the protection of wild plants.

2. Conservation of Habitats & Species Regulations 2010

Known as the 'Habitats Regulations', these transpose the *EU Habitats Directive* into UK national law. The most relevant parts deal with:

- the protection of habitats and species at a European level
- the formation of the 'Natura 2000' network of sites throughout the EU. In the UK these consist of Special Areas of Conservation [SACs], Special Protection Areas [SPAs] and 'Ramsar' sites (these are designated wetland sites of international importance). There are currently 251 SACs, 86 SPAs and 73 Ramsar sites in England.

3. The Countryside & Rights of Way Act 2000

As well as dealing with a wide range of issues outside the scope of this section, this Act also requires government to protect biodiversity by listing the most important species and habitat types, and then attempting to further their conservation. This duty falls on all government departments and the non-departmental government bodies they sponsor.

4. Natural Environment & Rural Communities Act 2006

Once again, this Act deals with a wide range of issues, but for the purposes of managers of public sites, Section 40 of it places a new duty on public authorities that *"Every public authority must, in exercising its functions, have regard, so far as is consistent with the proper exercise of those functions, to the purpose of conserving biodiversity"*.

5. Protection of Badgers Act 1992

Badgers are specially protected under their own legislation, not because they are rare, but in order to prevent 'badger baiting'. The Act was principally designed to make all aspects associated with this activity illegal and to make it easier to prosecute anyone associated with it. The Act makes it an offence to:

- kill, injure, capture or cruelly ill-treat a badger
- interfere with a badger sett: interference with a sett includes damaging, destroying or obstructing access to a sett, causing a dog to enter a sett, or disturbing a badger when it is occupying a sett.

Other legislation that needs to be considered when managing sites includes:

- *Town & Country Planning Act 1990*

  This Act allows the designation of Tree Preservation Orders [TPOs]. If a tree or trees are subject to a TPO, owners or managers must apply to the local planning authority for consent to carry out any work (including felling). The work may not be carried out unless this consent is received.

- *Planning (Listed Buildings & Conservation Areas) Act 1990*

  All trees with a diameter at breast height [DBH, taken at 1.2 m] of over 75 mm within conservation areas designated under this Act receive a basic level of protection. Owners or managers of these trees must give the local planning authority six weeks' notice of any intention to carry out work on the trees, including felling. The local authority has the option of consenting to the work, or placing a TPO on some or all of the trees.

- *Food & Environment Protection Act 1985*

  Under this Act, the *Control of Pesticides Regulations 1986 [CoPR] (as amended 1997)* were introduced, which, among many provisions to ensure the safe use of pesticides, require the certification of operators applying chemicals. Changes to regulations to implement the *Sustainable Use Directive* are expected in 2012. They are likely to tighten the rules on operators, bring in a requirement for advisors and specifiers to be qualified, and introduce CPD requirements for both groups.

- *Weeds Act 1959 (as amended)*

  This Act provides for the control of five 'noxious weeds': curled dock, broad-leaved dock, creeping or field thistle, spear thistle and common ragwort.

- *Ragwort Control Act 2003*

  This Act amended the 1959 *Weeds Act* in respect of Ragwort control, and provided for the introduction of a *Code of Practice on How to Prevent the Spread of Ragwort*, which contains useful advice and information (see **Bibliography**).

- *Environment Act 1995*

  Under this Act, the *Hedgerow Regulations 1997* was introduced, which provides a blanket protection for all hedgerows over 30 years of age. This is similar to the protection given to trees in conservation areas.

- *Environmental Protection Act 1990*

  This Act makes it an offence to spread by seed or vegetatively three invasive non-native plants: Japanese knotweed, giant hogweed and Himalayan balsam. Also, if the growth of these plants is controlled by other than chemical means (such as by cutting down or pulling up), the arisings become 'controlled waste' and are subject to controlled waste regulations, which require transport in sealed containers to licensed landfill sites, according to the EPA (Duty of Care) Regulations 1991.

Legislation, regulations and official guidance are inevitably subject to changes and revision. It is important to keep up-to-date with statutory requirements by checking the websites of the relevant government departments.

UNDERSTANDING & MANAGING ECOLOGICAL INTERESTS

Although ecological interest can be easily divided into flora and fauna, it should always be remembered that they do not operate in isolation, but interact with each other constantly to form a living ecosystem. For example, a whole host of small insects and animals feed on plants. These insects and animals are in turn fed upon by predators, and those predators by still other predators, and so on, until the 'climax predator' is reached. Even these animals, on death, become food for fungi and bacteria, thus returning to the soil as nutrients which then feed the plants. Even more complex relationships can exist where plants are reliant on animals to spread their seed to new sites. Some seeds even contain a droplet of attractive oil as an encouragement/reward for an insect (usually an ant) to move it.

A complete understanding of the ecology of a site can only be achieved over time, and by suitable survey and observation. However, there are some general points which can be considered for all sites to prevent owners and managers from unnecessarily harming the ecology of a site or breaking the law. These are considered in the following sections on fauna and flora.

Adult six-spot burnet moth (*Zygaena filipendulae*) on a common knapweed (*Centaurea nigra*) flower. The larval stage of this moth feeds mostly on common bird's-foot-trefoil (*Lotus corniculatus*), and both may be eaten by birds, small mammals and occasionally other insects or spiders.

The term 'fauna' basically covers all the animals on a site that can move around it. Much of the fauna on a site is rarely seen, but every site will have a range of fauna, including some, if not all, of the groups described below.

BIRDS

Often the best known and most frequently seen fauna on any site are birds. Many birds nest in trees and hedges, and can frequently be seen roosting on roofs. However, some have a much more intimate relationship with buildings. For instance, swallows, swifts, house martins and house sparrows are 'wall specialists'. As a replacement for natural cliffs or exposed tree trunks, they choose walls to build their nests. Birds, such as titmice, robins, wrens, blackbirds, thrushes and so on, commonly use any opportunity for a nest site and can often be found nesting on ruined walls, which offer some security. Occasionally, birds such as nuthatches will use a wall cavity in place of the hollow trees they normally prefer.

All wild birds are protected under Section 1 of the *Wildlife & Countryside Act 1981*. This makes it an offence to:

- kill, injure or take any wild bird
- take, damage or destroy the nest of any wild bird while that nest is in use
- take or destroy an egg of any wild bird.

House martins (*Delichon urbica*) are summer visitors, and construct nests made of mud on sheltered parts of buildings such as here in the gatehouse at Goodrich Castle. The nests are unlikely to damage the structure.

In most instances, this will not cause any problems for owners or managers of historic sites, although it is important to ensure that any planned works which might disturb nesting birds are timed for outside the nesting season (March to July), or are flexible enough to accommodate it.

Most birds do no harm to buildings, although some species, such as gulls, pigeons, and sometimes starlings, may cause problems, particularly where they are present in large numbers. Problems are associated mainly with droppings, which are both visually disfiguring and potentially damaging to surfaces, particularly on stonework and metalwork. Damage to buildings can also occur when droppings, old nests and dead birds block rainwater goods. Accumulations of droppings are also a health hazard to the occupants of a building, and to people carrying out maintenance or repair works (see **Managing Maintenance & Repair**).

The *Wildlife and Countryside Act 1981* allows specific licences to be issued for the control of certain birds, including gulls and pigeons, in specific circumstances. In addition, there are more general licences that permit bird control for a variety of reasons. Detailed guidance is available from the Department for Environment, Food and Rural Affairs [Defra] Wildlife Management website (see **Further Reading**).

More unusually, there have been increasing numbers of reports of woodpeckers damaging wooden (mainly cedar) shingles as they search for food and claim territory. The main culprit is the Great Spotted Woodpecker – the most adaptable and omnivorous of the family – which has become more common in residential areas.
⌐◔ROOFING

There are numerous methods for controlling the nuisance and damage caused by birds. These include physical deterrents, as well as a range of lethal (for particular species in certain circumstances) and non-lethal methods of control. The selection of the appropriate deterrent or control depends on the species of bird that is causing problems. It should be noted that lethal methods are never permissible in circumstances where deterrents and non-lethal methods have proved to be effective.

Bird Deterrents

Physical deterrents include netting, spikes, coil systems, wire, gels, electrical track systems and decoys. Deterrents are intended to prevent or discourage birds from perching, roosting and nesting on buildings, but they must not trap, injure or kill birds. All methods have a visual and physical impact on a building to some extent, and this should be considered when selecting a system. The service life and maintenance requirements of deterrent systems should also be taken into account. With the exception of sonic devices in some situations, deterrent measures do not conflict with wildlife protection legislation and no authorisation is required for their use. However, the installation of deterrents is likely to require Listed Building or Scheduled Monument Consent.

Top: Netting applied to the façade of a building to keep out birds.

Bottom: Bird spikes can be effective in preventing perching but periodic cleaning may be needed to remove trapped litter and leaves.

Netting

Netting is a very effective way of keeping birds off buildings. Architectural features, openings, entire façades, and even roofs, can be treated in this way. Netting is usually attached to steel perimeter wires or cables which are mechanically fixed to the building. The size of mesh and the space between the fixings of the perimeter cables depend on the bird species. However, care must be taken to minimise the risk of birds becoming trapped behind the netting. This may happen as a result of poor installation or inadequate maintenance. Careful design is also needed to minimise superficial damage to the building from fixings; ideally, these should be made into masonry joints wherever possible. Stainless steel should be used for the perimeter wires and fixings. Netting can be visually obtrusive, but in many situations its presence is barely detectable.

Spikes

Installations consist of arrays of spikes, usually of plastic or stainless steel, which are fixed to ledges and other features that birds could perch on. The spikes are normally fixed with plugs and screws, or a silicone adhesive, although in some instances, such as gutters, it may be possible to fasten them with plastic cable ties. Spikes are an effective deterrent, but they tend to be visually obtrusive, and care is needed to minimise damage to historically significant surfaces caused by fixings.

A typical tensioned-wire bird deterrent installation, here combined with netting.

Wire

Typically, systems consist of fine stainless-steel wires tensioned by springs, and supported by stainless-steel posts which are fixed at 1500 mm centres to the tops of sills, cornices, parapets and other ledges. The posts are either set in holes drilled in the masonry or, where there is a metal flashing, with adhesive. Although much less conspicuous than spikes, wire systems are not always entirely effective. Also, as the posts have to be located to suit the tensioning of the wires, there is little scope for adjusting their positions to minimise damage to historically significant building fabric.

Wire Coils

Although not as effective as spikes and less versatile, the wire coil system is useful for deterring pigeons from perching on exposed ledges and parapets. The coils are no less visually obtrusive than spikes, but require significantly fewer fixings.

Parallel Wires

This is a system used primarily to prevent seagulls landing and nesting on valley roofs. It consists of parallel steel wires installed horizontally at approximately 500 mm centres between the internal slopes of the roof. These are fixed each side to horizontal straining wires, running close to the roof slopes, which are fixed to the roof structure beneath the slates or tiles.

Gels

These are usually applied with a caulking gun to sills, parapets, ledges and other features that could provide perches for birds. The tacky surface of the gel discourages birds from landing, or staying long if they do. Repellent gels remain effective for periods of up to one year. Regulations governing the use of gels require them to be treated with a sealant to prevent harm to birds. However, constituents of the gel tend to leach into porous masonry, causing staining and leaving residues that are very difficult to remove. Installers normally recommend a sealant pre-treatment, but this too is likely to be harmful to porous masonry. For these reasons, and the short service life, gels are not recommended for use on most historic buildings.

In principle, this system works in the same way as electric fencing for managing livestock. The track is fixed to potential perches and delivers a mild electric shock to any birds that come into contact with it. It is visually unobtrusive, but fixings need to be carefully considered.

Decoys

Model birds of prey which are intended to scare off bird pests. These are sometimes used in conjunction with bio-acoustic devices, and are said to be effective in preventing roosting. All decoys need to be regularly moved to prevent birds becoming habituated to them, with the subsequent loss of deterrent effect. This needs particular consideration where access is not easy.

Bio-acoustic, Sonic & Ultrasonic Devices

Bio-acoustic devices replicate natural bird alarm and distress calls, as well as predator calls, to scare away the targeted bird species. Birds make alarm calls when they are in danger. Distress calls are made when a bird is in pain or has been attacked by a predator; predator calls are the sounds made by a predatory bird. Sonic devices produce a variety of electronically generated noises to scare birds. All are audible to the human ear. Ultrasonic devices emit sounds outside the normal range of adult human hearing. Although these systems have minimal impact on buildings, their effectiveness and acceptability in urban environments are limited, and ultrasonic devices cannot normally be used in localities inhabited by bats.

Top: Bird decoy in the form of an owl.

Bottom: Bio-acoustic bird deterrent devices mounted on brackets fixed to ridge tiles.

MAMMALS

Small Mammals

Small mammals are probably quite frequent on many sites but are often overlooked, and, as many are nocturnal, rarely seen. Small mammals in this group include voles, shrews, hedgehogs, and possibly stoats, weasels and mink. Some of these are protected, but even for those that are not, it is unlikely that control measures would ever be considered necessary. Many of these animals do no more than utilise existing holes in walls and occasionally enlarge them, but generally do no real harm.

Some small mammals may cause problems. Rats and mice, particularly in large numbers, can damage buildings, both inside and outside, and also carry a health risk. Neither species is protected and, where necessary, humane control methods may be used. Grey squirrels may also damage both inside and outside buildings, including gnawing cables, and may be similarly controlled using humane methods. However, particular care needs to be taken in areas which still contain protected red squirrels, to ensure that these are not harmed. Moles can be a nuisance in the gardens and landscaped areas, and are potentially harmful to sensitive archaeological sites, but are unlikely to damage buildings. Again, control is permissible; detailed advice is available from Natural England (see **Further Reading**).

Larger Mammals

Some larger mammals may also be damaging. Rabbits can certainly cause damage to landscape and archaeology, and, in extreme cases, may undermine foundations. Control is allowed, but not necessarily easy. A wide range of lethal and non-lethal methods is available; detailed advice on the management and control of rabbits can be obtained from Natural England (see **Further Reading**). Badgers can cause serious damage to landscape and archaeology, and their extensive digging within main setts could theoretically undermine foundations, although setts are not commonly found under buildings. Domesticated animals such as cattle, sheep, horses or occasionally goats are sometimes placed on a site to assist with management by grazing. However, their use needs to be carefully monitored, as damage can be inflicted to landscape features such as archaeological earthworks from surface trampling and scarring.

Extensive rabbit (*Oryctolagus cuniculus*) warrens are unattractive, can damage archaeology, may destabilise slopes, and can lead to surface erosion as the vegetation is killed by covering with soil or over-grazing.

Facing page: A decay cavity in an oak tie beam which has been enlarged by rodents.

Controlling Mammals

Protection of archaeological sites subject to damage from burrowing mammals is ideally carried out through a combination of removal and prevention of recolonisation through rigorous management. On scheduled sites, any intervention in the ground requires statutory consent, and archaeological guidance in all other cases. The means adopted should both be legal, specific to the site and humane. Detailed guidance is available from Natural England and Historic Scotland (see **Further Reading**).

Control of rabbits requires an initial period of intensive eradication, followed by permanent exclusion and regular monitoring. Legal methods are shooting, snaring, fumigation, destruction of warrens and trapping; a combination of approaches is usually required, preferably in the winter when colonies are smallest. Subsequent exclusion entails perimeter fencing of wire mesh, properly extended below ground to prevent burrowing under. Small sites can benefit from the placement of wire netting on the ground (including their outer perimeter), whilst clear of vegetation in the winter.

Badgers cannot be controlled, although it is possible in certain circumstances to obtain a licence to 'exclude' them from an area, although this is not always simple to achieve. Advice on the management of badgers can be obtained from Natural England (see **Further Reading**).

Moles can be eradicated by means of traps in tunnels, or by gassing within tunnels by professionals. Foxes are managed by shooting, trapping or snaring, and rats by shooting, trapping, fumigation, baits and proofing.

Damage from larger domesticated animals is more easily controlled through barriers, such as hedges, fences and dry stone walls. Management of pressure points is also essential, such as moving gates, fences and feeding troughs, or the creation of new sheltered tree cover away from vulnerable landscape features.

Bats

The one mammal which is common on historic sites and which needs special consideration is the bat. Bats frequently adopt old buildings for roosting. It is always safest to assume a historic building has bats, unless a survey has definitely shown that there are none and there are no signs of use. Bats will also use ruins, particularly if there are hollow cores or chimney features. There are 18 species of bats in Great Britain; all are protected (*Wildlife & Countryside Act 1981, as amended*). Furthermore, unlike birds' nests, which are only protected when in use, bat roosts are protected whether or not bats are present. This is because bats use different locations for their roosts at different times of year. It is most important to know if, how and when bats use a site, and it is vital to build up this information before any works are planned. Having this information at the very earliest stage will allow work to be programmed to either avoid committing an offence or, if unavoidable, to plan mitigation which will enable a licence to be obtained to carry out work which would otherwise cause an offence.

Living with Bats

Bats do not generally damage buildings, but their droppings and urine can be harmful. This is more likely to be a problem in buildings which are open to the roof, such as churches. Droppings and urine can cause pitting, etching and staining of porous and non-porous materials. Polished metalwork and stonework, polychrome surfaces, easel paintings and wall paintings are particularly vulnerable. Because bat urine is chemically more aggressive than the compounds found in faeces, it tends to cause comparatively more damage. Urine and faeces also present a human health risk, and appropriate precautions should be taken when removing droppings and cleaning.

Most bats can squeeze into very small spaces both for protection and to help regulate body temperature. This can make them very difficult to spot, even when they are actively being looked for.

Bat droppings are more obvious than urine, but both are likely to be found in the same area.

A variety of methods may be used to manage this problem. These range from simple low-cost techniques, such as organising effective cleaning routines, to extreme measures, such as excluding bats from the building altogether. Factors influencing the selection of appropriate methods include the scale of the problem, its impact on the use of the building, the cultural significance and sensitivity of the fabric vulnerable to damage, the conservation significance of the bats involved, the seasonal pattern of the problem, and the cost and practicality of the proposed actions.

Before deciding on what action may be appropriate and effective, information should be obtained about the nature and extent of damage to the building or furnishings, and the behaviour patterns of bats in the building.

CONSERVATION BASICS
ECOLOGICAL CONSIDERATIONS

Methods of management include:

- *Cleaning to remove droppings*
  Cleaning of significant and sensitive surfaces should be carried out only by professional conservators.

- *Moving vulnerable portable objects*
  Portable objects may be moved to locations in the building where the rate of droppings is lower. Statutory consent will be needed where the original position of the object is of historical significance or, in the case of churches, liturgical importance.

- *Providing temporary covers or screens*
  Temporary covers can provide protection to vulnerable surfaces or objects; vertically-hanging screens may be used to protect wall paintings and other artefacts on walls.

- *Installing removeable protection*
  'Protector boards' or canopies can be installed beneath areas used by bats to shelter sensitive areas or artefacts.

Covers and screens are visually obtrusive. In addition, they may harm vulnerable building fabric by creating adverse microclimates beneath them. However, they need only be used during the times of year when the bats are active, and can be removed at other times. The advice of professional conservators should always be sought because the selection of materials for covers and screens, and the methods for securing them, require careful consideration.

More interventionist methods of control include deterrents, relocation of roosts and, as the very last resort, exclusion of the bats. These methods should only ever be contemplated where the problem is extreme, where fabric of very high significance is being harmed, and after all other approaches have been tried and failed. It is very difficult to guarantee the success of these methods, all of which are illegal unless approved and licensed by Natural England, which should be contacted at an early stage if any of these methods are being considered.

English Heritage (now Historic England), The National Trust, and Natural England have published comprehensive and detailed practical guidance on living with bats in churches and traditional buildings (see **Further Reading**).

Top: This full-grown great-crested newt (*Triturus cristatus*) is around 14 cm long, and had spent the winter hibernating under the cover used to protect these medieval tiles at Cleeve Abbey.

Bottom: A juvenile smooth newt (*Lissotriton vulgaris*) is around 6 cm long and may reach 10 cm. This example is shown compared to the author's finger, to highlight the small size of cracks or crevices it could use for shelter or hibernation.

In England there are six native reptiles (three snakes and three lizards) and seven native amphibians (three newts, two toads and two frogs), and they are collectively known as herpetofauna (or more colloquially as 'herps'). All native reptiles and amphibians are protected under the *Wildlife and Countryside Act 1981*, and four species (sand lizard, smooth snake, natterjack toad and great-crested newt) are also protected by a European Directive. These may not be caught, handled or disturbed, nor their habitats damaged or disturbed.

All reptiles and amphibians may use walls, but lizards and newts are particularly fond of crevices and cracks in old walls. On sites with known herpetofauna interest, particular care needs to be taken if blocking cavities during the hibernation period, to ensure that hibernating animals do not become entombed.

ARTHROPODS & MOLLUSCS

Arthropods are undoubtedly the most numerous fauna on every site, and usually the most overlooked. This group consists of insects and arachnids (principally spiders). Molluscs include slugs and snails. Almost all arthropods and molluscs are under-recorded at sites simply because detailed surveys have never been carried out. A good example of this is Tynemouth Castle, where before 2005 nothing was known about the arthropod fauna on site. During the summer of that year, volunteers set a number of moth traps over several evenings, which resulted in the discovery that the site was used by 279 species of moth, of which 22 were included in the International Union for Conservation of Nature and Natural Resources [IUCN] *Red List of Threatened Species*.

INSECTS

Many insects will use walls, often as momentary resting places or as a place to warm up in the sun, but some insects are especially associated with walls because they feed on common wall plants. These include red admiral butterflies (*Vanessa atalanta*) and bloxworth snout moths (*Hypena obsitalis*), which feed on pellitory-of-the-wall (*Parietaria officinalis*); the swallow-tail moths (*Ourapteryx sambucaria*) and holly blue butterflies (*Celastrina argiolus*), which feed on ivy; and marbled green moths, which feed on lichens and whose scientific name – *Cryphia muralis muralis* – clearly show their connection with walls.

In all, there are around 21,000 species of insect, in several different families, in England. Many of these families have one or more species which use walls, though most of this use will be unnoticed unless specifically looked for. Occasionally, species such as wall mason bees (*Osmia sp.*), or the rarer mason wasps, cause noticeable damage by excavating tunnels into structures, usually into the mortar, and sometimes into the stone itself (usually only on soft sandstones or limestones). Generally, such damage is minimal, but in particularly good locations, large numbers of these insects can build up and damage can be substantial. In situations such as this, control measures may be necessary. Control with insecticides is undesirable and often not effective in the long term, as mason bees tend to reuse existing nest holes. The most reliable approach is to repair or replace the damaged mortar and masonry, to eradicate the nest sites and break the cycle of recolonisation. Alternatively, the use of temporary coverings of fine netting over the affected wall surfaces may be practical. This should be done during the spring when the bees are searching for nesting sites. Artificial 'nest boxes' made of clay can also be provided on or near walls to provide alternative nest sites. ⬡STONE

It should be noted that these bees/wasps are termed 'solitary' because individual females excavate the tunnels and provide for a handful of offspring before dying, and even in situations where there are large numbers using a site they are acting individually, not collectively. This is an important difference with honey bees or 'communal' wasps because with no hive or large food store to protect, these solitary bees/wasps are very unaggressive, and although capable of stinging are unlikely to do so unless actually picked up.

Several species of wood-boring insect may cause damage to timber in buildings. These, and measures for controlling them, are discussed in detail in the *Timber* volume of this series. ⬡TIMBER

ARACHNIDS (SPIDERS)

Out of 600 species of spiders found in England, 66 (over 10 %) are believed to be 'closely dependent on walls'. Like insects, spiders are plentiful on most structures, but are rarely seen, although the web-making species (not all spiders use webs) are probably more noticeable. Spiders cause no damage to structures.

Top: A night-flying privet hawkmoth (*Sphinx ligustri*), one of the largest English moths. Even such large insects as this may go unnoticed on a site.

Middle left: Wall mason bees making use of soft lime mortar for their tunnels.

Bottom left: On rare occasions a particularly favourable location may attract several hundred wall mason bees – all working as individuals – as in this case at Fordcroft Roman site in Croydon, South London.

Bottom right: A honey bee (*Apis mellifera*) swarm being removed from within the wall at Battle Abbey. Communal bees (and wasps) such as these can become aggressive if they perceive a threat to their hive, and may therefore need to be removed from public areas.

FLORA

Next to the historic fabric itself, plants are often the most noticeable feature of a site. For the purpose of managing flora on historic sites, it is convenient to divide plants into two main categories, and then subdivide these into:

Higher plants

- trees and shrubs
- woody herbaceous perennials
- soft herbaceous perennials
- annual and ephemeral plants
- climbers.

Lower plants

- lichen
- moss
- liverworts
- algae.

HIGHER PLANTS

Higher plants are often the most conspicuous, and generally provide scope for much wildlife to feed on, particularly when flowering. Long-lived plants such as trees and shrubs have increasing ecological potential as they age.

Trees & Shrubs

Trees and shrubs are always harmful to walls. As a tree or shrub grows, both stems and roots increase in girth by a process known as 'secondary thickening'. This growth is virtually impossible to contain. A tree or shrub established within a structure will eventually be capable of pushing apart stonework, brickwork, tarmac and concrete.

Except in very exceptional circumstances, trees and shrubs should always be removed: preferably at the earliest possible stage. Exceptions are where the tree and structure are already so intertwined that any action would eventually result in rebuilding.

Trees or shrubs growing close to, but not actually on, structures require more careful consideration. As they get larger, they may be able to impact on the structure, pushing it out of alignment and, in the worst case scenario, actually pushing it over. Tree roots can contribute to the shrinkage of clay soils during long spells of dry weather, which may result in subsidence. Wind throw, the uprooting of a tree in storm conditions, can also cause damage to an adjacent building.

Roots may also penetrate poorly maintained or defective drains, causing blockage and further leakage. The likelihood of damage, and the timescale, will vary depending on the species of tree (ultimate size and speed of growth) and its distance from the structure. Large growing and vigorous species include poplars, willows and oaks. Advice should be sought from an arborist or landscape manager.

Damage from trees and shrubs

Top left: Knowing the ultimate size to which these self-set sycamores (*Acer pseudoplatanus*) will grow, a prudent manager would remove them before they become large enough to push the wall over.

Top right: Secondary thickening of both stem and roots of a buddleia (*Buddleja davidii*) is causing serious damage to this brickwork.

Middle left: An old yew (*Taxus baccata*) pushing against the wall with its main trunk. When its branches move in the wind, they can dislodge stones.

Bottom left: A beech (*Fagus sylvatica*) growing in the masonry behind the high altar at Bayham Abbey. It has reached a size where any course of action would result in having to completely rebuild the wall. The best approach is to manage the tree, as far as possible, to prevent collapse or wind throw.

Bottom right: A partial collapse of the moat wall at the Tower of London was caused by the increasing girth of roots from a nearby London plane tree (*Platanus × acerifolia*).

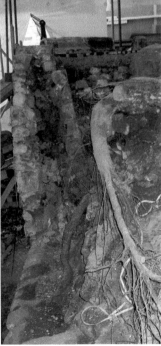

CONSERVATION BASICS
ECOLOGICAL CONSIDERATIONS

Individual branches from trees or shrubs may also affect structures, even where the main part of the plant is a safe distance away. Branches growing close to a structure can cause damage both directly, by pushing against a structure, or indirectly, by rubbing against it as they move in the wind. Roofs and guttering are particularly vulnerable. In such cases, it is usually sufficient to simply remove the offending branch. Injury and damage can also be caused by falling limbs or failure of the main trunk. It is advisable for mature trees growing near buildings or the public highway, to be professionally surveyed at regular intervals, to check on their general health, and assess the need for pruning or other work. Guidance on common sense management for tree safety is available from the National Tree Safety Group (see **Further Reading**).

The trees and shrubs most commonly found establishing on walls are those with seeds distributed by the wind such as sycamore (*Acer pseudoplatanus*), ash (*Fraxinus excelsior*) and buddleia (*Buddleja davidii*), or those eaten and then distributed by birds such as hawthorn (*Crataegus monogyna*) and yew (*Taxus baccata*).

Woody herbaceous perennials such as wallflowers (*Cheiranthus cheiri*) may cause damage as they get older.

Woody Herbaceous Perennials

Although not woody in the true sense of trees and shrubs, these plants can be damaging to stonework because their growth is strong and vigorous, capable of pushing stones apart and dislodging them. This often begins as the plant becomes older and larger, and produces a mass of stems or woody basal growth.

Examples of this type of plant are pellitory-of-the-wall (*Parietaria judaica*), valerian (*Centranthus ruber*), bramble (*Rubus fruticosus*) and wallflower (*Cheiranthus cheiri*).

It is frequently necessary to control these plants when they become older.

Left: Soft herbaceous perennials such as ivy-leaved toadflax (*Cymbalaria muralis*) are unlikely to cause damage unless walls are already in poor condition.

Right: Ephemeral plants such as the rue-leaved saxifrage (*Saxifraga tridactylites*) are small, short-lived and do not cause damage.

Soft Herbaceous Perennials

These plants are not generally damaging, although there is some possibility that they could grow sufficiently to dislodge stones on walls in poor condition.

Examples of this type of plant are ivy-leaved toadflax (*Cymbalaria muralis*), herb Robert (*Geranium robertianum*) and mind-your-own-business (*Soleirolia soleirolii*), and the common wall-growing ferns such as broad buckler fern (*Dryopteris dilatata*), polypody (*Polypodium vulgare*) and wall rue (*Asplenium ruta-muraria*).

There may occasionally be a need to control isolated plants of this group.

Annual & Ephemeral Plants

Small, short-lived plants that are quick growing, they normally germinate early in the spring, flower quickly, set seed and then die back as summer sets in. They are rarely very big and not known to be damaging. They often appear in cracks, but are merely taking advantage of opportunities which arise because cracks trap dust and moisture. They are never the primary cause of cracks and do not make them worse.

Examples of this type of plant are fern grass (*Desmazeria rigida*), wall bedstraw (*Galium parisiense*) and rue-leaved saxifrage (*Saxifraga tridactylites*). There is no reason to control these plants.

Climbers

A wide variety of plants are termed 'climbers'; these range from true climbers to what can be more accurately termed 'scramblers'. Trees and shrubs commonly grown against walls – such as fruit trees or flowering shrubs such as magnolia, ceanothus and similar – are not climbing plants.

Climbing is an evolutionary adaptation to get a plant to high light levels – increasing the chances of survival and procreation – without the high 'cost' of a rigid, self-supporting branch structure. Climbers generally have soft, flexible fast-growing stems.

Plants have developed a range of different methods to climb over inanimate objects such as cliffs, boulders and walls, or over their unfortunate living neighbours. These methods include:

- twining (such as honeysuckle)
- clinging tendrils (for example, sweet pea)
- downward facing thorns or hooks (such as climbing roses)
- self-clinging (such as ivy, Virginia creeper).

Many climbing plants will combine more than one method; for example, sweet peas have clinging tendrils but will also twine. However, with the exception of the last group (self-clinging), most of these plants can only climb vertical walls with assistance such as wires, trellis or hooks inserted for that purpose. Outside of 'gardened' sites, the only climber likely to be encountered on walls is ivy (*Hedera sp.*).

It has long been automatically assumed that ivy growing on a structure is damaging, but it is now known that this may not always be the case. In fact, there are some instances when it may even be protective. The two principal causes of erosion on exposed wall faces are the freeze/thaw cycle and the wet/dry cycle. A covering of ivy can create a microclimate at the wall face, moderating the upper and lower limits of both these cycles, thus creating a more stable environment. Additionally, in some locations, airborne pollutants and salts can prove harmful to stonework, and a covering of evergreen vegetation can reduce this problem.

Risk from ivy

Top left: Aerial rootlets produced all along the stem of ivy are only for holding on to the surface. They do not grow larger with time and do not penetrate into a structure.

Top right: In some circumstances, 'proper' roots may be produced at leaf nodes. These will increase in size and can penetrate a structure, causing increasing damage with time.

Bottom left: The heavy arboreal or flowering growth of ivy may cause problems for unsupported structures, especially in windy conditions.

Bottom right: Where ivy is clearly rooted into the wall and not the ground, it should be considered as damaging as any other tree or shrub, and removed at the earliest possible stage.

Removing Ivy

There are some very clear instances when ivy is damaging to a structure and removal should be considered.

These include:

- *When it has properly rooted into walls*
  In many cases, ivy simply uses a wall as a climbing frame, using small 'aerial rootlets' to cling on to the surface. These aerial rootlets do not grow bigger with time, and their only function is to adhere to a surface. In certain circumstances, including where the base has been cut, and possibly when a structure is very damp, proper roots are produced, usually only at leaf nodes (aerial rootlets are usually produced along the entire stem length). Under suitable conditions these roots will continue to increase in size and spread through the structure, causing increasing damage.

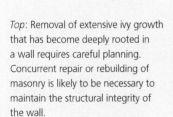

Top: Removal of extensive ivy growth that has become deeply rooted in a wall requires careful planning. Concurrent repair or rebuilding of masonry is likely to be necessary to maintain the structural integrity of the wall.

Bottom: A dense covering of ivy or other vegetation can make a condition survey of the structure difficult or impossible.

- *Where the entire plant is growing from the wall, with no connection to the ground*
  In this case, the ivy becomes like any other tree or shrub growing in a wall, and damage will escalate until it is removed. Although ivy may occasionally actually germinate in a wall, the most frequent cause of independent plants established in a structure is previous attempts at control by severing the stem at the base and leaving the foliage on the wall to die (which, inconveniently, it does not always do).

- *Where a stem of ivy, although rooted into the ground, has grown up through existing cracks or holes in the structure*
  In this case, the stem will make damage worse because it inevitably increases in girth each year.

- *Where ivy has reached the top of a structure, and has started to produce the 'arboreal' or flowering growth*

  This will become increasingly large and heavy, especially when wet, and presents a large 'sail' to the wind. On short or sound structures this may not be a problem, but on tall, unsupported sections of masonry, it may be seriously destabilising. Where funding is limited, cutting back this usually horizontal growth to the vertical stems can reduce the problem, although the flowering stems will regrow.

In addition to actual damage, there may be other good reasons for removing ivy. One is the need to carry out condition or structural surveys, which cannot be achieved where the structure is obscured by foliage, although complete removal for this purpose may not always be necessary. Another reason may be for presentational purposes. Leaving aside the arguments for romantic, ivy-clad ruins, there are frequently important reasons for having some or all of a structure visible, so that its construction, decoration and/or function can be properly understood.

Cut at the base over a year previously, this ivy is still flourishing and has clearly rooted into the structure.

A very well-established and deeply embedded root system that has displaced masonry in a wall has become, in effect, an integral part of the structure. In such cases it is important to carry out a structural assessment, and to determine the extent of consolidation and repair works that will be necessary to maintain the stability of the structure, before the plant is removed.

In summary, while there are a number of very good reasons for removing ivy from structures, each situation should be considered carefully and removal should not be automatic. If the ivy is not actually causing any damage, or the structure does not need to be uncovered, it might be better for the ivy to remain. Furthermore, removal might actually create worse problems, such as fabric erosion, in the long term.

One very clear piece of advice that has come from all research into ivy is that if removal is deemed necessary, it should be completed in one operation. Severing ivy at the base and leaving the foliage to die is likely, in many cases, to actually cause the plant to root into the structure.

LOWER PLANTS

On level surfaces, lower plants will often start the build up of soil, which allows natural 'soft caps' to form. Because it often takes a considerable time for lichen and moss to establish on vertical surfaces, the presence of these lower plants on the surface of stone can be an indication that it is in a relatively stable condition. Their absence might be an early indication that the surface of the stone is eroding. However, many materials are susceptible to varying degrees of biodeterioration, depending on their physical nature, chemical composition and environmental conditions. ⊖ENVIRONMENT

Lichen

Lichen is by far the most common lower plant found on walls. Lichens are a symbiotic organism consisting of a fungus, and an algae or a cynobacterium. In general, they do little harm to the surface of the stone on which they grow, although they may occasionally obscure carved detailing. Though they have the ability to dissolve this surface, their slow rate of growth is legendary, and there is strong evidence that a covering of lichen actually slows or prevents the erosion caused by rain, wind-borne pollutants, sand particles, and the freeze/thaw weathering of ice crystals.

Because of their intolerance of pollutants, lichen can also be important for monitoring air pollution, which in itself can be damaging to stonework. Superficial damage caused by lichen has been observed on certain types of sandstone in Scotland, although there are currently no recorded cases of this in England. If, however, damage is being caused, its removal or prevention is justified. ⊖STONE

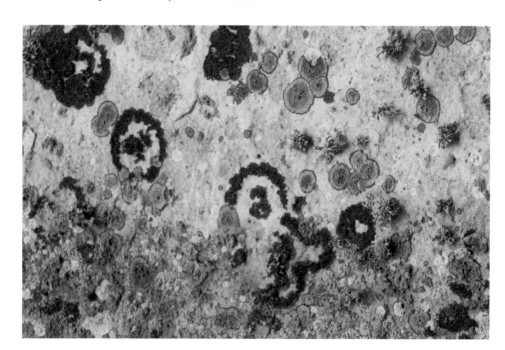

Lichens

Usually harmless, lichens can give stonework an extra patina of interest, as shown by this range of species growing together on one of the sarsen stones at Stonehenge.

Moss

Like lichen, moss seems to be a largely harmless and potentially beneficial covering, although on very soft fabric it may hold water, which can then cause damage. Careful inspection is recommended.

Corrosion of lead gutter linings is sometimes attributed to the colonisation of roof slopes by mosses and lichens. Typically, damage to the lead takes the form of narrow cleanly-cut grooves in the surface of the metal, where water drips from the slates or tiles. Although some lichens produce organic acids which have been shown to gradually affect stone, there is no evidence that they increase the acidity of rainwater running from colonised surfaces. Similar corrosion on lead flashings has been observed in situations where no lichen is present, and appears to have been caused solely by rainwater (which is naturally slightly acidic). Whatever the cause, the installation of a sacrificial lead flashing is an effective remedy in areas where this form of corrosion is taking place. ⊕ROOFING

Mosses

Top: Corrosion of this type, affecting lead gutter linings or flashings, can occur whether or not roof slopes have been colonised by vegetation.

Bottom: Moss is often a harmless covering and may be a good indicator of stability. The stonework on the right of this photograph is clearly eroding, but the moss-covered section on the left is in good condition.

Liverworts (*top*) and algae (*bottom*) are harmless indicators of constantly damp and shaded conditions.

Liverworts & Algae

Liverworts and algae are only found where walls are permanently damp and usually shaded. Damp conditions are often a problem for historic fabric and these plants are a useful indicator, but do not cause problems themselves: usually disappearing if the damp problem is solved. Typical conditions under which algae and liverworts establish are on walls where broken or missing drainpipes result in frequent saturation.

FUNGI

Several species of fungi are able to colonise buildings. Most do no direct damage to building fabric, but their presence may be symptomatic of environmental conditions that can impel other decay mechanisms. A small number of fungi species can cause serious damage to timber. These, and the methods for controlling them, are discussed in detail in the *Timber* volume of this series. ⊙ENVIRONMENT ⊙TIMBER

LEGAL & SAFETY IMPLICATIONS

FOOD & ENVIRONMENT PROTECTION ACT 1985

Under this legislation, the *Control of Pesticides Regulations 1986* (amended 1997) [CoPR] were introduced. 'Pesticides' are defined as *"chemical substances and certain micro-organisms prepared or used to destroy pests"*. The term therefore encompasses a number of products, including, amongst others, herbicides, fungicides, insecticides and masonry biocides. These Regulations are likely to be amended (or replaced) in a strengthened form when the *EU Sustainable Use Directive* is implemented in 2012. At present the most important aspects of the Regulations are:

- Only approved products may be sold, supplied, stored, advertised or used. The use of old chemicals which no longer have approval is not permitted.
- Users of pesticides must comply with the Conditions of Approval relating to use. These are clearly shown on the packaging and incorporated within the manufacturers' recommendations for use.
- A recognised Certificate of Competence is required by all contractors and all persons born after 31st December 1964, unless working under the direct supervision of a certificate holder. This 'Grandfather' rule only applies to people working on their own land or land owned by their direct employer. It does not exempt them from the need to be competent, only the need for the Certificate of Competence. This exemption is expected to be removed when the *Sustainable Use Directive* is implemented.

HEALTH & SAFETY AT WORK ACT 1974

These regulations are concerned with preventing or reducing the exposure of workers and other people to a wide range of hazardous substances. With regard to pesticides, the requirements of the *COSHH* regulations apply in addition to (not in place of) the *CoPR*. They require initially that consideration be given to whether pesticides are needed at all, or whether safer alternatives can be used. Where pesticides are to be used, organisations must ensure that:

- as a minimum, label precautions must be observed
- safety equipment and clothing must be properly maintained, and staff instructed in their use
- where necessary, exposure of workers is to be monitored and health checks carried out
- adequate records of all operations involving pesticide application must be made and retained for at least three years.

There are no specific requirements for people who advise on or specify the use of pesticides. However, they should be competent to offer such advice. The *Code of Practice for using Plant Protection Products* (produced by the Health and Safety Executive and Defra) states: *".....you should be sure that the person you ask for advice has the necessary skills, knowledge and experience. Also people who draft contracts should have suitable training and experience to do so....."* The safest option is to use someone with the British Agrochemicals Standards Inspection Scheme [BASIS] Certificate in Crop Protection and who is on the BASIS Professional Register (see **Further Reading**).

METHODS OF VEGETATION CONTROL ON STRUCTURES

ENVIRONMENTAL CONTROL

This may be useful under certain conditions and for certain types of vegetation: principally the lower plants like algae and liverworts. Removal of the damp and shaded conditions that favour their growth will usually cause their demise. This may be achievable by simple methods such as cutting back overhanging growth from adjacent trees, or repairing broken pipes and gutters.

PHYSICAL CONTROL

Most of the physical methods of vegetation control, such as hoeing or mulching, are not possible on vertical surfaces; the two methods available are hand removal and cutting. Stump treatment must be used in conjunction with cutting because almost all the common trees and shrubs found on walls will regenerate from cut stumps if left untreated. Stump treatment is a very targeted method of chemical control and there should be no non-target casualties. It is more effective on stumps over 50 mm diameter, possibly because of the amount of chemical which can be applied. If used on smaller stumps, it is better to cut these slightly longer (75–100 mm) and strip back the bark before treatment to provide a greater surface area for the chemical. Treatment is more effective if carried out immediately after cutting. When treating larger stumps, some thought needs to be given to what treatment the wall will require as the stump dies and rots away.

Failure to treat the cut stump on a previous occasion has caused this hazel (*Corylus avellana*) to regrow as a vertical coppice. Unless the cut ends are treated, this stump will grow a new set of branches, and the main trunk will continue to grow.

Top left: When stumps have been killed they will begin to rot. Thought needs to be given to what repairs may be necessary as this happens.

Right: When spraying vertical surfaces there is a strong chance of spray drift, and great care is needed to avoid damage to non-target species or areas.

Bottom left: The use of weed-wipers avoids the problem of spray drift and reduces chemical use.

CHEMICAL CONTROL

Spraying is only practical with hand-operated equipment because machinery-mounted equipment is designed for use on a horizontal surface. Conventional spraying has three major drawbacks: spraying on a vertical surface increases the risk of drift, making even spot treatment difficult to contain; there are increased health and safety concerns for operators if the lance is consistently being used above the head; and it is difficult to avoid some surplus chemical running down the face of the wall. Spraying is not recommended, except in a case where total vegetation control is required, both on the wall and the surrounding area.

Weed-wiping is a safer and more selective option for chemical treatment. It has several advantages: there is minimal risk of drift, control can be carefully targeted, there is reduced use of the chemical, and it is more environmentally acceptable. The disadvantage is that it is labour intensive and therefore more expensive.

Soft Cappings

'Soft capping' is a technique of using soil and vegetation (principally grasses) to protect exposed wall tops. Ruined walls that have remained untouched for decades slowly build up a natural 'soft' cap. When these have been removed during past consolidation works, little, if any, deterioration of masonry has been observed below this type of covering.

Research funded by English Heritage (now Historic England) since 2000, and carried out in conjunction with the University of Oxford School of Geography and the Environment, has shown very clearly that soft capping will protect the covered masonry from deterioration by freeze/thaw weathering. In laboratory and site trials, just 50 mm of soil under a turf capping prevented freezing at the wall head. Soft cappings will also reduce the potential damage to underlying stone from the thermal expansion of exposed wall tops. Because of the difficulties in accurately monitoring moisture within walls, the effect of soft caps on moisture penetration are less well understood. However, the data does suggest that soft capping helps to stabilise moisture levels within wall cores. Soft cappings also retain moisture during rain, thereby reducing the amount of water running down the face of walls and mitigating the harm incessant cycles of wetting and drying might cause to masonry surfaces.

This technique may be used on unconsolidated wall tops or those which have been previously 'hard capped', using either lime or cement mortar. Hairline cracks in existing hard caps need not be repaired before soft capping is undertaken, but any wider than 1 mm or any loose stones should be consolidated first.

There are various techniques for installing soft capping, but the simplest is probably using soil and turf. Turf is laid upside down on the edge of the wall top, with the turf hanging down the wall face. Soil is put on the top of the wall, covering the edge of the turf, to a firmed depth of 75–100 mm. The hanging turf is then folded back to cover the soil and wall top. When working from two sides, the opposing turfs can be trimmed to fit where they meet, or if the wall top is particularly wide, any gap filled with an additional piece of turf. Finally, the turf edges are pinned, using pegs of split bamboo cane to hold them in place until they have rooted into the soil.

The technique is flexible, and can be adapted to suit different types and shapes of wall top. Field trials have shown that work is more successful if a local turf containing a wide range of grass species can be used and cut slightly thicker (up to 25 mm of root) than is usual for commercial turf. Most soil is suitable, although any with a high clay content should be avoided, particularly in the warmer and drier areas of the country where cracking will occur if the soft cap dries out. It does not matter if the soil has a proportion of stone within it, although stones larger than 50 mm diameter should be avoided where possible. MORTARS

Facing page: Drawing showing the type of natural soft capping built up over decades or centuries (*top*).

Newly-installed soft cap, with the final pieces of turf waiting to be trimmed and folded into place. (*bottom left*).

The same soft cap after 18 months. The fresh 'lawn'-like appearance has matured into a more natural look (*bottom right*).

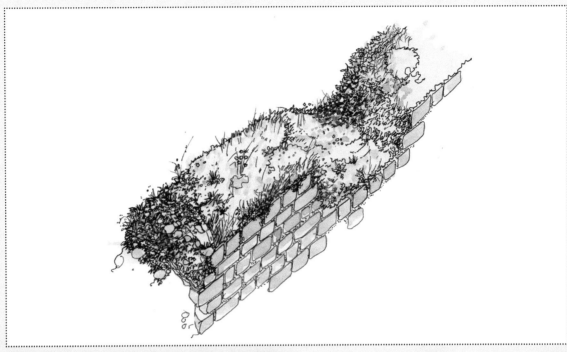

CONSERVATION BASICS
ECOLOGICAL CONSIDERATIONS

Further Reading

Cameron, S., Urquhart, D., Wakefield, R., Young, M. (1997); *Biological Growths on Sandstone Buildings: Control and Treatment (Historic Scotland Technical Advice Notes No.10)*; Edinburgh: Historic Scotland

Darlington, A. (1981); *Ecology of Walls*; London: Heinemann Educational

Dunwell, A. J., Trout, R. C. (1998); *Burrowing Animals and Archaeology (Historic Scotland Technical Advice Note No.16)*; Edinburgh: Historic Scotland

English Heritage, National Trust, Natural England (2009); *Bats in Traditional Buildings*; Swindon: English Heritage; available from *www.historicengland.org.uk/images-books/publications/bats-in-traditional-buildings/*

Gilbert, O. L. (1992); *Rooted in Stone: The Natural Flora of Urban Walls*; Peterborough: English Nature

National Tree Safety Group (2011); *Common Sense Risk Management of Trees: Guidance on Trees and Public Safety in the UK for Owners, Managers and Advisers*; Edinburgh: Forestry Commission; available from *www.forestresearch.gov.uk/research/common-sense-risk-management-of-trees/*

Paine, S. (1993); 'The effects of bat excreta on wall paintings'; in *The Conservator*; Vol.17, pp.3–10

Parsons, K. *et al.* (eds.) (2007); *Bat Surveys: Good Practice Guidelines*; London: Bat Conservation Trust

Pepper, H. (1998); *The Prevention of Rabbit Damage to Trees in Woodlands (Forestry Practice Note July 1998)*; Edinburgh: The Forestry Authority (Forestry Commission); available from *www.forestresearch.gov.uk/research/the-prevention-of-rabbit-damage-to-trees-in-woodland/*

Richardson, D. H. S. (1992); *Pollution Monitoring with Lichens (Naturalists' handbooks No.19)*; Slough: Richmond Publishing Co.

Segal, S. (1969); *Ecological Notes on Wall Vegetation*; The Hague: W. Junk

Stebbings, R. E. (1986); *Which Bat is it?: A Guide to Bat Identification in Great Britain and Ireland*; London: The Mammal Society and The Vincent Wildlife Trust

Watkins, J., Wright, T. (eds.) (2007); *The Management and Maintenance of Historic Parks, Gardens & Landscapes: The English Heritage Handbook*; London: Frances Lincoln

Useful Web Addresses

BASIS (Registration) Ltd (n.d.); Professional Register (web page); available from *www.basis-reg.co.uk/scheme-basis*

Department for Environment, Food and Rural Affairs [Defra] (2002); *MAGIC* [Multi-Agency Geographical Information for the Countryside] (web page); (updated 06/03/12); available from *www.magic.gov.uk/*

Natural England (2012); *Advisory Leaflets and Guidance Notes* (web documents: Advice on wildlife, problems that may occur and solutions); available from *webarchive.nationalarchives.gov.uk/20130301211844/http://www.naturalengland.org.uk/ourwork/regulation/wildlife/advice/advisoryleaflets.aspx*

MANAGING MAINTENANCE & REPAIR

The need for maintenance and repair arises primarily from the inevitable decay and deterioration of building fabric that occurs because of climatic conditions, wear and tear by building users, neglect or other threats. This chapter deals with the practicalities of organising and managing maintenance and repair programmes for heritage assets.

First, it considers maintenance: an activity essential for sustaining the heritage values and significance of historic buildings and places. The different types of maintenance are defined and the process of maintenance planning is explained. Other topics include: the management of information and records, the organisation, roles and responsibilities of people involved in maintenance, and, finally, maintenance procurement.

Next, it focusses on the design and management of programmes of planned repair, starting with the client brief; it addresses the skills and competencies required of the various project team members, and discusses the training and accreditation of professionals and craftspeople. The work stages of a project from inception to completion are outlined; in each stage, issues that have a bearing on the quality of the work eventually procured are highlighted. These include sourcing appropriate materials, carrying out site trials to evaluate the effectiveness of treatments and repairs, and the use of exemplars. Other topics include feasibility studies, cost planning, the preparation of contract documentation, and contact procurement methods.

In practice, the maintenance and planned repair of heritage assets are usually carried out concurrently, sometimes in conjunction with improvements or adaptations. The holistic approach described in **Conservation Planning for Maintenance & Repair** enables conservation needs and priorities to be identified, and works to be implemented effectively and economically. In larger or more complex places, estates or collections, a system of asset management planning may be adopted; the chapter concludes by examining this process.

Neglected buildings can quickly fall into a state of disrepair and dereliction. Once a building ceases to be wind and weathertight and secure, deterioration will occur at an increasing pace.

CONSERVATION BASICS
MANAGING MAINTENANCE & REPAIR

CARE & MAINTENANCE

It is perhaps natural to think that once a programme of conservation and repair has been completed, all the problems of deterioration will have been resolved, and no further work and expenditure will be needed for many decades. However, while there is an increasingly common expectation that all repair works and treatments should be guaranteed for a specific period, in practice this is unrealistic. Historic buildings respond to their environment and use, and will continue to deteriorate more quickly if not maintained. Thus recommendations and specifications for maintaining repaired elements should normally form part of any post-conservation report. Indeed, maintenance is considered so important by many grant-giving bodies that a priced maintenance plan must be submitted as a condition of funding. In some countries, the state has tried to encourage and support maintenance through schemes of subsidised maintenance inspections, such as Monumentenwacht in the Netherlands. In England, some private schemes are in operation which are aimed primarily at churches.

While it is true that grant funding tends to be focused on repair, and rarely if ever covers the cost of maintenance, this does not mean in any way that repair is more important. Rather, repair receives the largest amount of financial support simply because it is the most costly aspect of caring for heritage assets. By investing in maintenance, however, the need for larger, more invasive programmes of repair and conservation treatment will frequently be averted, reducing their impact on significance, maximising life expectancy and durability of building fabric, and saving money. Good maintenance ensures that the building fabric and services continue to function correctly, and slows down their rate of deterioration. It should be the primary focus of all property owners.

John Ruskin

"The principle of modern times is to neglect buildings first and to restore them afterwards. Take proper care of your monuments and you will not need to restore them. A few sheets of lead put in time upon the roof, a few dead leaves and sticks swept in time out of a water course, will save both roof and wall from ruin. Watch an old building with an anxious care: guard it as best you may, and at any cost, from every influence of dilapidation."

John Ruskin, *The Seven Lamps of Architecture*, 1848

Maintenance is defined as *"routine work necessary to keep the fabric of a place in good order"* (Historic England *Conservation Principles)*. It is distinct from repair, which is *"work beyond the scope of maintenance to remedy defects caused by decay, damage or use, including minor adaptation to achieve a sustainable outcome, but not involving restoration or alteration".* In other words, the main objective of maintenance is to limit deterioration, whereas the focus of repair is to remedy the harm caused by deterioration. Inspections carried out at regular intervals, coupled with prompt action to pre-empt or remedy problems, are the basis of effective maintenance.

Maintenance is cost-effective; the time and money spent on routine care, regular surveys and minor repairs protect the value of the building. In addition, maintenance helps to ensure the health and safety of building users and the general public. Although it is often seen as mundane, maintenance is invaluable in preserving the historic heritage. It forms a cornerstone of building conservation, as reflected in the opening policy in *Conservation Principles*: *"The conservation of significant places is founded on appropriate routine management and maintenance."*

Blocked gullies will back up and overflow during heavy rain, leading to water ingress to adjacent building fabric. They should be regularly checked and cleared if necessary.

MANAGING MAINTENANCE

Managing maintenance involves the following stages: determining the policies for maintenance; preparing the maintenance programme and allocating resources; implementing the programme; controlling the progress of the work and expenditure; and reviewing and assessing the effectiveness of the programme. Planning also enables maintenance to be integrated with other works such as repair or cleaning, and allows the best use to be made of access equipment that might be provided for such tasks. However, although maintenance is essential, it has the potential to be damaging if not carried out by personnel with the appropriate skills and understanding. The standards required for maintenance should be as high as for conservation project work. Both demand the appointment of suitably qualified, skilled and experienced personnel.

Periodic surveys, usually carried out at four- or five-yearly intervals, provide the basis for understanding what types of work are required, and how frequently they should be carried out. They also facilitate financial planning. Periodic surveys should always build on information gathered during previous building surveys and on the records of earlier maintenance work. In this way, an understanding will be gained about the behaviour and durability of the building fabric, and the life expectancy of past repair and maintenance interventions.

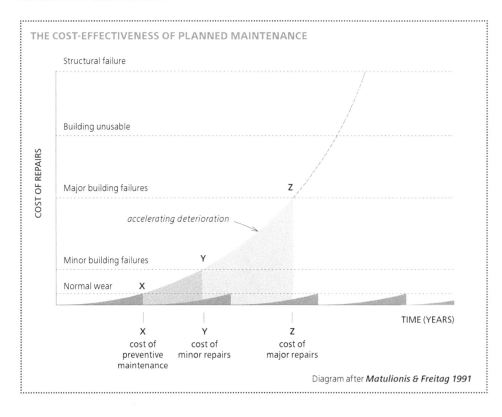

THE COST-EFFECTIVENESS OF PLANNED MAINTENANCE

Structural failure

Building unusable

Major building failures   Z

accelerating deterioration

Minor building failures   Y

Normal wear   X

COST OF REPAIRS

TIME (YEARS)

| X | Y | Z |
| cost of preventive maintenance | cost of minor repairs | cost of major repairs |

Diagram after *Matulionis & Freitag 1991*

Maintenance planning is dependent on three key sources of information:
- periodic building surveys that identify what work is required and when
- the building file or record, which provides a database of past maintenance work, including specifications, costs, materials, contractors and professionals employed
- a cost plan, including past budgets, actual costs and future budgets.

It is important for this record to be maintained regardless of changes in staff, so that past performance and future needs can be judged, and appropriate decisions made. The record should be readily available and accessible to future staff and external consultants and contractors. Where records are not maintained, there is no history upon which to base current and future decisions. This may result in over-repair or inappropriate intervention.

The evidence of past repairs may be buried within the structure and invisible to a surveyor carrying out a condition survey. This emphasises the importance of keeping detailed and accurate records of works carried out.

MAINTENANCE PLANS

According to *BS 8210:1986 Guide to building maintenance management*, maintenance should be *"organised and carried out with forethought, control and the use of records, to a predetermined plan based on the results of previous condition surveys"*.

Maintenance plans should be proportionate to the size and complexity of the building. For large buildings, or ones with complex uses, a maintenance plan may form part of a comprehensive asset management plan. For smaller buildings, such as privately-owned dwellings, it might consist simply of an inspection checklist (an example is given in **Appendix 1**). Although much of the advice in this chapter might, at first glance, appear irrelevant or unnecessarily complicated for the owner of a small or simple property, the basic requirements for regular inspection, linked with timely and appropriate action, are valid for everyone. Maintenance plans also provide a useful resource for new owners, and may go some way towards ensuring continuity of good maintenance practices.

While much maintenance is routine, there will be occasions when the need for action is triggered by unforeseen events, such as accidental damage or extreme weather. A good maintenance plan should make provision for dealing with such eventualities. The understanding of the building and its behaviour, gained through regular surveys, will help to identify areas that might be at risk. For example, if it is known that gutters and gullies are prone to blockage with leaves, they should be checked during and after autumn gales. The plan should highlight all areas particularly at risk, and assign responsibilities to individuals for *ad hoc* inspections and action.

Annual roof clearance

Clearing leaves from gutters each year after autumn will minimise the risk of harm caused by blockages.

"Organizations should ensure that suitable expertise is available for maintenance and its management at all levels within the organization. Where this expertise is not available in an organization, external resources should be employed instead."

BS 8210:2012 Guide to facilities maintenance management

The person responsible for maintenance – the maintenance manager – produces and implements programmes of periodic inspection and action, identifying who is to do what, and when, as well as ensuring that work is carried out at the right time and in the right way. Maintenance plans should be updated promptly if repairs or alterations to the structure, or changes in the way a building is used, bring about new or different conditions or requirements. For most privately-owned smaller properties, the maintenance manager will be the building owner. In other situations, such as larger residential, commercial or institutional buildings, the building owner or manager will normally appoint a suitably-qualified person or organisation to take responsibility for maintenance management.

Many routine maintenance tasks, particularly inspections and simple jobs, are often most effectively carried out by people familiar with the building – for example, the owner, occupier or, in the case of churches, the church warden – as they are best placed to recognise small changes in condition and be aware of potential trouble spots. It is helpful if inspections can be done by the same person each time. In that way, the maintenance manager will become accustomed to the building, developing a good understanding of its construction, use, and particular vulnerabilities. The maintenance manager will thus be likely to spot changes and faults more quickly, as well as comprehending what is possible and appropriate to prevent avoidable deterioration of historic fabric. A typical maintenance checklist is given in **Appendix 1**.

Tasks requiring high-level access should not be attempted without suitable safety equipment and may need to be carried out by appropriate contractors. Equally, some specialised tasks, such as the inspection and testing of electrical services and heating systems, should be left to suitably-qualified professionals.

Building users should know who is responsible for maintenance and be encouraged to report problems. However, there is little point in entrusting maintenance work to an individual unless there is an effective mechanism to enable reported defects to be dealt with appropriately and promptly.

DEVISING THE MAINTENANCE PLAN

UNDERSTANDING THE BUILDING

Understanding the design and construction of a historic building and how it behaves, particularly under challenging conditions such as snowfall, strong winds or heavy rain, is an essential prerequisite to devising a maintenance plan. Identifying weak points and anticipating where problems might occur will help in planning to deal with them promptly, or even avoid them altogether. Historic buildings were often designed with little thought about the practicalities of future maintenance. For example, in many older buildings there are internal gutters that are difficult to gain access to, and rainwater pipes concealed within roof spaces or built into walls. Careful examination of features such as these often indicates that problems have occurred frequently in the past, and that repeated attempts have been made to remedy them. When devising a maintenance plan, it is important to be aware of the existence of such features and make provisions for them.

Many historic buildings have internal gutters concealed in roof spaces, and rainwater pipes that are built into walls. In such cases, alterations to the rainwater disposal arrangements may be justified, to make them easier to gain access to and maintain.

Rainwater from the hall roofs at the Market Building, Covent Garden, London (*right*) originally passed through the supporting cast iron columns to underground drains. When the building was converted to its present use, the rainwater drainage system was redesigned to bypass the columns to reduce the risk of internal corrosion and to make maintenance easier.

The building as a whole, including its interior and the surrounding site, should be considered, and any factors that might conceivably cause problems, such as surface water drainage or the proximity of trees, taken into account. Orientation and location will also have an effect on the building. For example, the maintenance requirements for a building in an exposed upland area, or in close proximity to the sea, will be different to one on a sheltered inland site. The way the building is used will also have a bearing on its maintenance requirements. A rural church that is only used once a week might well suffer storm damage that could go undetected for several days until the next service, whereas damage to an occupied house is likely to be noticed almost immediately. The maintenance plan should also make provision for building services, both those that protect the building, such as fire detection systems and conservation heating, and those posing a potential risk to historic fabric, such as electrical and plumbing systems.

A maintenance plan should be devised by working methodically around the site, noting down areas of concern, and determining what needs to be done to prevent or manage problems, and at what intervals. The process is essentially a form of risk assessment that takes into account the likelihood and possible severity of any problems that might arise, and how the significance of the heritage asset might be affected. Some problems have more serious consequences than others, and should therefore take priority. For example, the implications of ignoring a blocked drain or a slipped slate will be much more serious than, say, failing to deal with a broken sash cord. The maintenance plan must be flexible enough to allow for such contingencies.

PREPARING MAINTENANCE PROGRAMMES

Having considered all aspects of the building, maintenance programmes or schedules can be drawn up that detail exactly what inspections and actions are required, and when they need to be done. There are two types of maintenance inspection:

- *Periodic inspections*

  These are carried out on a planned basis at defined intervals. The intervals might be weekly, monthly, seasonally, annually, or even quadrennially or quinquennially, depending on the nature of the building, the way it is used, and the part of the structure being considered. For example, ground drains and gutters may need to be checked weekly throughout the autumn and winter months, whereas roof voids would normally be inspected only once a year for signs of leakage. The condition of stonework might be checked every five years.

- *Occasional inspections*

  Occasional inspections are carried out following severe weather or unforeseen events. They should concentrate on areas of the external building envelope vulnerable to water penetration. For example, after violent storms it would be prudent to check roof coverings and rainwater disposal systems; following heavy snow, roof coverings, valley, and parapet gutters should be inspected.

BS 8210:2010 Guide to facilities maintenance management recommends that a full and detailed inspection of all aspects of the building fabric is undertaken by a suitably-qualified professional at least every five years. The Church of England, the National Trust and many other organisations have adopted this standard. Others, including English Heritage, carry out inspections on a four-yearly cycle.

Maintenance also falls into two categories:

- *Cyclical or planned preventive maintenance*

  This is carried out at predetermined intervals to reduce the likelihood of a building or its elements not meeting an acceptable condition in terms of function or significance. Tasks such as clearing gutters and rainwater pipes, painting and decorating, and testing and servicing of heating, plumbing, and electrical services – which includes fire protection systems – can all be forward-planned and incorporated into the maintenance programme at appropriate intervals, allowing the most economical collective use to be made of access arrangements.

- *Reactive maintenance*

  Reactive maintenance is carried out to correct defects arising from unforeseen causes, often in response to the findings of an occasional inspection or reports from building users. Some defects, especially those that pose significant risks to building users and the public, or to significant building fabric, will require urgent attention. Where expertise and access permit, they should be dealt with during or immediately after the inspection. Where this is not possible, appropriate contractors should be engaged to carry out the work as soon as possible. Ideally, a permanent repair should be carried out, but where this is likely to result in delay and, hence, increased risk of further damage, a temporary repair can be effected. Temporary repairs should be reversible, and not cause damage to significant fabric; they should not make permanent repairs more difficult to carry out. Some reactive maintenance tasks, such as repairing a broken latch or replacing a cracked pane of glass, may be less urgent. They can either be dealt with on a one-off basis, or 'saved up' for a programme of minor repair work at a later date.

RECORDING & MONITORING

All inspections and maintenance work should be recorded in a logbook, and include a synopsis of what was found during the inspection, a record of work undertaken and recommendations for any further work required. Using basic checklists that are completed during the inspection or routine maintenance makes this easy. The logbook should be supported by photographs, especially where access is difficult. There should be space to record the date and details of inspection and work undertaken, the weather, and any problems or issues noted at the time. Records of action should include what work was done and by whom, as well as the materials and methods used. An example of a simple maintenance log sheet is given in **Appendix 2**.

Some larger and more complex buildings may have parts that are difficult to gain access to for inspection and maintenance. In such cases, a computer-based building management system could be used to monitor, for example, the effectiveness of rainwater disposal, or to control systems such as trace heating in gutters which prevent blockage by snow.

As well as recording work undertaken, the logbook can be used to flag up problem areas that need to be looked at in subsequent inspections. Before each inspection, the latest logbook entry can be quickly checked to see whether there are any areas of particular concern. It can also be useful in monitoring the condition of various elements of the building, particularly if photographs are taken as part of the inspection. Reviewing past logbook entries can also help to highlight persistent areas of concern where some form of alteration or adaption might be considered to reduce problems in the future, such as providing bird deterrents or re-routing rainwater goods. The logbook should be archived alongside the major (typically quinquennial) inspection reports, and documentation of major repairs and alterations, to form a useful part of the building record.

MAINTENANCE PROCUREMENT

Maintenance work may vary in scope from large planned projects, such as the replacement of roof coverings, to routine periodic work, such as clearing gutters and rainwater pipes, or the inspection and testing and servicing of building services and equipment. Although the size of individual jobs may be quite small, they can be comparatively complex. Often, several workers with a variety of skills will be needed to fix a problem that on the face of it seems quite simple. For example, a plumber will be required to repair a leaking waste pipe. But if the pipe is concealed and has caused damage to wall linings, a carpenter and plasterer may also be needed to complete the work. Another important thing to consider is that maintenance works often have to be carried out while a building is occupied and in use. This may create further difficulties.

Procurement is the process by which a contractor's services are obtained. A maintenance manager will usually employ a number of different procurement methods to suit the range of requirements. Each method has respective advantages and disadvantages in terms of contract administration, cost control and flexibility. For instance, lump sum contracts predominate for larger planned packages of work, such as repainting external joinery and rainwater goods. The main advantage is that the financial commitment is known before the works commence. For private householders, this will also be the preferred method of procurement. Fixed-price contracts may also be let with utility companies for the maintenance of building services. For larger buildings in commercial or institutional uses, measured term or cost reimbursement contracts are commonly used for the procurement of both reactive and planned maintenance. These have the advantage of allowing a more rapid response to problems and more flexibility, but require greater diligence in controlling costs. Some large organisations – for example, health authorities and universities – have directly employed labour organisations to carry out maintenance works.

Whichever method is used, the objective is the same: to obtain work of the required quality at the time it is required and at reasonable cost. Clear instructions based on understanding, good contract documentation, and the selection of building operatives with the appropriate levels of skill and experience, are as important for maintenance works as they are for larger repair projects. More information on procurement methods and contract documentation is given in **Treatment & Repair**.

REVIEW & REVISION OF MAINTENANCE PLANS

Maintenance plans should be reviewed at least annually and revised if necessary. Where recurring problems cannot be alleviated by even the most meticulous programme of preventive maintenance, it may be necessary to consider making changes to the fabric of the building or, where possible, modifying the surrounding environment to help reduce future problems. For example, increasing the capacity of gutters; providing flashings to protect vulnerable elements and improve water-shedding; installing bird deterrents; modifying ground levels to improve surface water drainage.

PERMISSION FOR MAINTENANCE WORK

When planning maintenance on listed buildings, it is unlikely that most routine work will need consent. However, if changes are planned, such as the ones mentioned in the previous paragraph, Listed Building Consent may be required. Consent may also be required for certain maintenance works to Scheduled Ancient Monuments. It is advisable always to check with the controlling authority in good time so that work is not held up.

Building owners and maintenance managers should also be aware of the legislation protecting wildlife, particularly birds and bats, as this can affect the timing of maintenance work. It is an offence to destroy a nest being used by a wild bird, so removal of creepers and vegetation where birds might be nesting should not be carried out in spring or early summer. Similarly, it is an offence to wilfully disturb a bat, or to destroy or obstruct access to a bat roost, whether or not it is being used by bats at the time. If bats are using the roof space, the local authority ecologist or Natural England should be consulted. They will be able to advise on appropriate times for inspection and remedial work to roofs, and whether a licence for the work is needed. This subject is dealt with in detail in **Ecological Considerations**.

Mothballing Heritage Assets

The best way to protect a building is to keep it occupied, even if on a temporary or partial basis. However, in areas where the property market is weak and the opportunities for sale or re-use are limited, it can be very difficult to find uses for some historic buildings.

Unoccupied buildings are vulnerable to damage and dereliction. This threatens not only their heritage interest and significance, but can also jeopardise public safety. Damaged and derelict buildings are also more difficult to bring back into permanent use. However, owners and managers can take steps to reduce the risks. Factors to take into account include the following:

- *Review the options and plan strategically*
  Be realistic about how long the building is likely to remain vacant, and imaginative about ways to reuse the building, either permanently or on an interim basis. 'Mothballing' should be considered as a last resort.

- *Take stock of the building*
  Bring together all information about the building, including its layout, condition, heritage interest and significance, and features of special interest. An accredited architect or surveyor should be appointed to record the condition of the building, and assess any risks that make it vulnerable.

- *Make the building secure*
  Vacant buildings are more at risk from intruders who may cause deliberate or consequential damage. Measures to make the site and building secure should be implemented. Security lighting/CCTV and intruder alarms linked to an alarm centre should be considered.

- *Investigate the potential for temporary uses*
  The costs and risks of leaving the building unoccupied, compared with the benefits of a temporary use, should be considered. Leasing the building for a 'meanwhile use' provides income, discourages intruders and vandals, and may be beneficial in the longer term for the economic regeneration of the area.

- *Carry out urgent repairs*
  Prompt action minimises the escalation of repair costs.

- *Protect vulnerable features*
  Historic buildings usually contain architectural and decorative features which contribute to their heritage interest. These may be vulnerable to damage, loss or theft. Visitors and workers should be made aware of any vulnerable features, and how to avoid accidental damage. Physical protection, such as boxing-in, should be considered.

- *Protect the building from fire*
  Fire is more likely to develop in a vacant building and with greater risk of major damage. It is important to maintain fire-detection and intruder-alarm systems, and for these to be linked to a call centre.

- *Carry out regular inspections and maintenance*
  To ensure that potential threats are noted and acted upon quickly, there should be someone responsible for carrying out regular inspections. A weekly walk-through will swiftly pick up on changes to the building. A routine maintenance plan should be implemented and properly supervised.

- *Decide what to do about building services*
  A decision is necessary on whether to maintain, decommission or close down existing services, or to install new ones. A managed approach will help to avoid damage and unnecessary expense. Some systems may need to be kept running to protect the building.

- *Keep the building dry and well ventilated*
  To reduce the risk of decay it is important to control humidity levels by keeping water out and maintaining good ventilation. Where decay or infestation has occurred, the source of damp should be located and eradicated, and the building allowed to dry out. Specialist advice about remedial repairs may then be needed.

- *Prevent damage from vegetation*
  Gutters, rainwater goods and gullies should be kept clear of leaves and moss. Woody-stemmed plants should be removed where they are causing damage. Trees and climbing plants should be monitored and managed to minimise the risk of damage (see **Ecological Considerations**).

- *Insurance*
  Particular risks are associated with unoccupied historic buildings. Insurers should be notified as soon as a property becomes vacant to ensure that the required level of insurance cover is maintained.

- *Prevent damage and risks to health from wildlife*
  Empty buildings can provide a habitat for wildlife. This may pose a risk of damage to the building or a risk to human health. An ecologist can advise on whether species using the building are protected by legislation, and methods of control and removal. Natural England should be contacted as soon as possible if it is suspected that the building is used by bats or any other protected species.

- *Regulatory framework*
  Vacant properties are covered by a wide range of legislation and regulations in relation to building works, health and safety, and wildlife. It should be noted that the owner's legal responsibilities do not end when a building is vacated, and that the owner remains responsible for the safety of all visitors, both authorised and unauthorised, under the **Occupier's Liability Act 1984**.

- *Business rates and tax*
  Owners should ascertain whether they are required to pay business rates or council tax on an empty property, or VAT on building works. The local authority should be informed as soon as a building becomes vacant.

Detailed guidance on managing vacant historic buildings is available from Historic England (see **Further Reading**).

Health and safety are important aspects of maintenance (for relevant legislation, see **Current Law, Policy & Guidance**). Owners and building managers should be aware of their responsibilities regarding risks to those who undertake maintenance, as well as to users of the building and the general public. In all commercial situations, a risk assessment must be carried out. Building managers directing maintenance and repair work, and the contractors undertaking it, must comply with all statutory requirements regarding health and safety, particularly in relation to safe access and exposure to hazardous materials. There is less statutory control over safe access and exposure to hazardous substances when private householders undertake maintenance or repair work themselves, although it is strongly advisable to follow safe working practices. Care still needs to be taken to avoid risk to passers-by, as would be the case in the generation of dust during the sanding of paintwork.

PROVIDING SAFE ACCESS

Scaffolding

To gain access to historic building façades for treatment and repair, 'independent tied' scaffolds should normally be provided. These derive no vertical support from the building, but must be tied to it for horizontal stability. Ties can pass through window openings and 'hook' back against the internal surface of the external wall, or can grip the window reveals. In both cases, great care must be taken not to damage either the glass and glazing or any internal panelling.

Where independent tied scaffolds are not practicable, the scaffold may need to be fixed to the stone or brickwork (provided it is adequate for the purpose). This will usually mean setting anchor sleeves into the masonry to take fixing bolts. Sleeves and bolts that are to remain in place following completion of the works should always be made of stainless steel, to reduce the risk of further damage to the building. Transom poles must be fitted with end caps to protect masonry surfaces.

If fixing into the masonry is not permitted, then a freestanding scaffold will have to be used. As these are self-supporting, they are necessarily more substantial than scaffolds restrained by the building, and therefore are much more expensive. Mobile scaffold towers are cheaper and quicker to erect, but they do not adapt so easily to confined spaces or to steeply sloping ground, and there are strict limits on safe working height and load capacity.

Mobile Elevated Platforms

Hydraulic platforms provide an almost instant means of access. Like mobile scaffold towers, scissor lifts are difficult to use in confined spaces, or if the ground is rough or steeply sloping. 'Cherry pickers' – machines with the working platform on an articulated telescopic boom – can operate at greater heights, and their good reach gives access to elements set back from the front line of the building. However, they also require an access point sufficiently wide to accommodate the machine, and fairly flat and stable ground capable of bearing its weight. They are also comparatively expensive, although in some cases their cost may be offset by the working speed they permit.

Left: A mobile elevated platform in use. This illustrates the height and outreach that can be obtained provided there is adequate access and level ground.

Right: Rope access is a cost-effective and versatile way of checking the condition of a building façade, inspecting previous repairs, and carrying out tasks such as the removal of vegetation and minor repairs.

Rope Access

Rope access is increasingly common, and can provide rapid, flexible and cost-effective access to many parts of a building, not only for inspection but also for works such as cleaning, painting and repointing.

Workers are suspended from the building on both a working rope and an independently anchored back-up rope. Small tools are attached to the worker's harness by means of a lanyard, and larger tools and equipment are suspended on an independent rope. Some buildings (such as churches) may have architectural features that are suitable for anchoring; on other buildings, eyebolts must be fixed in place to take the ropes.

Rope access requires suitably qualified personnel, operating in accordance with the guidelines of the Industrial Rope Access Trade Association [IRATA] and the correct equipment; there are an increasing number of firms that specialise in rope access for historic building conservation.

Ladders

Ladders are comparatively cheap, easy to erect and can be used in many situations. On the whole they should be used only for reaching a working area rather than as a working platform. Ladders can be used for quick tasks (such as inspections and cleaning debris from gutters) if, after assessing the risks, the use of other equipment is not justified because of the low risk and short duration of the works. Ladders should be securely clipped or tied to prevent them slipping. Where regular access by means of a ladder is required (for example, to access a parapet gutter), it may be possible to fix an eyebolt to which the top of the ladder can be secured to stop it falling sideways or slipping down. Ladders should never be leant against delicate historic surfaces.

ASBESTOS

It is the duty of those responsible for a building to manage the asbestos risk within a non-domestic building. Current regulations oblige the 'duty-holder' of any building that might contain asbestos to:

- take reasonable steps to find out if materials containing asbestos have been used in the building, and, if so, where, how much and the condition of the materials
- presume materials contain asbestos, unless there is strong evidence that they do not
- make, and keep up-to-date, an 'asbestos risk register', which includes a record of the location and condition of the asbestos-containing materials, or materials which are presumed to contain asbestos
- assess the risk of anyone being exposed to fibres from the materials identified
- prepare a plan that sets out in detail how the risks from these materials will be managed, and take the necessary steps to put that plan into action
- periodically review and monitor the plan and the associated actions, so that they remain relevant and up-to-date
- provide information on the location and condition of the asbestos-containing materials to anyone who is liable to work on or disturb them.

The duty-holder and everyone else responsible for the building must comply with these requirements. The project manager of any proposed works on a building that might contain asbestos should ask for a copy of the asbestos risk register.

Where repair, alteration or demolition of a building (or part of it) are proposed, it will be necessary to appoint a trained specialist surveyor to carry out a refurbishment/demolition survey. The purpose of the survey is to locate and identify all asbestos-containing material [ACM] before any work is carried out. The survey involves opening-up and sampling of materials.

Due to the potential disturbance of asbestos, the area under survey has to be vacated until certified 'fit for reoccupation' after the survey. The survey report will state the presence, location and extent of asbestos-containing material and debris. Further information about refurbishment/demolition surveys can be obtained from the Health and Safety Executive (see **Further Reading**).

OTHER HAZARDOUS MATERIALS

Although many hazardous materials may accrete on building surfaces, white lead paint and bird droppings are encountered most often. Working with such materials is covered by health-and-safety legislation, and it is the responsibility of contractors to ensure that they comply with the latest guidance. Removal of hazardous materials must always be subject to a risk assessment, and all appropriate precautions must be taken to protect workers, building occupants and the general public.

In historic interiors, neglected coatings containing white lead paint that are loose, flaking and peeling are a hazard to health.

White Lead-based Paint

Until the 1960s, paints containing lead were common, and at the time of press could still be used in the UK for some listed buildings. It is always wise to assume that some proportion of lead will be present in pre-1960s paintwork (specialist analytical companies can assess the extent if necessary). If white lead paints are known or suspected, they must either be encapsulated under another coating, or removed in accordance with current guidelines.

Different methods of removal of lead paint pose different types of risk:

- dry sanding/scraping and mechanical abrasion techniques will generate airborne dust or debris that can be inhaled or ingested
- solvent-borne gels, poultices, as well as wet cleaning methods such as water blasting, wet sanding or water rinsing, generate contaminated slurry that can be ingested or contaminate water courses
- high-temperature paint-stripping methods (such as blow torches) are a fire hazard and release poisonous fumes that can be inhaled.

The safest way of dealing with white lead paint when it is in good condition is not to attempt to remove it, but rather to encapsulate it within a new coating. This also has the benefit of preserving historic paint layers. If its condition is too poor to over-coat, and removal is unavoidable, steps will need to be taken to reduce the risk of lead dust or fumes getting into the air. Blowtorches should never be used. If sanding, abrasive paper should be used wet, but gel or poultice paint strippers are usually a better choice.

Workers should wear disposable gloves and overalls, and a dust mask fitted with a P2 filter. Industrial-grade vacuum cleaners fitted with P2 cartridge filters should be used for clearing waste, and all surfaces where dust can settle wiped down. Removed paint and dust should be put in sealed bags, but they can be disposed of in a domestic dustbin. Commercial practice is covered by *The Control of Lead at Work Regulations 1998*, which require precautions to be taken for the prevention of lead hazards, personnel protection and hygiene, exposure control, and ventilation.

Bird Guano

Dust from infected droppings, feathers or nesting materials can cause lung infections if inhaled, particularly for people already suffering respiratory problems. Bird guano should first be wetted to eliminate dust, then scraped up using wooden spatulas and plastic dustpans, and sealed in plastic bags. Small quantities of bagged waste can be disposed of in domestic or commercial dustbins, but larger quantities must be handled by a Registered Waste Carrier.

Large quantities of guano can accumulate in unused parts of buildings – such as attics – that are accessible to birds.

USING HAZARDOUS MATERIALS

BIOCIDES

The Control of Pesticides Regulations 1986 [COPR] (as amended 1997) came into force in 1986, under the *Food and Environment Protection Act 1985*. This Act introduced statutory controls on biocides. Unless approval has been granted by Ministers, it is illegal to:

- make up laboratory formulations of biocides (only 'approved' formulations may be used)
- use biocides in a non-approved way (precautionary labels are now mandatory and must be adhered to)
- store, supply or use non-approved biocides such as arsenic, dieldrin or dichloro-diphenyltrichloroethane [DDT]
- employ unskilled or untrained staff to use biocides that have not been cleared for household use.

The regulations specify three categories of use: amateur, professional and industrial.

The definition of professional is *"a person who is trained in the use of the product and the equipment, and applies pesticides as part of his/her job"*. Preservatives cleared for amateur use may also be used by professionals, and those cleared for professional use may be used in industrial applications. In many cases, there is no difference between professional and amateur products except volume: amateur products are available in cans containing a maximum of 5 litres, and the professional application category covers bulk pre-treatments. In some cases, amateur products are ready to use, in order to avoid the handling of more toxic concentrates.

Further information on the use of pesticides for the control of vegetation on buildings and structures is provided in **Ecological Considerations**.

WHITE LEAD PAINTS

In the past, many oil-based paints contained quantities of lead carbonate and lead sulphate. However, the supply and use of lead paints was banned in the UK in 1988 for health reasons: lead is toxic if ingested or inhaled, and poses particular risks to pregnant women and children. Nevertheless, the European Parliament has permitted a 'derogation', which allows lead paints to be used for *"the restoration and maintenance of works of art and historic buildings and their interiors"*. In England this derogation is implemented through the *Environmental Protection (Controls on Injurious Substances) Regulations 1992*, and allows the manufacture and use of lead paint for Grade I or Grade II* listed buildings, Scheduled Ancient Monuments and works of art, under certain conditions. To apply to use lead paints in England, a declaration form must be obtained from the Office of Public Sector Information or commercial suppliers.

The person intending to apply the paint must sign the form, giving details about the building and a justification for the use of lead paint, and forward it to the paint supplier. The supplier then signs a second declaration before sending the form to the 'competent authority' (in England this is Historic England). If the intended use is not permissible, the authority must inform the user of this within three weeks.

Professionals should refer to *Lead at Work Regulations 1998 SI 543*, *Control of Lead at Work ACOP 1985*, *The Construction (Design and Management) Regulations 1994 SI 3140* and *The Environmental Protection (Controls on Injurious Substances) Regulations 1992 (Statutory Instrument 1992 No.31)*, which allow restricted use of lead paint in accordance with the *1989 European Marketing and Use Directive*.

VOLATILE ORGANIC COMPOUNDS [VOCS]

Most organic solvents present significant health-and-safety risks, and require COSHH assessments that must include the identification of the correct protective clothing and masks with appropriate filters for the materials being used. Those solvents that are particularly toxic, such as toluene and xylene, should only be used in well-ventilated areas or under fume extraction.

The use of paints with high VOC content is now limited by European Union Legislation (a derogation allows their use on listed historic buildings).

ABRASIVE CLEANING

Abrasive cleaning is hazardous. The onus is on the employer or contractor to assess the risk posed by using abrasives, and to implement appropriate controls in accordance with current legislation.

Noise & Dust

To protect operators from harmful noise and dust, personal protective equipment [PPE] will be required for all abrasive-cleaning systems. For dry abrasive cleaning, respiratory protective equipment [RPE] may also be needed. Dust may need to be controlled using sheeting or extraction systems to prevent nuisance to the surrounding area. The noise impact on the wider environment should also be considered.

Siliceous Abrasives

The use of siliceous aggregate (sand) for abrasive cleaning has been linked to silicosis. It is therefore banned for most blast cleaning. A loophole in the legislation permits its use for cleaning buildings, bridges and other similar structures. However, the *Control of Substances Hazardous to Health Regulations* [COSHH], covering containment of the silica dust produced, are so demanding that it is all but impossible to comply with them, so, in effect, the use of siliceous abrasives for cleaning buildings is outlawed. Even when used with water for wet abrasive cleaning, the risk of dust inhalation when cleaning up spent abrasive makes its use impracticable. There are a number of alternatives to silica sand that can be successfully used for abrasive cleaning.

Further material-specific information about health and safety issues is given in other volumes of the *Practical Building Conservation* series.

The purpose of repair is to intervene in the process of decay, to restrain it and correct damage. The objective is to conserve – and occasionally recover or reveal – significance, and enable a building to continue in, or be adapted to, beneficial use. In many cases, intervention may be needed because a building has been neglected or previous repairs have failed. To be effective, conservation treatment and repair strategies have to address the underlying causes of deterioration, not just treat symptoms.

Good conservation practice, considered in **Conservation Planning for Maintenance and Repair**, involves a systematic and analytical process in which heritage values and significance are defined, the condition of fabric assessed, threats and opportunities evaluated, clear conservation policies and objectives proposed and, finally, a strategy – a 'road map' – for reaching them drawn up. In this process, the available or required resources have to be considered in terms of finance, skills and materials. Then, the whole process, from inception to the completion of the works carried out, should be recorded, along with all assumptions made and conclusions reached at every stage. This record amounts to a 'medical history' that will assist in the diagnosis and treatment of future complaints. After conservation treatments and repairs have been carried out, continuing maintenance will be required if the values of the building – and the investments made in it – are to be sustained.

Provided that repairs are well-designed, specified and executed, an existing building element can perform as well as one that is renewed, and have a comparable service life. Repair itself, however, can introduce new hazards (*right*).

This section deals with the process of designing, specifying and procuring conservation treatments and repairs. There is a widely held perception that a repair is short-lived and inferior to a renewed element. There is no doubt that repairs will fail quickly if they are inappropriately selected, poorly designed, inadequately specified and badly executed. However, an element that has been repaired with care and forethought is not inferior to one that is renewed, particularly where the element is made from material of a quality no longer obtainable, as in the case of some timber species. Also, much aesthetic pleasure may be gained from a repair that is well-executed.

Decisions and actions taken at every stage of a repair project have a bearing on the quality of the works eventually carried out. The advice contained in this chapter is intended as a guide to current best practice.

ROLES & RESPONSIBILITIES IN REPAIR PROJECTS

The success of a repair project is dependent on several key factors: adequate financial, material and human resources (knowledge, skills and experience) should be available; the timescale for carrying out the project should be realistic; there should be commitment, enthusiasm and open-mindedness on the part of the client, the professional team, and the contractor. Good communication between the project team members is also important. These requirements are not unique to building conservation, but are the foundations for success in all industries.

Programmes of repair and maintenance to heritage assets should normally be led by an architect or chartered building surveyor who has a proven track record of experience in this field, and is formally accredited in building conservation. The only exceptions are limited programmes of minor works carried out by specialist contractors; for example, draught sealing of sash windows. Prospective companies or individuals should be interviewed to assess their knowledge and experience and to find out how they intend to approach the work. It is also an opportunity to gauge how they will interact with other members of the project team. References from previous clients should also be obtained. Speculative enquiries to companies or individuals of unknown experience will bring a high degree of risk to the client, and be less cost- and time-effective.

Projects largely consisting of specialised areas of work may be led by members of other professional disciplines: for instance, a professional architectural conservator in the case of sculpture, monument and wall-painting conservation, and a civil or structural engineer when repairing a bridge, crane, or in a project involving industrial archaeology. Larger projects may be led by other building professionals – project managers – who may not have direct experience in building conservation. In such cases, an experienced building conservation consultant will often be appointed to support the project manager, to ensure that the conservation objectives are addressed appropriately within the wider project. For instance, rail infrastructure projects with a conservation component have been led successfully by architects and structural and civil engineers with specialist rail experience, and supported by conservation consultants. In all cases, it is the type and range of work, and the context of the site, building or monument, that should determine the right professional for the lead job.

A logical and analytical approach is a key element in achieving a successful outcome. Programmes of repair should be based on an informed understanding that develops by continually questioning what is required, is achievable, and how; whether proposals are justifiable and in accordance with established conservation philosophy; and, finally, if there are any valid reasons for deviations from this philosophy.

Clients play an important part in a project. They are not only the commissioning party, but are responsible for financing the project, and their contribution can significantly influence its success. In the case of designated heritage assets – listed buildings and scheduled monuments – it is important that they have a basic understanding of their legal obligations so that any work commissioned is likely to be acceptable in principle. If in doubt, the advice of the local planning authority should be sought.

Guidance may also be obtained from Historic England, and organisations such as the Society for the Protection of Ancient Buildings [SPAB], the Georgian Group, the Victorian Society and the Twentieth Century Society. Clients should be able to prepare a brief for the appointment of a lead professional, and select an architect or lead consultant who has a proven track record of suitable experience. Where a project is receiving a grant, the client should be aware that the contract entered into with the grant-giving body will contain conditions concerning the procurement of the services of the lead and other professionals. It is important to ensure that these conditions are complied with, otherwise the grant may be put at risk.

Enthusiastic clients with a commitment to maintaining the historic environment can be a valuable asset to a project, and help ensure its success. They will be willing to cooperate with the design and construction team, thus enabling a high standard of work to be achieved without undue conflict. On the other hand, a client who is resentful of the legal obligations inherent in the care and conservation of a heritage asset may hinder success by being difficult to work with, and proceeding reluctantly. This emphasises the importance of promoting public engagement in decisions about change to significant places, and for the process, based on public policy, to be consistent and transparent.

THE PROFESSIONAL TEAM

The management of conservation repair projects (which may form part of a larger project encompassing alterations to a building) is similar to the management of any other building operation. There will typically be an architect or building surveyor as the lead professional with overall responsibility for the project, supported as necessary by other professionals, often a quantity surveyor (cost consultant), a structural engineer, a conservator, and, on projects where the work will last longer than 30 days or involve more than 500 person days of work, a CDM Coordinator. Larger projects tend to have a separate project manager leading the team, liaising with the client, and responsible for delivering the project to time, budget and quality, while smaller ones are often managed by the architect or building surveyor. In either case, an effective management role is fundamental for success. The professional team should have expertise in conservation work appropriate to the nature and scale of the project.

The main professions have established accreditation schemes:

- Register of Architects Accredited in Building Conservation [AABC], compiled and administered by Accon Ltd
- Royal Institute of British Architects [RIBA] Conservation Register
- Royal Institution of Chartered Surveyors [RICS] Building Conservation Accredited Scheme, which includes both building and quantity surveyors
- Institution of Civil Engineers [ICE] & Institution of Structural Engineer [IStructE] Conservation Accreditation Register for Engineers [CARE]
- Chartered Institute of Architectural Technicians [CIAT] Directory of Accredited Conservationists
- Institute of Conservation [ICON] Professional Accreditation of Conservator-Restorers [PACR].

The use of accredited lead professionals is normally a requirement for repairs undertaken with grant aid from Historic England and some other funding bodies.

CDM COORDINATOR

Where a project is notifiable under the *Construction (Design and Management) Regulations 2007* [CDM 2007], the client is required to appoint a CDM Coordinator to advise and assist them in meeting their responsibilities for health and safety. The CDM Coordinator's main duties include: notifying details of the project to the Health and Safety Executive; coordinating the health-and-safety aspects of the design process; facilitating good communication and cooperation between project team members; and preparing and updating the health-and-safety file. Detailed information about CDM 2007 is available from the Health and Safety Executive (see **Further Reading**).

LEAD PROFESSIONAL

The lead professional (also referred to as the 'designer' in the *Construction (Design and Management) Regulations*) is responsible for realising the client's objectives. The lead professional advises the client on what can be achieved, how works should be implemented, and what statutory consents and permissions are required. The lead professional should have appropriate skill, experience and accreditation; they will usually be an architect or chartered building surveyor. On more complex projects, the lead professional may be an architect supported by other building professionals and conservation consultants.

The lead professional is usually appointed at project inception, and should be retained through to project completion. Changing professional consultants, or dispensing with them, during the course of a project can result in duplication, increased costs and delays, and can adversely affect the quality of the completed work.

Conservation projects may also need to involve specialists knowledgeable about the evolution of the building. These might include building archaeologists, architectural historians and specialists in historic paint analysis. Building archaeologists who are members of the Institute of Field Archaeology [IFA] are bound by a code of professional conduct and practice. The IFA maintains a searchable database of registered archaeological organisations and companies. Specialist conservators familiar with the nature, defects and treatment of sculpture, wall paintings, glass, metalwork, ceramics, textiles, stone, plaster and timber may also be involved. The Institute of Conservation [ICON] manages the *Conservation Register*, a searchable database of people and firms within the Professional Accreditation of Conservator-Restorers Scheme, which includes architectural conservators.

Other specialist consultants may be required. For example, a fire engineer may be needed to devise an alternative method of meeting fire safety requirements. The Institution of Fire Engineers [IFE] registers suitably-qualified individuals.

Although not a form of accreditation, membership of the Institute of Historic Building Conservation [IHBC] indicates that an individual has recognised conservation expertise. Established originally as the Association of Conservation Officers, it is the 'umbrella' professional body for building conservation practitioners and historic environment experts. Members must be able to demonstrate that they have relevant skills, knowledge, experience and understanding in one or more aspects of historic environment conservation (although not necessarily the repair of historic buildings).

Analytical services should be obtained from individuals or companies with proven experience in the specific type of analysis or investigation required, and, wherever possible, familiarity with historic buildings and traditional materials. The United Kingdom Accreditation Service [UKAS] maintains a list of laboratories and other bodies accredited according to European and International standards.

PROJECT MANAGER

Traditionally, projects are led by an architect or chartered building surveyor, but with increasing project complexity, project management has become a specialist and distinct field. Project managers tend to be used on large projects that involve a multi-disciplined design team and complex site programmes. They are responsible for managing, monitoring and coordinating the necessary resources, meeting the budget and seeking appropriate advice when necessary, so that project constraints and objectives are realistic and achievable. There are a number of project managers who specialise in work to the historic environment.

Traditional Building Craft Skills

ASSESSING THE NEED

Traditional craft skills are essential for the continuing care of the six million or so buildings in the UK erected before 1919. This includes more than 500,000 listed buildings. In 2005 the National Heritage Training Group [NHTG] published *Traditional Building Craft Skills – Assessing the Need: Meeting the Challenge* that quantified the scale of skills shortages and gaps. This report, and further research carried out in 2008, showed that only around one-third of the workforce had the skills needed to work with traditional building materials. This skills gap is of particular concern, as the repair and maintenance of traditional buildings accounts for nearly half the current output of the construction industry.

Traditional building skills were passed down from generation to generation, but during the 20th century the quality and prevalence of many such skills declined as formal training focused on mainstream, 'new build' construction. There are important differences between the construction of traditional buildings and 'new build'. Traditional buildings were usually built with materials that could absorb and release moisture freely, while new buildings tend to be less permeable. Unlike their modern counterparts, traditional buildings were also subject to significant regional variations in materials and construction methods. This means that people repairing and maintaining traditional buildings require a different – and often wider – range of skills than those working on new buildings. Furthermore, in some regions, specialised skills, such as earth and cob wall building and thatching, will be needed.

A masterclass in the consolidation of ruined masonry in progress at West Dean College, Sussex.

Historic buildings can be harmed when people lacking the appropriate skills, knowledge and experience carry out the work. For example, poorly executed repointing is not only disfiguring, but can cause damage in the longer term if the ability of the masonry to 'breathe' is inhibited by the use of inappropriate cement mortars. Also, there are potential technical and human health risks in refurbishment and 'retrofitting' works aimed at improving the energy performance of traditional buildings. For instance, some insulation materials and systems, if not properly installed, may increase the risks of surface and interstitial condensation that could lead to mould growth and timber decay.

MEETING THE CHALLENGE

Historic England works with CITB–Construction Skills and the NHTG to ensure that the right traditional building craft and professional skills exist to care for traditional buildings. NHTG works on a strategic UK-wide remit supported by a network of Regional Skills Action groups. This regional expertise coordinates action and partnerships to deliver training matched to local materials and local needs.

Training programmes and qualifications developed specifically for this sector ensure that those working with traditional buildings can demonstrate the required skills and competency. The *Construction Skills Certification Scheme* [CSCS] Heritage Skills card demonstrates that an individual has reached appropriate standards of skill, qualification and safety awareness in trades relevant to traditional buildings. A flexible and evolving framework of education, training and qualifications exists to provide a range of options to suit people with different levels of experience. This includes:

- *NVQ Diploma in Heritage Skills*
  A work-based qualification, gained through on-site assessment via a portfolio of evidence that demonstrates attainment of NVQ Level 3 skills in a range of building craft trades.

- *Heritage Specialist Apprenticeship Programme*
  For craftspeople already experienced in mainstream construction who wish to develop their skills in order to work on heritage buildings. Training is tailored to fit each person's individual needs, and consists of a combination of college-based training, on-the-job work experience and assessment for an NVQ *Diploma in Heritage Skills*. Led by sector practitioners, it is available for brick work, specialist lead work, plastering, roofing, stone masonry, painting and decorating, and carpentry and joinery trades.

- *Mastercraft qualification*
  Allows highly-experienced practitioners to demonstrate excellence to NVQ Level 4 through work-based assessment by accredited assessors.

- *Foundation Degrees in Building Conservation*
  Work-related higher education qualifications aimed at craftspeople who want to progress on an academic route. They combine academic study – lectures and tutorials – with practical work-based learning in either full-time or part-time courses.

For more experienced craftspeople from a mainstream construction background, there are modular programmes of training to enhance their skills. For example, *Understanding the Principles of Repair and Maintenance of Traditional (pre-1919) Buildings* is an NVQ Level 3 that covers the fundamental principles and approaches to appropriate methods and materials for repair and maintenance. In this two-day course, held at a historic building or site, experienced built heritage practitioners and trainers give lectures, and give a guided tour to illustrate how principles can be applied in practice.

The availability of a sufficiently large and appropriately-skilled workforce is essential for the continuing care of the built heritage. Incentives for training include NHTG *Skills for the Future*. Building on the success of the *Traditional Building Skills Bursary Scheme*, this project offers a wide range of placements, training and qualification opportunities for new entrants and highly experienced practitioners alike in a range of traditional building crafts throughout England. In addition, both schemes focus on encouraging people who are currently under-represented in the sector, especially women and people from minority groups, to enter the workforce (detailed information is available from the website listed in **Further Reading**).

QUANTITY SURVEYOR

The quantity surveyor is responsible for estimating the cost of the work, and setting a budget, monitoring and controlling costs throughout the duration of the project, and producing the final account. Detailed design information and site measurements are converted into a bill of quantities, or alternatively a quantified schedule of works as part of the specification. These documents are used as the basis for tendering and post-contract cost control. The quantity surveyor will also advise on forms of contract and procurement. In common with other members of the professional team, the quantity surveyor appointed should have a sound knowledge and experience of working with historic buildings.

CLERK OF WORKS

In larger, more complex projects, a Clerk of Works may be appointed by the client to monitor the quality and progress of work on their behalf. The main responsibility of the Clerk of Works is to ensure that works are carried out in accordance with the specification and contract programme. The Clerk of Works, who may be on site full-time or make regular visits, will keep detailed records and report on various aspects of the work. These include, among other things: progress and delays; weather conditions; the number and type of workers employed; deliveries; the receipt of drawings and instructions; and details of significant events that affect the works, including deficiencies in provisions for health and safety. As with other members of the project team, it is important that the Clerk of Works has experience in conservation contracts.

CONTRACTORS

The contractor is the company that implements the detailed design by carrying out the programme of conservation and repair, or construction and alteration work on site. As with professionals, selecting one with appropriate skill and experience is crucial to achieving success. A contractor is appointed on the advice of the lead professional, following the tendering or price negotiation process. The company, normally referred to as the 'principal contractor', is employed as part of the project team, but has a direct contractual relationship with the client. The principal contractor has overall responsibility for the construction work supervised by the lead professional or project manager, and primary responsibility for health and safety on the construction site. An experienced contractor will be able to contribute valuable specialist practical knowledge to a project.

There is no formal register of accredited 'conservation contractors', although the National Training Heritage Group, under the aegis of the Sector Skills Council and Industry Training Board, has proposed the development of an Accredited Heritage Building Contractors Register. The National Federation of Roofing Contractors approves, vets and maintains a list of specialist heritage roofing contractors. The Lead Contractors Association [LCA] also trains, grades and vets members listed in its annual *Directory of Specialist Leadwork Contractors.* In addition, the LCA provides insurance backing for work carried out by its members. A useful source of information on contractors with experience in the conservation field is *The Building Conservation Directory*, published annually. It provides designers and specifiers of works with information about appropriate products, services and expert advice. No form of 'vetting' is applied to entries in the Directory, however, and it is up to the designer or specifier to seek evidence of competence and appropriate experience. The Institute of Conservation [ICON] maintains a register of professionally accredited conservators.

SUB-CONTRACTORS

Sub-contractors are employed on many projects to supplement the skills and experience of the principal contractor. In most instances, the sub-contractors are selected by and solely responsible to the principal contractor, and are referred to as 'domestic sub-contractors'. In some circumstances, the sub-contractor may be selected by the lead professional and employed by the contractor. There are two types of sub-contractors: 'named' and 'nominated'. 'Named sub-contractors' are similar contractually to domestic sub-contractors, except that the client provides a list of firms that are acceptable without taking contractual responsibility for selection; the principal contractor appoints a sub-contractor selected from the list. 'Nominated sub-contractors' are selected and appointed directly by the client.

SUPPLIERS

Material suppliers are involved in all projects. In most instances, the contractor will choose the suppliers, but in some cases, where specific design elements require specific materials, the supplier can either be named or nominated in a similar way to sub-contractors. Where long procurement periods are required, the supplier is frequently chosen before the contractor.

PROJECT ORGANISATION & PLANNING

Managing a programme of conservation and repair is no different, in principle, to managing any building construction or civil engineering project. The *RIBA Plan of Work* is a useful guide for the type and amount of work to be completed at each key stage. The main work stages are as follows.

PREPARATION: PROJECT APPRAISAL & DESIGN BRIEF

Initially, the brief should be a clear, logical and concise statement of requirements setting out the aims, objectives and timescale for the project for the purpose of appointing a lead consultant. Contracts for the appointment of consultants include the *RIBA Conditions of Engagement for the Appointment of an Architect* or the *NEC 3 Professional Services Contract.* Where a Conservation Plan or Conservation Management Plan has been produced, the conservation policies and strategy will form part of the brief. The lead professional will develop the brief to identify the range of professional skills that will be required, organisational structures and the procurement strategy. The procurement strategy should take into account:

- the size and complexity of the project
- the likely value of the works
- the proposed programme
- the availability of the necessary skills
- any constraints, such as the *EU Procurement Regulations*.

Based on this information, decisions can then be made concerning:

- the procurement method (type of contract)
- the documentation needed to specify the works.

DESIGN: CONCEPT & DESIGN DEVELOPMENT

The starting point for any scheme of repair, alteration or development is translating the requirements and objectives set out in the brief into practical actions. The necessary skills and information must be available at appropriate stages so that the project can progress efficiently and effectively. An important difference between conservation projects and new build is that there is usually a much greater degree of uncertainty, and a larger number of unknowns when dealing with an existing building. Thus, the more investigations that can be carried out before a contract is let, and a contractor is on site, the less risk there will be of uncovering some unexpected form of construction or defect that may by that stage be difficult to deal with in a cost-effective way. Limited opening up on site is often advisable during the design development stage to increase the level of understanding, and enable decisions to be made about detailed technical design, specification and budget. To minimise uncertainty, it may be necessary in some large or complex buildings to phase the works, or to let a separate contract for exploratory works prior to completing the technical design for the main contract.

A historic building should not be repaired, conserved, altered or adapted to a new use unless it is properly understood. The importance of the assessment and fact-finding stage in conservation planning has already been emphasised (see **Conservation Planning for Maintenance & Repair**). This should include a condition survey to investigate and establish the condition, causes of decay and deterioration, assess threats, and identify priorities for repair. This stage may also include investigation work to determine the original construction and materials, or archaeological analysis to determine the historical development of a building, structure or site.

The range of survey techniques available has been reviewed in **Surveys & Investigation Methods**. These should be proportionate to the significance of the building and the nature of the repairs envisaged, and commissioned to develop an informed understanding of the heritage asset that is the subject of the project. It is important that clear and comprehensive briefs are provided when commissioning surveys and other investigations. A sample brief for a condition survey is given in the **Appendices**.

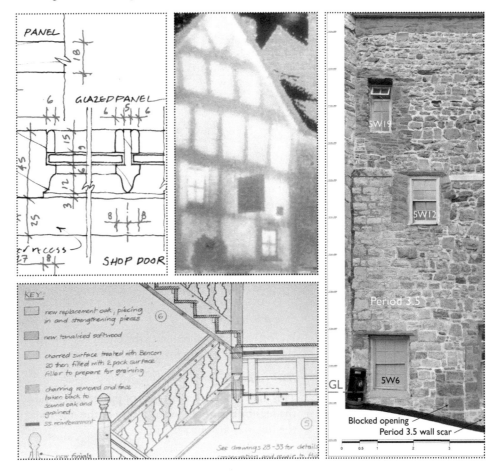

FEASIBILITY

Feasibility studies in conservation repair projects may form part of the preparation or design stages. This is where ideas are investigated and developed, and various options considered. It corresponds to the 'Devising strategies to safeguard the significance of a building' stage of the conservation planning process, described in **Conservation Planning for Maintenance & Repair**. At this stage the following questions should be addressed:

- What is to be achieved?

- What options are available?

- What will the impact be on significance of the available options?

- What resources in terms of materials and skills are required?

- What funding is available?

- When should the project start and be completed by?

- What are the implications of other limiting factors; for example, access or restrictions on the timing of works?

The feasibility stage may be as simple as a meeting between key participants, who will discuss ideas in order to come up with a series of statements about the scope, intention and available funding of the project. Alternatively, it may be a piece of commissioned work by an external party, which will investigate what is possible and advise on a budget.

Feasibility Studies

A feasibility study is a useful tool for exploring the viability of proposals and the options available. Feasibility studies may be carried out for a wide variety of projects, including programmes of repair, alteration and redevelopment, as well as representation or interpretation of a site. Feasibility studies would normally be based on information about the site obtained from ground level; however, closer inspections may be more useful. A feasibility study would normally include some element of background research before site inspection, but may also simply be desk-based analysis of existing documents. A typical feasibility study report will include:

- *Summary*

- *Client brief*
  It is often useful to reiterate the client brief at the start of the report, since it places the report in context for future readers who may be unfamiliar with the project.

- *Background research*
  This may include a summary of past reports and recommendations, as a context for the feasibility report. A summary of the historical development and significance of the site may also be relevant.

- *Condition assessment*
  A broad assessment of condition usually based on ground-level inspection and assessment of threats.

- *Options appraisal*
  Discussion of the various options for project implementation, including their impact on heritage values and significance, based on the desired client objectives.

- *Conclusion*
  Summarising conclusions that can be drawn from the appraisal.

- *Recommendations*
  Recommendations for the way forward.

The feasibility stage concludes by providing a framework of requirements and objectives, practical proposals for meeting them, and a budget estimate.

Prioritised Recommendations

Projects can seldom afford to complete all recommended repair and conservation work, and varying rates of decay often require intervention at different periods of time if the life cycle of original materials is to be maximised. It is therefore desirable that all condition surveys include prioritised recommendations based on need and timescales for completion.

PRIORITIES FOR CONDITION SURVEYS		
PRIORITY	**CRITERIA**	**TIMESCALE FOR COMPLETION (YEARS)**
1	Essential repairs which should be carried out urgently to prevent risks to health and safety	<1
2	Essential repairs which should be carried out to prevent ongoing accelerated decay or significant loss of architectural detail	5
3	Desirable repairs which should be carried out within 10 years to prevent decay and loss	10
4	Major elements of work where replacement is expected within 10 to 20 years	20
5	Desirable repairs to reinstate missing or heavily-decayed architectural detail	20
6	Cosmetic defects that do not adversely affect durability	—

COST PLANNING (BUDGETING)

Cost planning or budgeting is carried out and reviewed as the project develops. At the project concept stage, the preliminary budget might simply consist of the amount of available funds, or the funds considered necessary for the project. The preliminary budget may be determined simply by judgement, rather than a calculation based on quantities of work and costs. It is then reviewed during the feasibility stage. This may lead to an increase or decrease in the total funding, or result in an increase or decrease in the scope of work. The budget is the overriding constraint to a project. Work should not be commissioned unless it is within the financial constraints.

Following completion of the survey phase, the budget should be reassessed with the additional information, and a more detailed understanding of what is required and can be achieved. The budget can now be calculated based on quantity of work and the cost of the necessary resources. It may still include estimates or provisional sums if the detailed design has still to be completed. Reassessment may cause a change in direction for the project, or may reveal that less money is required than previously estimated. Projects should not proceed if they are 'over budget'; the amount of work should either be tailored to suit the available funding, or additional funding must be found so the budget can be increased. If the original amount cannot be increased, the project is no longer feasible, or the scope of work has to be reduced or split into phases.

The budget is a cost plan for the life of a project, and should be reviewed periodically during the design phase and monthly during the construction phase. As information increases and understanding evolves, budgets can be refined. Work elements that are defined by provisional sums can be accurately priced once the scope of work is known. The budget is concluded once the 'final account' (the final agreed payment) is made for all elements of work completed.

REFINING PROPOSALS

As a project evolves, understanding of the heritage asset may increase or client requirements change. It may then become necessary to refine proposals for works. If mistakes are recognised early enough to be corrected, the proposal should be amended. Changes should not be avoided solely because of concerns about wasted time or resources when the refinement of ideas and proposals will bring about tangible benefits. Although refinement is not a distinct stage in project development, proposals for works to heritage assets frequently benefit from reassessment and review; this can happen at any stage of the project.

Detailed technical design follows completion of the survey stage and reassessment of the budget. It consists of design drawings, specifications, schedules of work, and any other information necessary for the work to be implemented. The proposals cannot be vague at this stage: there must be a high degree of certainty about what exactly is required and how it should be achieved. It is more economic and effective to address questions and uncertainties at the design development or feasibility stages. Attempting to find the answers to design questions during the project construction phase is likely to increase costs, and cause disruption and delays, simply because more people and resources are involved at this stage.

The detailed design should provide at least 90 % of the information needed. Where it is not possible to complete the design (for example, due to access limitations), these deficiencies should be clearly identified, and provisional sums (of money) and a programme period (duration of time) allowed. Alternatively, the works can be covered by provisionally quantified items of work based upon an assessment of the scope of the works necessary. Provisionally quantified items should be used wherever possible where the scope of the works is not fully known, as they ensure that the works are competitively priced and also provide rates which can then be used *pro rata* to price variations. It is far easier to manage a programme of work when deficiencies are known.

There will be exceptional cases where programmes of work are very complex or specialised, and the technical design cannot be completed without the skills and resources of a specialist contractor or manufacturer. In these circumstances, the contractor or manufacturer should be employed during the design phase; the design should not be left to the construction phase. Detailed technical design is a distinct phase during a project, and should be completed before the tendering and construction phases.

FURTHER INVESTIGATIVE WORK

Investigation work is normally undertaken as part of survey work, but can also be needed at the technical design stage to provide targeted information to help answer specific questions. For example, building fabric might be opened up to investigate concealed construction or defects, or mortars and paints analysed. Most building conservation projects will involve investigative work of this type as part of the development of an informed understanding. Statutory consent may be required when opening-up is proposed in a listed building and is essential for Scheduled Ancient Monuments (see **Survey & Investigation Methods**).

STATUTORY CONSENTS

Formal consent is required for all work that affects the heritage interest and significance of the listed building, though consent may not be required for repair and maintenance work using materials and techniques which match the original. In all cases, the local planning authority should be consulted to determine the need for Listed Building Consent. Almost all work to a scheduled monument requires formal consent from the Secretary of State for the Department of Digital, Culture, Media & Sport; Historic England administers the application process. It is advisable to consult Historic England on proposed work as soon as possible. Works materially affecting the external appearance of a listed building or scheduled monument will additionally require planning permission from the local planning authority, unless they are 'permitted development'. Approval under the Building Regulations may also be required; for example, where alteration, extension or change of use are proposed (see **The Building Regulations & Related Provisions**).

Applications for Listed Building and Scheduled Monument Consent and planning permission are usually made following completion of the detailed design. In practice, investigation, trials and exemplar work often require Listed Building and Scheduled Monument Consents, and the relevant authority should always be consulted.

MATERIAL PROCUREMENT

Material procurement is not always identified as a distinct stage within project development. In conservation projects, however, matching historic materials is often the key to achieving successful repairs. This process can take time. Furthermore, it is important to remember that stone quarrying and brick production may be reduced during winter months, and material procurement in the spring and summer months is more desirable to provide programme certainty.

There are many reasons why procurement is so time-consuming. For example, research and analysis may identify the original types of stones used in a building. However, further investigation may reveal that original stone sources are no longer in production. A quarry or delve may need to be located and perhaps reopened, or suitable alternatives have to be found. Finding matching bricks can also be time-consuming. Brick making is often believed to be a straightforward process, yet much modern brick production differs significantly from historic methods. In order to produce new bricks that are similar to the originals, similar clay types and production methods may have to be used. It might be necessary, therefore, to have bricks specially made by a small-scale traditional brick manufacturer. Before they produce a larger order, the manufacturers will need to produce sample bricks that demonstrate that a close match in size, colour, shape, texture and consistency with the original can be achieved.

Material procurement

Appropriate materials may be difficult to source or have long lead-times for procurement. This should be taken into account when planning a programme of works.

Top: Air drying of timber: oak stacked for seasoning in a timber yard.

Middle: A 'delve' in Shropshire re-opened to provide tile stones for a roofing repair project.

Bottom: Production of handmade bricks at a small-scale traditional brickworks.

Brick procurement will take at least four months, but periods of 6–12 months should be expected. Stone matching and procurement frequently take between three and nine months before material is ready to use on site. Terracotta normally takes at least six months to obtain, and a year should be allowed to manufacture sufficient materials before site works commence. Heavy oak timbers can also be difficult to source quickly, particularly if very large sizes or particular shapes are required. If a tree has to be felled specially, the timber will need to be seasoned for at least two or, preferably, three years before it can be used.

Obtaining stone slates and special tiles for roofing can also involve significant lead times, and bad harvests may affect the availability of thatching straws. These constraints on material procurement often mean that bricks, stone, timber, terracotta and roofing materials, for example, must be procured during the year before project work commences on site. This may require the client to place orders for the supply of materials, because the contractor will seldom have been appointed at this stage.

SITE TRIALS

Trials are required whenever there is uncertainty over how a piece of work will be carried out. A trial may explore production methods or finishing techniques. It might help decide, for example, whether replacement stones should be worked with a stone axe or broad chisel, or how a patch repair in a concrete surface should be finished. Trials may also test and refine repair methods that are not widely practised or understood, or explore how work will appear once complete. They can help determine whether repairs should be finished to match original tooled surfaces or 'distressed', so they appear more compatible with surrounding weathered surfaces. Stains and colour washes may be trialled to assess their effectiveness in toning down new work. With each approach, conservation philosophy should be reviewed, and decisions made on the basis of an informed understanding which can be substantiated and reasoned.

Site trials are also a useful method for assessing the competence and performance of a contractor. Whenever possible, it should be stipulated that the person carrying out the trial will be involved in the work, either as a foreman or tradesperson. Site trials are also a useful means of managing expectations, since everyone involved will understand the likely outcome of the intervention from the outset.

Trials are commonly used to assist in formulating repair mortars; the trial investigates sources and suitability of aggregates, mortar ratios and compositions, and is used to create mortar tablets for matching to original mortars and masonry substrates. The mortar tablets should be approved and kept as reference samples for the project duration, and ideally will form part of the ongoing building file. ⊃MORTARS

Samples and site trials

Left: Detail of a record of treatment site trials carried out at Howden Minster, Yorkshire, to compare the behaviour of mortars made with different binders.

Top right and bottom right: A range of cured mortar sample 'patties', based on different aggregates and binders, is being used to help select a repair mortar that has the desired colour and texture when compared with the host material.

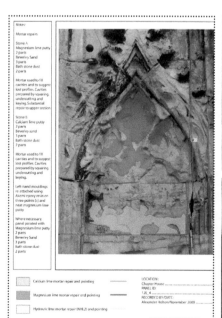

Left: *In-situ* pointing exemplars on 18th-century brickwork. The sample on the left is a poor match for the surviving historic pointing, whereas the one on the right is much more successful.

Right: This sample panel of masonry, prepared to demonstrate the materials for an extension to a listed building, is bedded and pointed in a bland grey mortar, which bears no resemblance to the birds-beak profiled white pointing that is an essential part of the character of the original building.

Exemplars

All projects should include exemplars to establish minimum standards or benchmarks, to be achieved in terms of materials and workmanship. Exemplars are distinct from trials and are completed following them. They may be produced during the design phase to provide examples of completed work for the client and tendering contractors to view, and can be particularly useful for tendering contractors to clearly demonstrate what is required. Exemplars should be retained for reference throughout the project.

PROTECTING BUILDING FABRIC DURING TREATMENT & REPAIR

Conservation treatments and repairs in a particular location may be potentially harmful to other adjacent materials. Therefore, it is important to assess the likely effect of proposed works, including the way access is provided, on adjacent areas and materials.

Protection should be designed to prevent inadvertent damage. For example, it will probably be necessary to protect:

- any handmade glass in the building (this is especially vulnerable to all types of building works)
- nearby decorative surfaces, particularly plasterwork and wall paintings
- other glazing, timber and ironwork, especially during cleaning
- floors, wall surfaces, special architectural features and roof coverings, especially during erection and dismantling of scaffolding
- fragile areas requiring consolidation or surface treatment.

Where necessary, specialist conservators should be engaged to advise on suitable protection. This should be specified and costed as part of the repair programme. The materials used (for example, plastic sheeting, hardboard or oriented strand board) should be sufficiently robust to last for the duration of the works. Allowance should be made for modifying or replacing protection as work progresses.

Areas of the building requiring protection should be identified, and the method of protection clearly specified. Similarly, any special requirements for protection of flora and microflora, or for cutting back of shrubs to enable access, should also be included in the specification.

INTEGRATING BUILDING SERVICES

The installation of new building services and equipment has the potential to cause serious harm to the significance of historic buildings and places. Even an apparently routine operation, such as lifting and relaying floorboards when installing pipe work or cabling, can be damaging if not properly specified or carelessly executed. The routing of services should be devised to minimise disturbance and destruction of significant fabric. Ideally, they should remain fully accessible to allow for maintenance and upgrading in the future. Existing voids, holes and slots should be reused wherever possible. A pre-installation desk study and site survey of the building will ensure that this can occur. Where cable trays or conduits are provided, consideration should be given to oversizing them to provide capacity for future enhancements. Services such as water supply or heating distribution pipes should be routed to avoid areas where leaks would damage significant building fabric or contents, and leak monitoring equipment considered. Surface-mounted cables, pipe work and equipment can be visually obtrusive, and historic surfaces may be damaged by their fixings.

The key to the successful integration of building services installations in historic buildings is to plan for them from an early stage. This enables the least obtrusive and disruptive service layouts to be devised, and for the associated builders' work to be anticipated and properly designed and specified. English Heritage (now Historic England) has produced a number of advisory publications on building services – both old and new – in historic buildings. These are listed in **Further Reading** at the end of this chapter.

Programmes of conservation treatment and repair are normally procured by means of a building contact. This is an agreement between the 'employer' (that is, the client) and a contractor to carry an agreed amount of work to an agreed standard, in an agreed period of time for an agreed amount of money. Formal contracts set out the obligations of the employer, contract administrator and contractor. The contract also states how disagreements between the parties may be resolved.

There are various types of building contract: these include the JCT forms of contract (Joint Contracts Tribunal) for building related projects; the ICE (Institution of Civil Engineers), and NEC 2 or 3, for civil engineering and infrastructure projects. Each of these contract providers publishes a variety of contract types to suit various project types.

The procurement method adopted can have a considerable bearing on the outcome of a project. Most standard forms of building contract can be successfully used for building conservation, providing the necessary skill and experience are available. However, traditional forms of contract tend to be the preferred option, as they are tried and tested. They also give the client a greater degree of control over the design process, as this is separated contractually from construction. However, sufficient time has to be allowed for these processes to be carried out in sequence. Traditional forms of contract for larger projects include measured works contracts with bills of quantity. For smaller projects, or where the work cannot be measured precisely, fixed-price contracts without bills of quantity, based on specification, provisionally quantified schedule of works and drawings, are appropriate. Prime cost contracts or contracts based on a schedule of rates may sometimes be used where work has to be carried out urgently, and the full scope is not known. However, it can be difficult to control costs with these types of contract. Other methods of procurement, developed primarily for new build projects, such as design and build contracts, management or construction management contracts, do not usually lend themselves to conservation projects, as they may not offer sufficient control or certainty for the employer.

It is important to recognise that formal contracts are only a framework for administering the project defined by the contract documentation, including drawings, schedules of work, specification, preliminaries and any appropriate specialist reports. The contract can allow for the provision of site-based investigations, trials and exemplars, as long as requirements are identified at the appropriate stage in the project, and sufficient time and funding is allowed for them.

CONTRACT DOCUMENTATION

READERSHIP

When producing contract documentation, it is important to know and understand for whom the documents are written. The specification and schedule of work must be written clearly and concisely, and contain relevant information so that the conservators and craftspeople on site can use them.

Documents may also have a wider readership, including local authority conservation officers and Historic England advisors. The primary readership and function of the documents must be prioritised, however, if the documents are to remain effective. If the documents become too complex, or use language that is too technical, the value of the documents as a practical tool to convey information for site conservation and repair will be diminished.

Clarity can be achieved by using a combination of concise relevant text, short paragraph structure and formatting that divides the documents into clear sections, and can be quickly sifted through to find the appropriate information.

SPECIFICATIONS

Specifications are divided into two parts: the preliminaries and the works specification. The *Preliminaries* is an essential document that provides the background details that a contractor needs in order to price the works:

- project particulars, which sets out the name and location of the site, and the names and addresses of the employer and the consultants
- a list of all the drawings and any reports which will accompany the tender
- details of site boundaries, existing buildings, existing mains and services
- tendering requirements, and details of how qualifications to, and errors in, tenders will be adjusted, and arrangements for site visits
- requirements for selection of sub-contractors
- restrictions on the management of the works; for example, whether a contract manager or foreman is to be on site at all times during the works
- the description and amount of any work to be carried out under a provisional sum.

The *Works Specification* contains design information about the actual work, and includes information about materials, suppliers, manufacturers and methods of work. The specification conveys design intention and quality. Works specifications may be prepared using either a prescriptive or a performance approach, or possibly a hybrid of the two. A prescriptive specification is more detailed and, therefore, lengthier than a performance specification. It describes materials, work methods, and procedures and equipment, in detail. In essence, it is a precise method statement, and is not intended to be open to interpretation.

The advantage of a prescriptive approach in the context of building conservation is **285** that the document describes exactly how work is to be achieved, and can clearly convey this to a wider readership, including the conservator, the client and statutory bodies such as the local planning authority. The disadvantage is that the specification tends to be lengthy, and is heavily reliant on the skills and experience of the designer. It is considered by some practitioners to be unnecessary, since an experienced conservation contractor should know and understand conservation repair techniques well enough not to need detailed descriptions of work methods and processes.

A performance specification sets out criteria for completing work, and defines the minimum standards to be achieved. This is the approach adopted by the National Building Specification [NBS]. Only limited information about work methods and processes is given; it is intended to allow choice and interpretation by the contractor. A performance specification will state the type and extent of repair or replacement techniques to be used – for example, mortar repairs, piecing-in, cleaning or replacement – but will not describe the process or techniques. Instead, it will simply set out the performance criteria for the repair, such as minimum depth of repair, mortar types and materials to be used. This allows contractors to implement the required work by relying on their own experience, judgement and skill, as well as employing methods that might be unfamiliar to the designer. Unfortunately, performance specifications allow for differing interpretation by the parties directly or indirectly involved in the project, and may not be easily understood by the client and statutory bodies. However, when well written, and used in appropriate circumstances, such as in situations where work is to be carried out by an experienced conservator, performance specifications can be useful and effective.

A hybrid specification using a combination of qualities from a performance and prescriptive approach can often be appropriate in conservation projects. For example, the design and installation of lead weatherings, roofs and cladding is well described by the Lead Sheet Association's design guide, which is widely known and available. Therefore, for this type of work, a performance specification is ideal. However, materials and techniques for the repair of many historic building materials tend to be less well understood, and sources of information more widely dispersed in books and journals that may not be easily accessible. Furthermore, the designer or specifier may decide that although a variety of approaches might be followed, in the circumstances, only one particular technique is suitable. In these cases, a prescriptive approach will be more appropriate.

Irrespective of the approach, specifications in England tend to be written on a works trade system, which groups work into trade types like masonry, joinery and plastering. This approach was refined and formalised into the *Common Arrangement of Works Sections* [CAWS] by the Coordinated Project Information Committee, and is used across all project documents, including drawings, specifications, schedules of work and bills of quantities. The CAWS system is widely used by construction professionals, and software packages are available with linked information databases that are periodically updated on a subscription basis.

There are circumstances where conservation work is highly specialised and only practised by a limited number of highly trained individuals, as is the case in wall-painting conservation. In these circumstances, the designer may not produce a works specification for this element of work, but may commission reports and proposals, including method statements, from the conservators who will, with other suitably-qualified specialists, tender for the works.

COMMON ARRANGEMENT OF WORKS SECTIONS [CAWS]

A	Preliminaries; General conditions	**N**	Furniture; Equipment
B	Complete buildings / structures / units	**P**	Building fabric sundries
C	Existing site / buildings / services	**Q**	Paving; Planting; Fencing; Site furniture
D	Groundwork	**R**	Disposal systems
E	*In-situ* concrete; Large precast concrete	**S**	Piped supply systems
F	Masonry	**T**	Mechanical heating / cooling / refrigeration systems
G	Structural; Carcassing metal / timber	**U**	Ventilation; Air conditioning systems
H	Cladding; Covering	**V**	Electrical supply / power / lighting systems
J	Waterproofing	**W**	Communications; Security; Control systems
K	Linings; Sheathing; Dry partitioning	**X**	Transport systems
L	Windows; Doors; Stairs	**Y**	Services reference specification
M	Surface finishes	**Z**	Building fabric reference specification

Specifications frequently include information that is irrelevant, resulting in large documents that take a long time to sift through to find the information that is required.

Document structure is very important. The specification should follow the CAWS system as relevant to ensure familiarity and integration with other project documents. The specification will include a contents page listing out each work section. Each work section is then subdivided into:

- Scope generally
- General/Preparation
- Scope of work/Conservation/Repair
- Workmanship generally
- Materials/Production/Accessories

These broad subdivisions vary across each work section, but are intended to give consistency and standard content from one project to another. Specifications incorporate standard text that will be common to all jobs, but this should only be included where it is relevant.

Scope Generally

At the start of each section it is useful to set out the scope of the work and list out techniques. This list refers to the specific clauses that describe the element of work or techniques, as well as providing quick access to information.

General/Preparation

This section includes several relatively generic clauses that describe general requirements such as preparation work. Although these clauses are described as generic, and are likely to be commonly seen on many projects, they must be reviewed and amended, if required, to ensure relevance. For example, in a masonry repair specification, a clause might be included describing the requirement to remove plant growth from walls before carrying out any repair work. This section should include all work activities that are logically carried out before commencing specific repairs.

Scope of Work/Conservation/Repair

This section describes the various elements of work, conservation treatments and repair techniques to be carried out. For example, it might include clauses describing the installation of lead coverings to stone cornices, the repair of a plain clay tile roof, or scarfed repairs to timber windows. In a performance specification, the clause would refer only to the materials to be used, and information related to performance criteria and quality, such as the depth of cutting out masonry joints in preparation for pointing.

Workmanship Generally

This section describes the materials to be used and standards of workmanship required. There are standard documents, such as the *National Building Specification*, that provide comprehensive libraries of clauses. It is important that clauses are chosen and edited to be project specific.

Materials/Production/Accessories

This section lists and describes the materials to be used, together with product and supplier details. Clauses are also included about application, quality and minimum standards for materials.

Specification Clauses

COMPARING PRESCRIPTIVE, PERFORMANCE & HYBRID SPECIFICATION CLAUSES

The examples show how the same item of work – the piecing-in of stone ashlar – is described in the three forms of specification.

EXAMPLE A

PRESCRIPTIVE SPECIFICATION CLAUSE

PIECING-IN OF NATURAL STONE ASHLAR: REPAIR TECHNIQUE P1

1	Refer to drawings
2	Mark out size of repair on the face of stone, as described within the schedule of works and agreed on site with the Contract Administrator. The intention is to minimise the size of each repair to the area of spalling, decay or previous repair. However, where appropriate, the repair may be extended to the edge of a moulding or joint to prevent a patchwork appearance. Each repair is to be marked out square.
3	All cutting out to be undertaken using either mason's fire sharp or tungsten tipped chisels and electric drills, unless otherwise directed. Electric cutters will only be permitted with the written consent of the Contract Administrator, and only if it can be demonstrated that no over-cutting or damage will occur.
4	Carefully cut out the existing decayed stone or previous defective repair within the areas marked to the required depth. The sides of the mortice are to be undercut to form a key for the repair. The top and bottom faces of the mortice are to be square cut to allow accurate location of the natural stone indent.
5	Where a corroded fixing cramp is exposed, carry out further repair works in accordance with Repair Technique CR.
6	Ensure that the perimeter of the mortice is square, with a sharp arris. No chips or spalling to this edge will be accepted.
7	The new natural stone indents are to be cut from stone scant or sawn six sides, sourced in accordance with specification clause C41.3a/215. The sides of each indent are to be coated with a lime slurry mortar: NHL 3.5 mixed with water, the consistency of single cream, before fitting into the mortice. Remove excess mortar and smearing with clean water and a sponge. Where the repair bridges a joint, the indent is to be made in separate pieces, with the joint width and position maintained. All indents are to be laid on their natural bed unless standard masonry practice would dictate otherwise.
8	Mortar joints to be filled with mortar composed of 1 part NHL 3.5 to 2½ parts blended aggregate in accordance with Repair Technique RP-1.

EXAMPLE B

PERFORMANCE SPECIFICATION CLAUSE

PIECING-IN STONE: REPAIR TECHNIQUE P1

1	Drawing
2	Stone: Portland Stone clause 240A
3	Mortar joints: Mortar: 1 part NHL 3.5 to 2½ parts blended aggregate. Aggregate blended from 3 parts crushed Portland Stone and 2 parts sharp white sand. Aggregates to be well graded from 3 mm down. Refer to specification section Z21.
4	Mortar slurry: NHL 3.5 mixed with water to form consistency of single cream. Refer to specification section Z21.
5	Wherever possible, the piecing-in repair should follow the irregular shape of the existing defect rather than being 'squared up'. Refer to drawings.
6	Mortice to receive stone piece: • Cut out accurately. Undercut sides of mortice where necessary to provide space for grout. • Clean out thoroughly.
7	Stone repair piece: • Cut to the smallest rectangular shape necessary to repair the defect and provide a firm seating. Install accurately and securely. • Ensure the new piece is accurately cut and shaped before fixing. Only carry out minor trimming *in situ*. Do not re-work the face of the original stone. All repairs should be hand worked on site. • The perimeter joint around the repair piece should not exceed 1 mm in width.
8	Exposed faces: keep clean.
9	Existing iron cramps: retain and treat or remove as clause C41/450.

EXAMPLE C

'HYBRID' SPECIFICATION CLAUSE

PIECING-IN STONE: REPAIR TECHNIQUE P1

1	Drawing
2	Stone: Portland Stone clause 240A
3	Existing iron cramps: retain, and treat or remove as clause C41/450
4	Mortar joints: • 1 part NHL 3.5 to 2½ parts blended aggregate. Aggregate blended from 3 parts crushed Portland Stone and 2 parts sharp white sand. Aggregates to be well graded from 3 mm down. Refer to specification section Z21. • Finish: slightly recessed to maintain original joint width. • Mortar slurry: NHL 3.5 mixed with water to form consistency of single cream. Refer to specification section Z21.
5	METHOD: • Keep the repair size as small as possible to cover the area of spalling, decay or previous repair. Where appropriate, the repair may be extended to the edge of a moulding or joint to provide a more discreet repair. Each repair is to be marked out square. Repairs to stair treads should be dovetailed on plan. • Cut out the repair area using sharp masonry chisels. Percussive rotary drills fitted with depth gauges may be used where appropriate for 'chain drilling' a series of holes to speed up the cutting out process for piecing-in repairs. Disc cutters are not permitted. • Ensure that the perimeter of the mortice is square, with a sharp arris. No chips or spalling to this edge will be accepted. • Make the repair piece or 'indent' to accurately match size and shape of the repair mortice. Work moulding to match the existing shape and profile. Where moulding varies in size of profile, seek instructions before proceeding. • Where the repair bridges a joint, the indent is to be made in separate pieces with the joint width and position maintained. All indents are to be laid on their natural bed unless standard masonry practice would dictate otherwise. The indent is to be carved on site from one piece of stone, and must match the original size and profile of the moulded detail. • The exposed face of the indent is to be finished with a sand-rubbed face to achieve a slightly open texture, to match the surrounding weathered stone. • Each indent is to be cut square and slightly oversized to the repair mortice. The edges of each indent are to be rubbed down by hand to form fine rubbed joints to the perimeter of the repair. This joint must not exceed 1 mm in width. No chips or spalling to this edge will be accepted. • Clean out the back of the mortice using clean water to remove all loose dust and debris. Ensure that the repair aperture has been sufficiently wetted to control mortar curing. • Coat the sides of the indent with a lime slurry mortar and fit each indent into the mortice. Lightly wash the face of the repair with clean water and a sponge to remove excess mortar and smearing.

DRAWINGS

Drawings are an essential part of all projects, and can include elevations, plans, sections and details, to convey the location, extent and type of work. Drawings can also include sketch details to illustrate specific elements. Drawings and sketches are invaluable for communicating information that would otherwise be difficult to describe clearly or concisely in words. Photographs can also be used as a template for conveying information about the location, scope and types of treatment, and repairs to be carried out. Sketches can often be used very effectively to show repair techniques, and are often more memorable to conservators and craftspeople.

Wherever possible, drawings should be produced to standardised scales, and using standard drawing conventions, so that a wide readership may understand the information being conveyed. When issuing drawings via email, it is advisable to mark on the drawing the intended printing sheet size, as well as the scale (for example, 1:20 @ A1) to avoid scaling errors.

Annotated sketches and photographs are helpful in communicating information that would be difficult to describe clearly or accurately in writing.

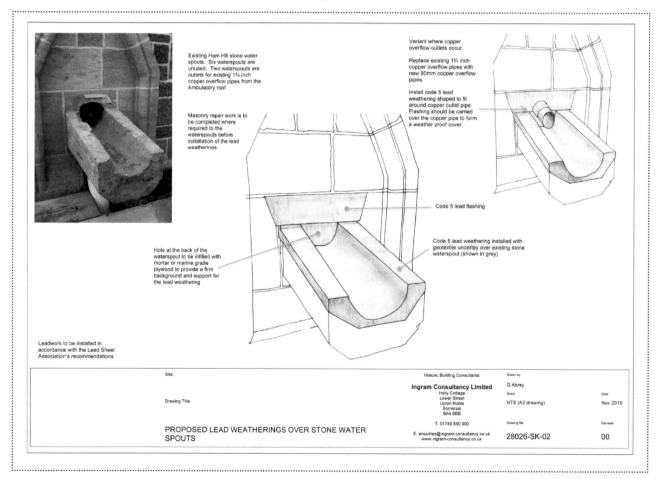

Existing Ham Hill stone water spouts. Six waterspouts are unused. Two waterspouts are outlets for existing 1¾ inch copper overflow pipes from the Ambulatory roof

Masonry repair work is to be completed where required to the waterspouts before installation of the lead weatherings

Hole at the back of the waterspout to be infilled with mortar or marine grade plywood to provide a firm background and support for the lead weathering

Leadwork to be installed in accordance with the Lead Sheet Association's recommendations

Variant where copper overflow outlets occur.

Replace existing 1¾ inch copper overflow pipes with new 90mm copper overflow pipes.

Install code 5 lead weathering shaped to fit around copper outlet pipe. Flashing should be carried over the copper pipe to form a weather proof cover.

Code 5 lead flashing

Code 5 lead weathering installed with geotextile underlay over existing stone waterspout (shown in grey)

Site:

Drawing Title:

PROPOSED LEAD WEATHERINGS OVER STONE WATER SPOUTS

Historic Building Consultants:

Ingram Consultancy Limited
Holly Cottage
Lower Street
Upton Noble
Somerset
BA4 6BB

T. 01749 850 900
E. enquiries@ingram-consultancy.co.uk
www.ingram-consultancy.co.uk

Drawn by:
G Abrey

Scale:
NTS (A3 drawing)

Date:
Nov 2010

Drawing No
28026-SK-02

Revision
00

rebed copes.

reinstate fallen finial.

record and remove memorial fragments to store.

reinstate fallen pediment.

carefully take down and rebuild leaning memorial on new footing.

Type and scope of repairs indicated on a photographic base. The ease with which digital images can be edited and imported into CAD applications makes this a versatile and cost-effective method of conveying information.

SCHEDULES OF WORK

Schedules of work are normally prepared by the lead professional, a quantity surveyor or a specialist conservation consultant, and are used in conjunction with the drawings and the specification to provide clear concise descriptions of items of work for pricing by the contractor. A condition survey report may sometimes include a schedule of works for use in planning future project work. Schedules of work tend to be used on small to medium-size projects. They are also very useful on larger conservation projects for specific elements of work. Schedules of work differ from bills of quantities in that they list the individual items of work to be carried out in a given location. As such, they are helpful to contractors, conservators and craftspeople on site in determining what works need to be carried out and where. A typical schedule of works describes each work item and gives its location, often by means of survey grid coordinates that are cross-referenced to drawings. Further cross-references to specification sections or clauses, as appropriate, may also be provided.

Where the full extent of the work is known, a quantified schedule of work should be produced. This includes measured quantities and clearly defines the scope of the work. By leaving none of it open to interpretation, it thereby ensures that contractors all tender on the same basis. This is by far the most efficient method of obtaining realistic competitive tenders in the first instance, and also provides the most effective way of controlling costs during the construction phase. Specialist advice on preparing a quantified schedule of works may be obtained by appointing a quantity surveyor who is experienced in working with historic buildings.

Provisionally quantified items may be included in the schedule of works where the full extent of the work is not known. In this case, an approximate quantity of work is included (based on an assessment of the likely scope of the necessary works), and the contractor is required to include a rate for such work in the tender. This rate can then be used for future price adjustment. Where the need for repairs is anticipated, but the scope of the work is completely unknown, provisional sums are normally allocated. Where this is necessary, the contractor must provide rates for the items of work that will most likely be needed to enable subsequent measurement and valuation.

BILLS OF QUANTITIES

A bill of quantities is prepared by the quantity surveyor. It sets out the terms and conditions of the works contract, and itemises, clearly and concisely, all the works to be carried out. A bill of quantities provides a strong basis for budgetary control and financial risk management. It enables contractors to prepare tenders accurately and efficiently, and allows tenders to be appraised prior to awarding a contract. Once a contract has been entered into, it provides the means for valuing work executed for the purpose of certifying payments, and is the basis for calculating and negotiating the cost of variations.

Where there is uncertainty about the scope of some parts of the work, a bill of quantities may include provisional quantities; these items are re-measured as the work is carried out. In cases where the nature of the works is known but not their precise extent, an approximate bill of quantities may be used. This will be subject to re-assessment and re-measurement during the contract, as detailed information becomes available. For example, it may not be possible to determine the full extent of masonry repair until scaffolding has been erected that allows a close inspection to be carried out.

Bills of quantities are usually prepared on a trade or elemental basis. In this form they are not helpful to conservators and craftspeople on site, as insufficient information is provided about the location of the work to be carried out. However, the form of bills of quantities can be adapted to suit the project in hand. For example, the bill for the surface repair of a building might be prepared on an elevation-by-elevation basis, with the work items set out so they can be clearly related to drawings.

PROGRAMME

The programme is an essential project-planning and management tool. It gives the contract commencement date, duration and completion date, as well as the sequence of work activities. Programmes are normally presented in the form of critical path or Gantt charts. Programmes for specific work elements can be presented as flow charts. They should be developed and maintained from project inception to completion, so that all activities can be planned for and progress monitored. Programmes should include a progress line to indicate the progression of individual activities. Delayed events will be highlighted, and their impact on subsequent or linked activities can be assessed.

During the design stages, the lead professional should provide the client with a programme and give regular updates on progress. In the construction phase, the contractor will provide a detailed programme and report on progress at regular intervals.

Programming works to historic buildings may be subject to restrictions arising from the use of the building. For example, there may be a requirement to avoid working on historic buildings opened to the public during the peak tourist season. Also, the need to protect wildlife habitats, such as bat roosts (see **Ecological Considerations**), may have programming implications. Because of these kinds of restrictions, and the timing of annual financial cycles, works may sometimes have to be carried out during parts of the year when conditions are least favourable for traditional materials such as lime-based mortars and renders. This may have a bearing on the technical specification, and the way in which works are organised and carried out. For example, temporary covers may have to be provided to protect newly-placed mortar. Protection may be required for several months, and this might mean that scaffolding has to remain in place for much longer than would be the case during summer working. This will have financial implications, and should be taken into account at the technical design stage.

TENDERING (PRE-CONSTRUCTION STAGE)

It is usually desirable to obtain several quotations or 'tenders' for a programme of work. This process is begun when a letter – a 'tender invitation' – is sent to the contractor, asking him or her to provide a quote for the work. Tender invitations should always include as much information as possible, to ensure that fair and reasonable accurate quotations are received. Wherever possible, invitations to tender should include the detailed design (drawings, specifications and schedules of work) and, for larger projects, a bill of quantities (see **Contract Documentation**).

The clarity and adequacy of tender documentation is a major factor in determining whether or not a repair project is delivered satisfactorily. Good-quality tender documents provide contractors with a level playing field for pricing, planning and controlling the works. They form the basis for allocating and managing risk, controlling costs, and determining how much the contractor appointed is ultimately paid for the completed project. Tender documentation should identify the contract information, including bills of quantities, drawings and schedules, to be referred to. The scope and standard of the works, and the roles and responsibilities of the parties to the contract, should also be clearly defined. If it fails to address any of these issues, the project will be at risk of not delivering what the client wanted in the time required or at the cost anticipated. Poor-quality tender documents lead to inaccurate pricing, and increase the likelihood of claims and disputes. Although clear, concise tender documentation should result in realistic competitive tenders being submitted, market conditions prevalent at the time of tendering will also have a bearing on the level and spread of the tenders.

Only contractors with proven skills and experience should be invited to tender. Public authorities and large organisations often maintain lists of 'approved' contractors and consultants, but these may not necessarily include companies with the required skill and experience for a conservation project. Therefore, approved lists should always be reviewed on a project-specific basis; standard procedures and requirements of a client organisation should not disqualify companies with the appropriate skills and experience. They should certainly not favour those with insufficient skill, simply because they meet the generic requirements of an approved list. Alternatively, a pre-qualification process may be carried out in order to select the list of contractors and consultants with the appropriate skills and experience to tender for the works. This has advantages for both employers and tendering contractors. For employers it can help avoid the problem of low tenders submitted by companies of doubtful competence. It allows contractors lacking sufficient qualifications to avoid the expense of tendering, and reassures qualified contractors that they will be competing against firms that also meet the appropriate standards of competence. A formal pre-qualification process will often be a requirement where a project is receiving grant funding. Tenders for public sector projects that are valued above a certain financial threshold must be published in the *Official Journal of the European Union* [OJEU] to comply with EU legislation.

In most cases, three or four tenders should be sought, although for larger projects up to six may be required. There should be no reason to obtain a greater number of quotations if good quality documents are produced. Tendering is often a very time-consuming and expensive process for contractors and building professionals to undertake. A 1-in-3 or 1-in-4 chance of success is reasonable. Increasing the number of contractors tendering reduces their chances of success and increases their overhead costs. Tendering is part of a company's overheads, and will ultimately be passed on to clients in future project costs.

Tender documentation and invitations to tender should clearly state the procedure that is being used to conduct the tendering process. One of the construction industry standard codes for tendering should be adopted. This will set out the way the tendering process is to be carried out and clarify issues such as qualifications inserted by tendering contractors, and the correction of errors. A tender report is then prepared for the client.

Contracts should always be awarded on the basis of value for money. Value for money may include other factors, such as reduced programme durations, less inconvenience for the client or a substantiated 'better service'. In the ideal situation, where tender contractors are all comparable in terms of size, experience and availability, the lowest tender can represent the best value. However, in practice, tender lists are not always comparable, and the lowest price may not necessarily represent the best value. In some cases, where the size or sensitivity of the project warrants it, interviews may be conducted with the 'short-listed' tenderers to assess their approach and attitude to the project. Once a tender has been accepted, and the dates for possession of the site, completion of the works and the first site meeting have been agreed upon, the contract is signed.

In grant-funded projects the procedure for awarding contracts is usually prescribed by the grant-giving body, due to requirements for tendering processes to be open, transparent and non-discriminatory. This allows the client less discretion in the choice of tender for acceptance.

NEGOTIATED CONTRACTOR & CONSULTANT APPOINTMENTS

As an alternative to competitive tendering, it can be advantageous to obtain a negotiated price from a contractor or consultant for programmes of conservation work. In many instances, this will achieve better value for money, since specialist companies obtain much of their work from repeat business, referral and reputation, and will charge a fair cost that generates reasonable profit for reinvestment. Charging unreasonable costs will damage their reputation and harm their ability to obtain further work. Negotiated appointments may also be used where a contractor and/or designer plays a significant part in the technical design development of a project, and it is desirable to retain their knowledge and experience for the construction phase. It should be noted that the conditions imposed by grant-giving bodies may preclude the use of negotiated contracts for the procurement of the lead and other professionals, and the works contractor.

PROJECT IMPLEMENTATION

CONSTRUCTION PHASE

Project implementation is referred to as the construction phase, the stage when conservation, repair, alteration and construction work are completed on site. With appropriate investment of time and resources during earlier phases of the project, the construction phase should be completed within budget and on programme.

Conservation projects generally require a greater amount of professional time on site than new build projects. As the work proceeds, there will inevitably be the need for further or revised instructions to be issued to the contractor when the construction or condition of building elements prove not to be as anticipated, or when a specified technique requires adjustment or modification to suit site conditions.

It is important to maintain adequate levels of supervision to ensure that the required quality of work is obtained. The degree of supervision necessary will depend to a large extent on the experience of the contractor. The regular presence of conservation professionals on site communicates commitment and enthusiasm, and can help ensure that building operatives at all levels understand the value and significance of the building, and the elements they are working on. Also, by spending time on site and observing the work as it proceeds, the building professional gains valuable practical experience and insight into the way theory translates into practice. In some large and complex projects, the client may appoint a full- or part-time Clerk of Works to inspect and monitor the progress of the works on his behalf.

COMPLETION

Where projects employ a formal construction contract – for example, the JCT, NEC or ICE forms of contract – the project is not complete when construction work concludes on site. The project is only complete once the 'rectification period' ends. This period follows 'practical completion', when all of the work carried out under the contract has been completed, and provides the contractor with the opportunity to remedy any defects in the work that have become evident. For most contracts, this period is either six or 12 months. A sum of money is retained by the client until the defects period elapses. The amount of retention is usually 3 % or 5 % of the contract sum. Half is released at practical completion; the other half is paid once the rectification period is complete and all defects have been remedied.

LEARNING FROM EXPERIENCE

'Documenting and learning from decisions is essential' (Historic England *Conservation Principles*, Principle 6). The value of a cumulative record of what is known about the form or design, construction and services of a historic building (particularly a large, complex or especially sensitive one) cannot be underestimated. Organised archives can help save the time and cost of duplicating surveys and investigations. They provide the basis for an incremental database of knowledge about the building fabric and services, and are a source of information about what lies beneath finishes. This information enables a rapid and intelligent response to emerging problems, and is an important tool in assessing the feasibility of potential interventions. Increasingly, such records are held in digital form, and so can be made easily available to all members of a project team. The building record should be kept in a safe place with appropriate back-ups, and a designated individual or post-holder nominated to be responsible for maintaining and updating it. Unfortunately, such records are often discarded or lost, and with them the data that could help in planning future interventions.

The management of historic buildings should include regular monitoring and evaluation of the effects of change, whether passive or as a result of intervention. When building works are carried out, it is important to record not just what has been done (the 'as built' or 'as installed' drawings and specifications), but also the reasons for doing so and the expected outcomes. These can include not only technical performance (including the medium- to long-term performance of materials), but also the perceived effects on the significance of the building. Outcomes of decisions can subsequently be compared with expectations, and often reveal unanticipated consequences; they also inform future policy and decisions. Unfortunately, 'as built' records are infrequently produced because of their cost and the common desire to cut costs towards the end of projects; this can often prove to be a false economy in the long term.

Facing page: Example of an 'as built' record drawing. This forms part of the building archive, an incremental database of information about the building fabric and services. 'As built' records enable the effectiveness of repairs to be assessed over time and provide valuable information to guide future interventions.

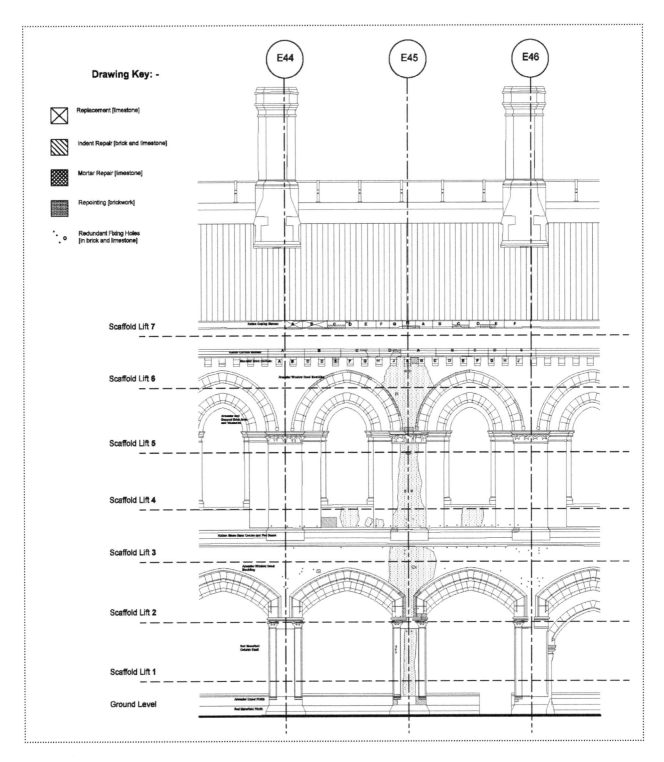

Drawing Key: -

- ⊠ Replacement [limestone]
- ⊠ Indent Repair [brick and limestone]
- ▦ Mortar Repair [limestone]
- ▦ Repointing [brickwork]
- ⊙ Redundant Fixing Holes [in brick and limestone]

E44 E45 E46

Scaffold Lift 7
Scaffold Lift 6
Scaffold Lift 5
Scaffold Lift 4
Scaffold Lift 3
Scaffold Lift 2
Scaffold Lift 1
Ground Level

A building record can be established at any time, preferably immediately, if one does not exist. Historic data can be added later, but even if the opportunity to do so does not arise, the record will still grow in value as information from subsequent surveys and interventions is added to it. As discussed in the preceding section, **Care & Maintenance**, recording begins by keeping a log of all maintenance inspections and minor works (see **Appendix 2**). Providing basic checklists to be completed as these activities are carried out makes this easy and a matter of routine. The log should be archived alongside major periodic (typically quadrennial or quinquennial) surveys, specialist inspection reports, and the documentation of major repairs and alterations.

The choices we make about what, and how, to conserve – and our changing opinions about conservation – reflect current attitudes to the past and the values we ascribe to its material remains. These choices become part of the story that we pass on to future generations who will reinterpret it from their perspective, as we have done in our time, then add further chapters of their own.

Asset Management Planning

Organisations responsible for the care of large numbers of historic buildings or places, or for complex estates and collections, will normally implement a systematic and holistic approach to the management of heritage assets. The allocation of resources must be planned and provided for long term. This is accomplished through understanding the baseline condition of heritage assets, whatever immediate repair work is needed, and the maintenance and repair work that will be necessary in the future.

A large number of properties may be assessed by a rolling programme of condition surveys, perhaps managed and implemented on a regional basis. Surveys are typically undertaken on a quadrennial or quinquennial cycle, and carried out to a consistent standard, usually using a template to optimise consistency.

The maintenance and repair requirements of heritage assets and their elements are assessed individually. The assessment takes account of a range of factors that will include statutory and legal requirements; significance; condition (fragility and variability of deterioration, cause and extent of decay); original design and construction (materials, workmanship and specification), and any subsequent alterations; location (orientation, vulnerability and severity of exposure); function; and frequency and intensity of use (past, present and future). It is therefore important that the survey process is able to capture all the necessary information in a consistent way, so that work across the estate can be prioritised. Alongside the survey and assessment of individual sites, a further essential requirement for an asset-management planning system is the ability to store and manage data on current condition and required work. This enables detailed work plans to be quantified and costed, and facilitates effective procurement and delivery of maintenance and planned repair works.

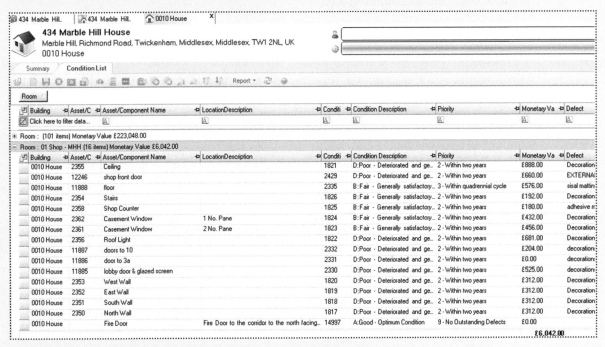

Asset management planning software applications allow the information gathered in condition surveys to be processed and presented in a variety of ways, taking into account a wide range of factors.

Further Reading

Ashurst, J. (ed.) (2007); *Conservation of Ruins*; Oxford: Butterworth-Heinemann

Ashurst, N. (1994); *Cleaning Historic Buildings* (Volumes 1&2); Shaftesbury: Donhead

Brereton, C. (1995); *The Repair of Historic Buildings: Advice on Principles and Methods* (2nd edition); London: English Heritage

Collings, J. (2002); *Old House Care and Repair;* Shaftesbury: Donhead

Earl, J. (2003); *Building Conservation Philosophy*; Shaftesbury: Donhead

English Heritage (2010); *Caring for Places of Worship*; available from *www.historicengland.org.uk/images-books/publications/caring-for-places-of-worship-report/*

Historic England (2018); *Vacant Historic Buildings: Guidelines on Managing Risks*; Swindon: Historic England; available from *www.historicengland.org.uk/images-books/publications/vacanthistoricbuildings/*

Hunt, R., Suhr, M. (2008); The Society for the Protection of Ancient Buildings [SPAB]; *Old House Handbook: A Practical Guide to Care and Repair*; London: Frances Lincoln

Insall, D. W. (2008); *Living Buildings: Architectural Conservation, Philosophy, Principles and Practice*; Mulgrave (Australia): The Images Publishing Group

Klemisch, J. (2011); *Maintenance of Historic Buildings: A Practical Handbook*; Shaftesbury: Donhead

Lithgow, K. (2006); 'Building work: planning and protection'; in *The National Trust Manual of Housekeeping: The Care of Collections in Historic Houses Open to the Public*; Oxford: Butterworth-Heinemann; pp.710–21

National Heritage Training Group (n.d.); *NHTG Initiatives* (web page); available from *www.nhtg.org.uk/*

Normandin, K. C., Slaton, D. (2005); *Cleaning Techniques in Conservation Practice*; Shaftesbury: Donhead

Robson, P. (1999); *Structural Repair of Traditional Buildings*; Shaftesbury: Donhead

Seeley, N. (2006); 'Commissioning conservation work'; in *The National Trust Manual of Housekeeping: The Care of Collections in Historic Houses Open to the Public*; Oxford: Butterworth-Heinemann; pp.774–83

Wrightson, D. (2002); *A Stitch in Time: Maintaining Your Property Makes Good Sense and Saves Money*; Tisbury: The Institute of Historic Building Conservation in Association with the Society for the Protection of Ancient Buildings; available from *www.ihbc.org.uk/stitch/Stitch%20in%20Time.pdf*

EMERGENCY PLANNING FOR HERITAGE BUILDINGS & COLLECTIONS

This chapter discusses the principles and practicalities of planning for an emergency, and dealing with its immediate consequences.

Since the unexpected is an expected part of life, all building owners and managers should be prepared for emergencies. However, those responsible for heritage assets have a particular duty of care. The definition of 'emergency' is the subject of continuing debate. For some, a tiny leak in the ceiling that allows water to drip onto a priceless painting is an emergency. Others would regard this simply as an 'incident'. Whatever the point of view, an emergency is a sudden, urgent, usually unexpected occurrence or occasion requiring immediate action. It is clear that creating a workable and proportionate Emergency Plan is an important part of heritage management.

Emergency planning is a process that involves:

- identifying hazards
- assessing risks
- implementing policies and appropriate precautionary or preventive measures to alleviate risks
- ensuring that adequate arrangements are provided to cope with eventualities.

Plans must also be made to cover the actions necessary during and after an emergency, to minimise damage and long-term deterioration of the building and its contents.

EMERGENCY PLANNING

Heritage assets differ in size, structure, age, materials, site, use and vulnerability, and thus face different risks. However, the basic tenet of emergency planning remains the same: that prevention is better than cure. Emergency plans will vary in scope and detail, according to the size and complexity of the premises concerned. While smaller buildings might have only a simple Contact List and Emergency Strategy Document (see **Appendix 4**), larger premises such as museums, art galleries and historic houses (some also containing valuable collections) will require comprehensive and detailed written plans, designed to meet as many eventualities as possible. Typically, these plans will include:

- vulnerability and risk assessments
- routine housekeeping policies and actions to mitigate hazards
- fire safety measures to minimise risks
- emergency procedures in case of fire, flood, storm damage or chemical spillage
- contact and facilities lists
- prioritised salvage plans, for rescuing contents during an emergency
- management and training arrangements for a salvage squad
- business continuity plan
- key insurance arrangements.

It is also advisable to fully record heritage assets, both fixed and moveable, that may be damaged. The existence of detailed photographs and drawings will be of immense value in the aftermath of an emergency (see **Survey & Investigation Methods**).

COMPILING AN EMERGENCY PLAN

Site staff and management will be the best people to compile an emergency plan, as they will already have an idea of the hazards and threats to the building, and the value of its contents. They will know that the hidden valley on the roof will leak after a snowstorm, or that leaves will block the gullies in the autumn. They will also know where the concentrations of combustibles are kept, and where stock is temporarily stored when too much is delivered.

The advantage of producing a plan in-house is that it can easily be updated. In fact, it should be continuously reviewed, perhaps quarterly or when new collections arrive, so that it is never 'signed off' and filed.

The plan should be easy to understand and accessible to any authorised person who needs to use it. Important information such as contact lists and salvage priorities may need to be read in adverse conditions, so plans should be in large print and preferably laminated, so that they remain readable should they get wet. Diagrams, plans and pictures are a concise way to convey messages.

One copy of the salvage plan should be kept in the most secure location on site, and another securely held offsite by a responsible person.

Fire and flood are arguably the two biggest threats to heritage assets. Lapses in security arrangements – both routine and precautionary, or during salvage operations – can also pose threats, such as arson or theft. Although the examples given in this chapter focus on fire and flood, other threats, such as storm damage and chemical or oil spillages, should be considered and included in risk assessments.

FIRE

Heritage assets presented to the public, such as museums, art galleries and historic houses, may not appear at first to be high fire-risk premises. Most have a very low fire load (that is, the quantity of fuel for the fire), few ignition sources, high ceilings, large open spaces, and a high level of supervision and management. Although these observations may apply to the public spaces, there are other smaller rooms, often hidden away in basements or upper floors, which might include offices, archives, workshops, kitchens, boiler rooms and storage areas, where real dangers exist. This is because:

- the highest concentrations of combustible materials are kept here
- more ignition sources exist in these areas than in the public rooms
- ceilings are lower, which will hasten the spread of fire
- housekeeping is often less rigorous in these areas because they are out of the public eye
- valuable collections may be stored here when not on public view
- these areas are more likely to have had holes made in fire-resisting walls for the passage of services.

There are also other risk factors relating to firefighting operations. Hidden voids can lead to unseen fire spread to areas remote from the origin of the fire, and a fire in these spaces is extremely difficult to fight. Upper floors are more difficult to access to fight fires, and hoses may need to be laid up stairs and along corridors. This will lead to delays before a fire can be tackled.

The basement is often very difficult for firefighters to access because they may need to fight their way through the heat layer to tackle a fire, and subsequent ventilation is not always easy. Roof spaces are often used for the storage of combustibles; these, together with the roof timbers, provide fuel for a serious fire, which may be almost impossible to tackle successfully. This is because fire-fighting water applied from outside the building may be prevented from reaching the fire, as the roof is designed to keep water out and internal access is usually restricted.

Even if access is gained, the conditions inside the roof space are likely to be extremely hazardous because of the lack of ventilation. In many of the large fires witnessed in historic buildings, the lower floors became involved only after the roofs collapsed.

Facing page: Fire threatens all buildings and its effects can be disastrous in both human and economic terms. In the case of a heritage asset, part of a finite and irreplaceable cultural resource is also lost.

CAUSES OF FIRE

For a fire to start there has to be a source of ignition and combustible materials to provide fuel, and oxygen. Some of the more common causes of fire are described below.

Arson

Historic buildings are often the target of deliberate fire-raising by arsonists. Measures to minimise the risk of arson include physical security, the installation of security systems, and management policies and procedures, to ensure that appropriate levels of vigilance and security are maintained. A plan should be devised based on a risk assessment of the building in question. Advice can be obtained from the police, insurers and the Loss Prevention Council.

Electrical Installations

Electrical faults are a common cause of fire in all types of building, and are more likely to occur in older installations not protected by residual current circuit-breakers, or where sockets are overloaded. Electrical services and equipment should be inspected and tested regularly, and the results recorded in the maintenance logbook.

Level of fire risk

Top left: A large room in a public building with high ceilings, having a high level of supervision, few sources of ignition and containing only a small amount of combustible materials represents a low fire risk.

Top right: Poor housekeeping: overloaded electrical sockets are a potential cause of fire.

Bottom left: A poorly managed store with large amounts of combustible materials, ignition sources (portable standard lamps) and restricted access is a high fire risk.

Bottom right: A basement archive, having a low ceiling and containing a large quantity of combustible material, may also represent a high fire risk.

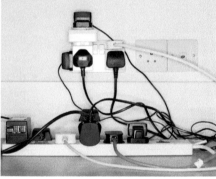

Fires can be caused by defective or poorly-maintained flues. Faults in flues may permit heat to be transferred into roof spaces or floor voids, igniting timbers. Thatch roofs are particularly at risk when sparks or burning brands from wood-burning stoves are emitted from the flue, or heat is transferred through a defective or inadequate flue. Where open fires or stoves are used, it is essential to ensure that flues and linings are fit for purpose and in good condition, and are inspected and cleaned at least once a year, or more frequently if the circumstances warrant it. Inspections and cleaning should be recorded in the maintenance logbook (see **Managing Maintenance & Repair**).

OTHER FIRE HAZARDS

New situations bring new risks. For example, building works, catering for an event or displaying a new collection can all create hazards, and affect risk levels. Therefore, a new risk assessment should be undertaken to address each situation.

Catering

Where caterers are used for events, the contract should be clear about what operations can and cannot be carried out, what areas can be used, and who has the authority (preferably site staff) to make sure that the conditions of contract are being adhered to. Menus should be agreed in advance so that additional hazards such as deep-fat frying or the use of blowtorches can be anticipated. Additional safeguards can then be put into place or, if necessary, the menu changed to eliminate the hazard. By making these conditions clear, problems arising during an event can be avoided.

Lightning Protection

The purpose of lightning protection systems is to reduce injuries to people, physical damage to buildings and contents, and the failure of electrical installations caused by lightning strikes. *BS EN 62305:2011 Protection against lightning* provides detailed guidance (in four parts) on threats from lightning, risk assessment procedures and protection systems. Risk assessments, advice on new installations, and periodic inspection and testing of existing installations, should be obtained from a suitably-qualified and experienced chartered electrical engineer, or specialist lightning protection contractor, such as a member of the Association of Technical Lightning & Access Specialists [ATLAS].

When considering the installation of a new lightning protection system for a heritage building, it is important to take into account the impact that the system will have on the appearance of the building, also the archaeological implications of ground-disturbing work. Historic England has published guidance on lightning protection for churches and on the installation of electrical surge protection equipment (see **Further Reading**).

Building Works

Threats posed by building works can include:

- loss of fire separation caused by the removal of doors, or repair of partitions or ceilings
- temporary isolation of fire detectors to avoid false alarms caused by dust
- increased fire loading caused by the temporary storage of building materials and packaging
- additional sources of ignition caused by temporary lighting, plumbing works, sparks from cutting gear, burning paint and lead welding; these ignition sources should be controlled by a system of Hot Work Permits or, better still, banning hot work altogether
- poor water supplies, because hydrants have been covered or have not yet been fitted
- poor fire service access, because of temporary hoarding or site huts.

'Hot Works'

Hot work is any process that generates heat or sparks (such as welding, or using angle grinders, cutting discs, blowtorches and hot air paint strippers). Such processes are an enormous fire risk and should only be allowed in historic buildings where there is no safer alternative (such as hand sawing or removing an item to a workshop for treatment).

If hot works are contemplated there should be a full risk assessment, and they should be controlled by means of a Hot Work Permit (indeed, many insurance companies insist on these as a condition of insurance). This should be issued by an 'authorised person' (someone who has sufficient technical knowledge of the hazards of hot works to undertake a risk assessment, such as a fire officer or architect). The permit is issued for a set period of time, and specifies the particular job to be carried out and the exact location of the work, as well as listing any special conditions. It includes checklists to ensure that all potential risks have been identified and mitigated: these will relate to site preparation, safe working practices, site-checking and clear-up.

Hot work should be undertaken only by a competent person, trained and experienced in the type of work proposed. All equipment to be used should be serviced before work begins, so that everything is in good working order. Thorough site preparation is vital. Fire detection and protection systems, such as alarms and sprinklers, must be checked and working properly, and fire-fighting equipment, such as a hose reel or fire extinguishers, should be readily available. If the work might generate hazardous fumes, adequate ventilation must be provided for personnel, and all people on site should be familiar with emergency procedures and fire escape routes. There must be no combustible material left within 10 m of the work site; if it is not possible to remove a building element at risk, it should be protected with non-combustible drapes or boards.

While hot work is in progress, a trained person, not directly involved with the work, should provide a continuous fire watch during and for at least one hour following each period of work. For this reason, hot work should not be scheduled less than two hours before the site is due to close for the evening. When work is complete, any waste, such as paint and ends of welding rods, should be collected and removed from the site, and equipment should be stored in a secure area.

Gaining an understanding of how a fire in a particular building might develop and spread is an important part of emergency planning. One way of determining the likely effect of fire on a building and its contents is to use the ***Building Fire Performance Evaluation Methodology*** ('The Method'). This begins by looking for the room or rooms that appear to be the highest fire risk, and the most likely place for a fire to start.

Fire Development in the Room of Origin

It is then assumed that if a fire occurs, it will involve the whole room; the point at which this happens is called 'flashover'. Whether flashover happens, or the fire just burns out, depends on a number of factors: the fire load (the amount of combustible material); how close the combustible materials are to each other; the available ventilation; the volume of the room; and the ceiling height. These factors, and the flammability of the contents of the room, will also give an indication of how quickly the fire will reach full room involvement.

Subsequent Fire Spread

A judgement is then made as to the probability that the fire will extend to an adjacent space, and then perhaps to the whole floor and then the whole building. This can be judged by considering the fire resistance of the partitions, ceilings and doors which separate the spaces within the building. The time taken for fire to spread from the room of origin to adjacent areas will depend on the fire resistance of these elements.

Away from the seat of the fire, heat and smoke damage typically affects the upper part of a room.

Time Taken Until the Fire is Fought

The time taken before an intervention, such as first-aid firefighting or tackling by the fire service, should then be assessed by looking at the following factors:

- How quickly will the fire be detected?
- Is automatic detection provided, and is it heat or smoke activated?
- How long will it take before first-aid firefighting commences?
- Is there a 24-hour presence so that first-aid firefighting can be instigated?
- If there is not a 24-hour presence, is the fire alarm monitored so the fire service can be called automatically?
- How long will it take for the fire service to attend? This will vary according to the time of day, whether the fire service are full-time or retained, the distance from the fire station, the traffic conditions and the ease of access.
- When the fire service arrives, how much time is taken before water can be applied to the fire? This will depend on how close the fire engines can get to the building, the available water supplies, the distance that hoses will need to be laid between the water supplies and the fire engine, and between the fire engine and the fire.

If the length of time between a fire starting and it spreading beyond the room of origin is less than the time taken for the fire service to respond and start fighting the fire, then some remedial action is necessary.

The availability of access for the fire service will affect the time taken before the fire can be fought.

There is a range of actions that might be taken to improve building fire performance. These include passive measures to increase the fire resistance of building elements including doors, partitions and floors to limit the spread of fire, and active measures such as fire detection and suppression systems. Standards of management and housekeeping also play an important part in achieving a satisfactory level of fire safety. It might not be possible to undertake all potential improvements because of their impact on heritage values and significance, cost or desirability. The following questions may help decide which course of action to take:

- What factors have been identified as the biggest threats to the building and contents?

- Can these threats be reduced to an acceptable level that does not involve any upgrading, such as reducing the fire load, or changing the use of the building or parts of the building?

- If improvements are necessary, what impact will they have on the significance of the building? Will they be reversible?

- Will the improvements be effective? For instance, a fire alarm and detection system that is not remotely monitored will not provide any protection during periods when a building is unoccupied.

- The provision of an automatic detection system may reduce the time taken to discover a fire. But will discovery be early enough to ensure that effective action can be taken before the fire spreads to adjoining spaces? If not, another layer of improvements, such as a sprinkler system or local water mist system, may be necessary.

- Will the improvements be affordable? If not, is there a more cost-effective alternative?

The importance of maintaining good standards of housekeeping, and implementing effective management policies to minimise fire risks, cannot be over-emphasised. Ensuring that all fire safety systems and equipment are properly maintained, and that adequate training is provided for building occupants, are important aspects of fire risk management (see **Further Reading**).

FLOODING

Heritage assets can be subject to flooding from natural causes, such as tidal surges, bursting river banks, runoff from hillsides during rainstorms or melting snow, or, indeed, faulty water services within or around a property.

ESTABLISHING NATURAL FLOOD RISK

When historic buildings or places are subject to the risk of flood, knowledge of the degree of risk is essential in planning appropriate mitigation or responses. The risk assessment should consider both local topography and the history of flooding in the area. This may require consultation of a range of sources, such as the Environment Agency flood map (which only includes risks from river and coastal floods). Risk is classified into three levels: significant, moderate and low. Historic England and other organisations provide a range of guidance (see **Further Reading**).

PREPARATION FOR NATURAL FLOODING

Forward planning and implementation of sensible preventive measures can avoid or minimise flood damage. These should encompass three stages: preparation (assessing, understanding and managing risk); reaction (coping with a flood in progress); and recovery (limiting damage after flooding).

FLOOD PROTECTION SURVEY

In addition to the flood risk assessment, a flood protection survey is needed to determine areas of significance and vulnerability, and the nature of protection required. This would normally be executed by an architect or surveyor with appropriate experience. A thorough photographic record of the building (both interior and exterior) should also be undertaken by building occupants, as this can be a valuable resource in the event of damage for insurance compensation.

PROTECTION AGAINST FLOODING

Good maintenance is the vanguard against flooding, since blocked drains or watercourses impede the evacuation of flood waters (see **Managing Maintenance & Repair**). These may be on adjacent public property and require local authority intervention.

Facing page: Flood protection defences in place beside a swollen tidal river.

Flood protection barriers around a property boundary.

There are two forms of flood protection works:

- *Flood-resistance or flood-proofing works*
  Works that reduce the amount of water entering a property.

- *Flood-resilient works*
  Works that reduce damage from water which enters a property.

In historic buildings these need to be applied with due sensitivity to significant fabric, and may require statutory consent.

Flood-resistance or flood-proofing measures can consist of temporary flood barriers around the property (bagged barriers) or fitted to the building itself (window or door barriers, covers for air bricks). Permanent site barriers can be bunding, walls or additional drainage ditches.

Flood-resilient measures are appropriate for properties subject to repeated flooding. These are modifications to building services, fixtures and fittings to limit damage, such as moving utility infrastructure above probable flood level; installation of a pump in vulnerable cellars; and installation of backflow valves in plumbing to prevent water ingress from drains and sewers. Removal of valuable items to higher levels of the building is also a sound precaution.

Flooding from Building Services

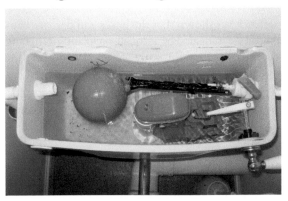

A survey of building services (see **Survey & Investigation Methods**) should record the route of water pipes, position of tanks, cisterns, ballcocks, valves, hose reels and overflow pipes, as well as the position of stopcocks. These identify locations of potential flooding. Maintaining infrastructure so that it is fit-for-purpose is obviously the basis for preventing damage.

Other preventive measures include moving valuable objects away from these places, if possible. An alternative approach might be needed where, for example, a valuable feature (such as a wall painting) is located below a bathroom. In this case, it might be prudent to decommission the bathroom. When work is being carried out on water or heating systems there is a higher risk of leaks occurring, so it would be wise to carry out checks at the end of the working day, before the premises are vacated.

Runoff Water from Firefighting

Fire services try to cause as little water damage as possible when fighting fires, but at serious high-level incidents, the water from their hoses will pour into the rooms below. The weight of water may bring about further structural damage, and will certainly cause debris and contamination to cascade onto lower floors.

Top: Plumbing leaks are a very common cause of water damage within buildings.

Bottom: A large fire may typically require the use of six jets, each discharging over 500 litres of water per minute. A hydraulic platform can discharge 1800 litres a minute. This makes a total of 4.8 tonnes per minute.

PLANNING FOR SALVAGE

PREPARING A SALVAGE PLAN

Experience has shown that the losses and damage caused by emergencies to heritage assets can be mitigated by adequate pre-planning. In buildings where there are collections of works of art, books or furniture of high significance, a written emergency plan should be prepared that identifies:

- the personnel responsible for salvage operations, including the Salvage Officer and their deputy (contact list for management teams and service providers)
- site and building plans (with keys)
- salvage priorities
- salvage procedures for the removal of items
- specialist facilities for the temporary storage and emergency first-aid conservation of objects
- arrangements for the longer-term storage or treatment of salvaged material.

The plan should also prompt the procurement of response equipment, and identify a schedule for its maintenance. The fire and rescue service should be made aware of the existence and content of the plan.

Once a plan is written, it must remain fit for purpose, so it needs to be tested and maintained. Regular training exercises aid this process and ensure that the people using the plan are familiar with it. The steps for testing a plan are as follows:

- undertake a full-scale exercise
- update the plan
- full-scale exercise
- update the plan again.

The contact lists for the management teams, members of salvage teams and equipment suppliers are often difficult to keep up-to-date. Review contact lists regularly and ensure the escalation ('telephone tree') process works regardless of the time of day.

When a disaster occurs, contacting the staff may take a long time, and occupy the person who is first on the scene and has many other tasks to perform. An alternative solution is to contract-out the task to a third party, such as a call-receiving centre. The third party would be required to periodically check the lists and make test calls.

All incidents, even if they are relatively minor, should be reported to management, so that a record can be made of their nature, size and potential impact. These reports can then be used to take action before the incident is repeated, which could be more serious the second time around.

Training should include how to undertake a risk assessment and working safely in adverse conditions. The most likely risks in order of magnitude are: manual handling, trips and slips, falling objects (tiles, overhead working), burns and scalds, electric shock, smoke inhalation, structural collapse, and drowning.

Practical aspects of training include reading plans, identifying objects on the salvage list, removing paintings from their fixings, handling objects and carrying out first-aid treatment of damaged items. All these should be practised in simulated conditions. These practices should periodically include joint exercises with the fire brigade. A log book should be provided so that a record of training can be kept.

Training of salvage teams

Top left: Participants on an emergency training course at a historic property discuss access issues with the fire service.

Top right: Check in/out procedure in operation to ensure everyone is accounted for in the event of an evacuation.

Bottom left: A 'human chain' is an effective method for moving a large number of objects quickly.

Bottom right: Personal protective clothing and equipment for the salvage team, stored ready for use in an accessible location.

Training should address how to move large numbers of objects out of a building. One method is to form a human chain. This involves a line of people who pass objects from one to the other. To be effective, the distance between each person in the chain should be such that the objects can be passed down the line without anyone needing to move. The disadvantage of a chain is the number of times the objects are handled, but this should be weighed up against the speed of operations. To allow access, the salvage teams should have personal protective equipment, which includes identification, hard hats, fluorescent jackets, safety boots or shoes, and torches.

All participants should be aware of the evacuation procedure. Training should also include a check in-and-out exercise that would account for everyone during a potential evacuation.

Salvage training exercise for staff of historic houses and museums

Top left: Organic and inorganic objects, representative of the contents of a historic house, are immersed in water and ash to mimic post-fire or flood conditions. Objects are initially moved to a triage area, where they are recorded, examined and sorted according to treatment requirements.

Right: Wet objects are transferred to the 'wet treatment team', as shown here. As items are removed from crates, the conservator decides on the form of treatment. Some may be air dried, on blotting paper (*bottom right*). Air-drying is labour intensive and requires a great deal of space. It should be done in a controlled way to dry the objects slowly. Although freezing followed by freeze-drying is quicker at the time of the emergency, the process of drying will take longer overall. This method may be appropriate if objects are very wet, structurally unstable or contain water-soluble dyes.

Bottom left: Sound and dry objects being packed by the 'dry team', before removal to a storage facility.

Salvage lists ensure that objects are rescued in the correct order, with, if possible, those of the highest significance rescued first. The lists should consist of photographs of the items to be rescued, their position in the room and building, and any special measures needed to remove them. This may be the manual handling requirements, removal techniques or equipment required. The financial value of the exhibits should not be included for security reasons. To help salvage teams identify objects quickly, a simple description is sometimes more useful than the proper title; for example, 'Lady's head in marble'. If a room is completely filled with items of similar value, it is still worth sorting them into an order of removal. This could perhaps be done by order of rarity, historic significance, ownership or simply ease of removal, rather than simply giving them all a 'Priority 1' rating. The example below shows the front and reverse of a typical two-sided salvage priority sheet, which is often laminated to make it more hard-wearing.

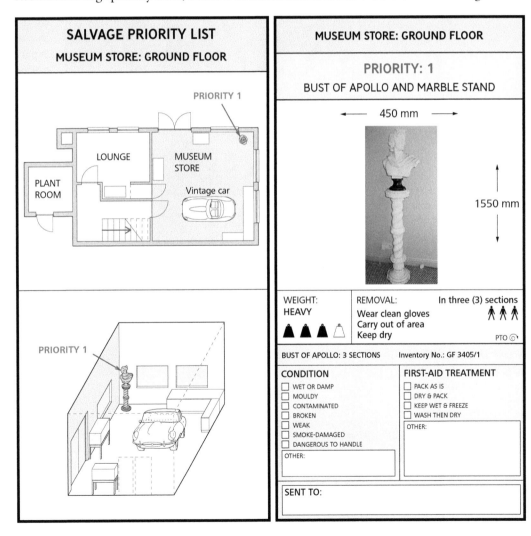

ACTIONS DURING AN EMERGENCY

During a real emergency situation, the officer in charge of the emergency services will decide which areas are safe to enter, and then give permission to begin salvage. It may be possible to enter:

- when a fire has been extinguished
- when a fire is remote from collection rooms
- when a fire is at a high level (entry into rooms below may still be possible)
- when a fire is in an adjacent room, and the 'fire is surrounded'
- after the electrical and gas services have been shut off to a flooded property
- after a flood when there is no risk of contamination from sewage or stored chemicals.

In some cases, inspection by a structural engineer expert in a post-emergency assessment will be necessary before salvage of contents can be undertaken.

In a fire, the scope and sequence of salvage operations will be determined by the damage inflicted by the fire, the water to extinguish it, and ultimately by the harm incurred to the stability of the structure itself.

The procedures for salvage will vary according to the scale of the incident, but it makes sense to plan for the worst case scenario and try to remove all the objects.

Top: An example of building debris collapsed onto a lower floor, partly as a result of water damage. This also illustrates the type of hazardous conditions that need to be addressed before safe access for salvage can be permitted, and the challenging context in which the salvage of objects must take place.

Bottom: In a fire it may be possible to cover objects to minimise water damage. The best course of action would be to divert as much as possible to the outside, using waterproof sheets and hoppers if available. Removing objects before the water reaches them is an alternative course of action. However, this depends on there being sufficient people and enough time available to do this safely.

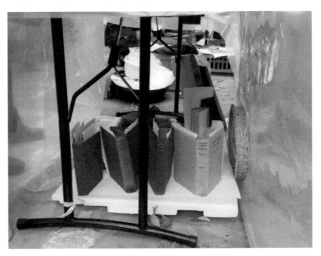

The first few hours after a disaster are critical to the long-term survival of fragile historic artefacts. If the condition of the objects can be stabilised quickly, the risk of long-term damage – for example, from mould or decay – can be minimised. The salvage plan should include the provision of first aid equipment and a suitable place, either permanent or temporary, for treatment.

SECURITY OF OBJECTS

The degree to which security needs to be considered in an emergency depends upon the value of items which will be removed. Providing for the security of valuable objects, while they are being moved and temporarily stored, should be pre-planned. Storage areas which are easily supervised, or take advantage of existing security arrangements (such as an outbuilding on site), should be identified.

In addition, the need to involve as many volunteers as possible to assist with the salvage operation has to be balanced against the need to control access to temporary stores, and parts of the building containing valuables that might be unaffected by the incident.

It is important to consider and plan the techniques for removing items which have been secured to protect them from theft. Methods for securing objects should be designed to facilitate removal in an emergency. Instruction on removal techniques should be provided in training exercises and salvage snatch sheets, and if tools are required they should be included in the salvage equipment. Of equal importance, the Salvage Priority List needs to be immediately accessible in an emergency. However, because of its sensitive content it also needs to be held securely. Therefore, its location needs careful consideration to meet both these requirements.

Water-damaged books, otherwise in robust condition, being dried in a provisional wind tunnel on site. This is formed of polythene sheeting over a table, with one or two fans at the end to provide a constant air flow. Blotting paper within the books also absorbs moisture.

Once the immediate salvage operation is complete, an inventory of items recovered should be made. If objects are removed from the site to alternative storage facilities, the inventory will help ensure all items are accounted for throughout their transit and eventual return. Where building security measures have been damaged as a consequence of the incident, the longer-term security of the premises needs to be considered. A manned guarding presence, temporary alarms and hoardings might be required.

DEALING WITH LONG-TERM CONSEQUENCES

Emergencies such as fires and floods present particular challenges that will need to be addressed by conservation professionals. The building may well have developed structural problems. Even areas not directly affected may be damaged by smoke, chemicals and other contaminants, and water.

In the first instance, specialist cleaning and decontamination is likely to be required. Trapped moisture can result in mould growth and decay, and measures to promote progressive drying of the fabric will be required to reduce the risk of further deterioration. Timber is particularly susceptible to damage caused by moisture-induced dimensional changes and organisms. On the other hand, stripping out wet timber or lime plaster, or incorrect storage, may cause more damage than the flood itself. These issues are discussed in detail in the other volumes of this series. ⊖MORTARS ⊖METALS ⊖E B & T ⊖ROOFING ⊖GLASS ⊖STONE ⊖TIMBER ⊖CONCRETE ⊖ENVIRONMENT

Moisture problems are a typical consequence of both fire and flood emergencies. Here a fan is being used to increase the rate of drying but careful monitoring is required to ensure that materials are not damaged by trying to dry too quickly.

Left: After the fire in the Long Gallery at Hampton Court Palace, the timbers were removed and stored in special drying racks. Drying timber will need to be checked regularly, and moved at intervals to prevent distortion.

Right: Drying after an emergency requires close monitoring. It may be possible to limit the potential damage of forced drying by using plastic sheeting to protect more fragile areas or materials.

Drying after a flood or fire emergency can be a particular challenge: trying to dry too rapidly can cause severe damage to materials, and may even slow the long-term removal of moisture from the building fabric. On the other hand, scaffolding and other post-emergency sealing of the building will constrain natural ventilation, often making some form of forced drying desirable.

It is important to be aware that some of the post-emergency responses promoted by loss adjusters and insurance contractors may not only be highly damaging to historic fabric, but may also breach listed-building legislation. If in any doubt, the local planning authority's conservation officer should be consulted before works are begun.

Further Reading

Forsyth, M. (ed.) (2007); *Structures and Construction in Historic Building Conservation (Historic Building Conservation series)*; Oxford: Blackwell

Historic England (2015); *Flooding and Historic Buildings: Revised 2nd Edition*; Historic England; available from *www.historicengland.org.uk/images-books/publications/flooding-and-historic-buildings-2ednrev/*

Historic England (2017); *Metal Theft from Historic Buildings: Prevention, Response and Recovery*; Swindon: Historic England; available from *www.historicengland.org.uk/images-books/publications/metal-theft-from-historic-buildings/*

Historic England (2019); *Lightning Protection: Design and Installation for Historic Buildings*; Swindon: Historic England; available from *www.historicengland.org.uk/images-books/publications/lightning-protection/*

Ingham, N. (2004); *Practical Conservation Guidelines for Successful Hospitality Events in Historic Houses*; available from *www.english-heritage.org.uk/learn/conservation/collections-advice-and-guidance/*

Kidd, S. (ed.) (2006); *Heritage Under Fire: A Guide to the Protection of Historic Buildings* (2nd edition); London: Fire Protection Association

Kidd, S. (2010); *Guide for Practitioners 7: Fire Safety Management in Traditional Buildings: Part 1: Principles and Practice*; Edinburgh: Historic Scotland

Kidd, S. (2010); *Guide for Practitioners 7: Fire Safety Management in Traditional Buildings: Part 2: Technical Applications and Management Solutions*; Edinburgh: Historic Scotland

Matthews, G., Smith, Y., Knowles, G. (2009); *Disaster Management in Archives, Libraries and Museums*; Burlington: Ashgate

Maxwell, I. (ed.) (2008); *COST Action C17: Fire Loss to Historic Buildings Built Heritage – Final Report*; Edinburgh: Historic Scotland

O'Farrell, G. (2007); *External Lighting for Historic Buildings*; London: English Heritage; available from *www.historicengland.org.uk/images-books/publications/external-lighting-for-historic-buildings/*

Roberts, B. (2009); *Historic Building Engineering Systems & Equipment: Heating & Ventilation*; London: English Heritage; available from *www.historicengland.org.uk/images-books/publications/heating-ventilation*

Rowell, C., Martin, J. (1996); *Uppark Restored*; London: National Trust

Stovel, H. (1998); *Risk Preparedness: A Management Manual for World Cultural Heritage*; Rome: ICCROM; available from *www.iccrom.org/sites/default/files/ICCROM_17_RiskPreparedness_en.pdf*

Xavier-Rowe, A. *et al.* (2010); *Historic House Collections: Drawing up a Collections Management Plan: Guidance Notes*; Swindon: English Heritage; available from *www.english-heritage.org.uk/siteassets/home/learn/conservation/collections-advice--guidance/drawing-up-collections-management-plan.pdf*

APPENDICES

APPENDICES

1. Typical Maintenance Checklist

2. Building Management Records

3. Sample Brief for a Condition Survey

4. Sample Contact List & Emergency Strategy Document

TYPICAL MAINTENANCE CHECKLIST: OCCASIONAL, REGULAR & CYCLICAL MAINTENANCE TASKS

Checklists provide a systematic and consistent basis for gathering information and ensuring that essential maintenance tasks are not forgotten. They also allow the condition of building elements to be monitored over time, and faults to be quickly identified.

MAINTENANCE CHECKLIST		
ELEMENT	MAINTENANCE TASK	COMMENTS
AFTER STORM WEATHER		
ROOF AREAS & RAINWATER GOODS GENERALLY	Inspect roof coverings and flashings from ground level, and accessible high points; engage a contractor to remedy any loss or damage	
CHIMNEYS	Inspect from the ground and accessible high points; engage a contractor to repair any damage	Where structural problems are suspected, obtain professional advice from a conservation structural engineer on monitoring, investigation and repair
EXTERNAL WALLS GENERALLY	Inspect from ground level, and arrange for any damage and signs of movement to be investigated and made good	
SURFACE WATER DRAINAGE CHANNELS & GULLIES	Inspect and note if ineffective or blocked; clear channels of vegetation and debris; arrange for blocked gullies and drains to be cleared	
INTERNAL SPACES GENERALLY	Inspect roof void and internal spaces, particularly below gutters; arrange for evidence of water penetration to be investigated and remedied	
DAILY		
FIRE PRECAUTIONS	Check that: • fire escape routes are unobstructed and that fire doors operate correctly • combustible materials and rubbish are stored in a safe area • the building is secure from unauthorised entry • no faults are indicated on fire alarm system	
WEEKLY		
FIRE ALARM & AUTOMATIC FIRE DETECTION SYSTEMS	Make operational test; engage a qualified specialist contractor to remedy any defects found	Batteries should be replaced in single-station smoke alarms, if required
FIRE EXTINGUISHER SYSTEMS	Make visual inspection; engage a qualified specialist contractor to remedy any defects found	
MONTHLY		
SURFACE WATER DRAINAGE CHANNELS & GULLIES	Clear channels of vegetation and debris	From May to November
EMERGENCY LIGHTING	Make operational test; engage a qualified specialist contractor to remedy any defects found	

MAINTENANCE CHECKLIST

ELEMENT	MAINTENANCE TASK	COMMENTS
TWICE YEARLY		
CHIMNEYS & FLUES	Engage a specialist to sweep chimneys where fireplaces or stoves are in use Check condition of parging or flue liner Clean tar and debris from spark arrestors, where fitted	Depending on the frequency of use and type of fuel burned, chimneys may need sweeping less or more frequently Spark arresters can become clogged with tar and debris, and this might increase the fire risk for thatched roofs
ROOFS RAINWATER GOODS: Valleys Gutters Hoppers Rainwater Pipes	Engage a contractor to clear roofs and rainwater goods of leaves and debris, and ensure overflows are clear; rod if necessary Check: • that guards and wire balloons are in place • falls and ensure that water is not pooling • support brackets Adjust gutter brackets to ensure appropriate fall Inspect rainwater goods for cracks and leaks Repair or replace cracked or damaged sections with matching items; if this cannot be done immediately, consider a temporary repair using plastic sections and plan for permanent repairs to be carried out within an appropriate timescale Check that water is discharging correctly into drains and gullies	Regular clearance of gutter and hoppers will reduce the risk of downpipe blockages Blockages in downpipes will result in overflow higher up the pipe When replacing downpipes the joints should not be sealed When re-fixing rainwater pipes, consider using spacers to set the pipes away from the wall; this will enable the back to be painted and reduce the rate of corrosion
DRAINS: Surface Water Drainage Gullies Below-Ground Drains	Engage a contractor to open up inspection chambers Check that all gullies and gratings are free from silt and debris, and that water discharges freely to an appropriate drain or soakaway Use drain rods or high-pressure water jetting to clear blockages Consider cutting back or removing trees and shrubs growing close to the line of a drain	Inspect gullies during heavy rainfall to see whether they back up: this might indicate blockages Most instances of 'rising damp' in walls are attributable to defective drainage
VENTILATION GRILLES & AIR BRICKS	Ensure that ventilation grilles, air bricks, louvres, and so on, are free from obstruction	
DOORS & WINDOWS	Check operation of opening parts Ensure that locks and security devices are operating properly Lubricate hinges, bolts and locks as required	
FIRE ALARM & AUTOMATIC FIRE DETECTION SYSTEMS	Engage a qualified alarm engineer to inspect and test fire alarm and detection systems, and to maintain them	

MAINTENANCE CHECKLIST

ELEMENT	MAINTENANCE TASK	COMMENTS
YEARLY: EXTERNAL FABRIC		
CHIMNEYS	Engage a contractor to inspect chimney pots, flaunchings and chimney stacks for defects Carry out localised repointing and repairs; replace broken pots Consider installing ventilated caps in pots on disused chimneys to prevent rain penetration	If the chimney stack is cracked or leaning significantly, obtain professional advice from a conservation architect or a building surveyor
OVERLAPPING COVERINGS TO PITCHED ROOFS VERTICAL CLADDING	Engage a contractor to inspect for missing, displaced, defective slates and tiles or shingles; insert matching replacements Note heavy accumulations of moss	Roofing maintenance tasks should ideally be carried out concurrently with late spring roof clearance (*described above*) Moss holds moisture; for clay and stone tiles it may increase the risk of frost damage; obtain professional advice from a conservation architect or building surveyor about removal
RIDGE & HIP TILES	Engage a contractor to inspect bedding and jointing of ridge and hip tiles, and make good defects Re-bed and repoint as required	
THATCHED ROOFS	Engage a thatcher to inspect for defects such as: holes, grooves of decay, thinness, heavy accumulations of moss, bird damage etc. Carry out patch repairs as necessary using appropriate materials	Consider bird netting if damage is a persistent problem Obtain professional advice from a conservation architect, building surveyor or thatcher concerning moss removal Assess the adequacy of the fire-prevention measures
SHEET METAL ROOFS & CLADDING	Engage a contractor to inspect condition of panels, joints and clips Make temporary repairs to cracks and splits where permanent repairs cannot be carried out immediately Plan for permanent repairs within an appropriate timescale	Theft of lead has become an increasing problem in recent years: lead roofs should be checked regularly, particularly on unoccupied buildings Consider installing security devices
ASPHALT ROOFS	Engage a contractor to inspect condition of flat areas and upstands; look for evidence of ponding and blistering Where permanent repairs are not to be carried out immediately, make temporary repairs to splits and holes Plan for permanent repairs within an appropriate timescale	Bumps and blisters require immediate attention only where there is evidence that water penetration is occurring
LEAD WEATHERINGS & FLASHINGS	Inspect condition of lead flashings and weatherings: make minor repairs, for example, refixing and redressing dislodged or displaced sections, and replace missing or damaged sections Make temporary repairs using adhesive flashing material where permanent repairs are not to be carried out immediately; plan for permanent repairs to be carried out within an appropriate timescale Make good any defective mortar fillets	Splitting of lead flashings may indicate they are oversized or over-fixed; obtain professional advice before renewing to the same specification
VALLEY & PARAPET GUTTERS	Engage a contractor to inspect condition Make temporary repairs using adhesive flashing where permanent repairs are not to be carried out immediately; plan for permanent repairs within an appropriate timescale	Consider installing permanent safe access aids to allow regular inspection of hidden valley and parapet gutters Lead valleys on thatched roofs are particularly liable to become choked with debris Splitting of lead gutter bays may indicate that they are oversized; obtain professional advice from a conservation architect, building surveyor or accredited leadwork contractor, before renewing to same specification

MAINTENANCE CHECKLIST

ELEMENT	MAINTENANCE TASK	COMMENTS
YEARLY: EXTERNAL FABRIC		
COPINGS	Engage a contractor to inspect condition of copings, including the mortar bedding and joints Remove any vegetation beneath and between copings Repoint and re-bed as required using appropriate lime-based mortar	Staining, algal growth and vegetation growth in the area of wall below a coping joint is a sure sign that a joint is leaking
BRICK OR STONE WALLS	From ground level, inspect for: • signs of structural movement, cracking and bulging • areas of surface deterioration including open joints, and cracked or deeply eroded pointing • evidence of failed or ineffective water-shedding features (for example, drips and string courses) • areas of dampness Plan for defects to be investigated more closely and remedied within an appropriate timescale	Where structural problems are suspected, obtain professional advice from a conservation structural engineer for advice on monitoring, investigation and repair Comparing signs of damage with photographs of previous inspections will help to determine rates of decay and highlight new structural problems Note colonisation by algae, lichen and higher plants (*see below*); also note any accumulations of bird guano Consider installation of bird deterrents if guano is a persistent problem Obtain professional advice from a conservation architect or building surveyor about investigation and remedial treatments for moisture problems Obtain professional advice from a conservation architect or building surveyor about surface cleaning and repairs
TIMBER-FRAMED WALLS	From ground level, inspect for: • signs of structural movement or timber decay • open joints between timbers and infill panels, and other defects that might allow water penetration • evidence of failed or ineffective water-shedding features or flashings • infill panel defects • areas of dampness Plan for defects to be investigated more closely and remedied within an appropriate timescale	
EARTHEN WALLS	From ground level, inspect for: • signs of structural movement and cracking • holes and rat runs • areas of dampness Plan for defects to be investigated more closely and remedied within an appropriate timescale	
CONCRETE WALLS	From ground level, inspect for: • cracking and spalling • rust stains • water seepage Plan for defects to be investigated more closely and remedied within an appropriate timescale	
SURFACE FINISHES	From ground level, inspect for: • bulging, detachment, cracking, blistering or spalling of render • blistering or flaking paint finishes Plan for defects to be investigated more closely and remedied within an appropriate timescale	Defects in renders and paint finishes may indicate high levels of moisture, and soluble salts Obtain professional advice from a conservation architect or building surveyor about appropriate remedial work

MAINTENANCE CHECKLIST

ELEMENT	MAINTENANCE TASK	COMMENTS
YEARLY: EXTERNAL FABRIC		
EXTERNAL JOINERY: Doors & Doorcases Windows Shutters Blind Boxes Canopies & Porches Shopfronts Weatherboarding Cornices Dormer Windows Cupolas	Inspect for: • decay and insect attack • open or loose joints, cracking or splitting • cracked, loose or missing putty • defective fixings, fastenings and stays • broken sash cords • loose, blistering and flaking paint • defective flashings • blocked weep holes or condensation drainage channels Engage a contractor to ease and adjust opening components and carry out essential minor repairs Plan for defects to be investigated more closely and remedied within an appropriate timescale	Maintaining the integrity of paint finishes is essential to prevent deterioration of external joinery Consider measures to improve the thermal performance of windows and external doors
TRADITIONAL GLAZING	Inspect for cracked or missing panes In leaded-light windows, inspect lead cames, glass, wire ties and saddle bars; engage a specialist contractor to carry out essential minor repairs Plan for defects to be investigated more closely and remedied within an appropriate timescale	Avoid loss of historic glass: cracked panes should be retained, and may be repaired if necessary
MODERN GLAZING	Inspect for failure of gaskets and sealants Check integrity of fixings, flashings and protective coatings; ensure drainage routes are clear Engage a specialist contractor to carry out minor remedial works, and lubricate moving parts and operating equipment	Inspection, maintenance and repair of large areas of modern glazing should normally be carried out on an annual rolling cycle
EXTERNAL IRONWORK: Windows Balustrades Porches Gates & Railings	Inspect for: • corrosion • cracking or buckling • damaged or failed fixings • cracked, loose or flaking paint Engage a specialist contractor to adjust and lubricate moving parts and carry out essential minor repairs Plan for defects to be investigated more closely and remedied within an appropriate timescale	Maintaining the integrity of paint finishes is essential to prevent deterioration of most external ironwork
VEGETATION	Ensure that climbing plants are not growing into rainwater goods, under copings or into roof spaces; remove unwanted growth and trim climbing plants as required Check nearby trees for dead branches and other signs of ill health; engage a tree surgeon to remove dangerous limbs of nearby trees Check for evidence for root damage to the building or below ground drains	Avoid cutting back climbing plants, shrubs and trees in late spring or early summer when birds are nesting Algal growth may be a symptom of high levels of moisture: there is little point in removing algae until the source of the water has been addressed
BIRD DETERRENTS	Check that window ledges, belfry openings, roofs and other sensitive areas are bird-proofed before the nesting season If necessary, engage a specialist contractor to clean and repair	When guano is rinsed off, ensure that wall below is protected (runoff contains high levels of nutrients and can encourage biological growth in previously clean areas) Do not disturb bats

MAINTENANCE CHECKLIST

ELEMENT	MAINTENANCE TASK	COMMENTS
YEARLY: INTERNAL FABRIC		
INTERNAL STRUCTURE & FABRIC GENERALLY	Inspect for: • signs of structural movement such as cracks and bulges • area of dampness • mould and fungal growth • condition of decorative finishes (in particular. flaking or peeling white lead-based paint) • signs of infestation by vermin Plan for defects to be more closely investigated and remedied within an appropriate timescale	Where structural problems are suspected, obtain professional advice from a conservation structural engineer concerning monitoring, investigation and repair Obtain professional advice from a conservation architect or building surveyor concerning investigation and treatment of moisture problems and timber decay Consider measures to control vermin
ROOF VOIDS	Inspect for: • evidence of roof or gutter leaks • signs of active decay or insect attack in structural timbers and battens • defective tile nibs, pegs and other fixings • signs of damage by vermin (for example, chewed timber and shredded insulation materials) • evidence of use by birds or bats • displaced insulation materials Engage a contractor to remedy roof leaks Replace missing or displaced insulation	Where active decay or insect infestation is suspected, obtain professional advice from a conservation architect or building surveyor Consider measures to control vermin (avoid using poisons where the roof void is used by bats) Obtain professional advice from an ecologist where works are proposed that may affect bats (such as repairs or measures to control bird access) Consider increasing the thickness of insulation
FLOOR VOIDS	Inspect for: • signs of active decay or insect attack in timbers • signs of damage by vermin Plan for defects to be investigated and remedied within an appropriate timescale	Where active decay or insect infestation is suspected, obtain professional advice from a conservation architect or building surveyor Consider measures to control vermin
WOODWORK: Doors Panelling Staircases, Floorboards	Inspect for: • evidence of active decay or insect infestation • splitting, cracking or warping • defects caused by wear and tear • defects that affect fitness for purpose or safety affecting doors, staircases, floors and so on Check: • condition of decorative and protective finishes such as paint, varnish or polish • that doors open and close smoothly; operation of hinges, latches and bolts • that fire doors are in satisfactory condition, and that hinges, latches and self-closing devices are in good working order Engage a contractor to carry out essential minor repairs	Where active decay or insect infestation is suspected, obtain professional advice from a conservation architect or building surveyor
SOLID FLOORS	Inspect for: • areas of moisture • wear of surface finishes Engage a contractor to carry out essential minor repairs to areas that might present a hazard	Where active decay or insect infestation is suspected, obtain professional advice from a conservation architect or building surveyor Obtain professional advice from a conservation architect or building surveyor for advice on investigation and remedial treatments for moisture problems

MAINTENANCE CHECKLIST

ELEMENT	MAINTENANCE TASK	COMMENTS
YEARLY: BUILDING SERVICES		
LIGHTNING PROTECTION	Engage a specialist contractor to inspect components, fixings, connection and earth rods; service and repair as necessary	
ELECTRICAL SYSTEM	Carry out routine check of fixed electrical installation Check for breakages, wear and tear, signs of overheating, missing parts, obstruction of switches preventing operation, loose fixings etc. Engage a registered electrician to undertake remedial work as required	
HEATING & BOILER SYSTEMS	Engage a registered contractor to inspect, test and maintain heating and boiler systems and controls Where applicable, carry out gas-safety inspection	
PLUMBING SYSTEMS	Check that all sinks, washbasins and sanitary equipment are functioning properly, securely fixed, undamaged, and free from leaks Clean shower heads Inspect cold water storage tank for signs of leaks; check ball valve, and overflow and vent pipes Check inside tank for cleanliness, debris etc, and that tank cover is adequate and in place Engage a registered plumber to carry out maintenance and repairs as required	
FIRE EXTINGUISHER SYSTEMS	Engage a specialist contractor to inspect, test and maintain all fire-extinguisher systems	
INTRUDER ALARM SYSTEM	Engage a qualified alarm engineer to inspect, test and maintain all intruder alarm systems	
EVERY 4–5 YEARS		
ALL ELEMENTS OF BUILDING FABRIC & CURTILAGE	Engage a conservation architect or building surveyor to carry out a 'quadrennial' or 'quinquennial' inspection, and report on condition, maintenance and repair requirements, timescales and planning	*BS 8210:1986* recommends that a full and detailed inspection of all aspects of the building fabric is undertaken by a suitably-qualified professional at least once every five years
EVERY 5 YEARS		
PREVIOUSLY PAINTED OR COATED SURFACES	Carry out pre-decoration repairs Prepare and redecorate previously painted surfaces	Maintaining protective coatings is essential to prevent the deterioration of external softwood joinery, most metalwork, and some masonry or rendered surfaces
ELECTRICAL SYSTEM	Engage a registered electrician to fully test and inspect electrical installations	

Note: All high level inspection and work will require safe access to be provided

APPENDIX 2

BUILDING MANAGEMENT RECORDS

The following example illustrates a typical building management record, and the type of information that should be recorded in it.

BUILDING MANAGEMENT RECORD					
PREMISES:	*The Hall House* *23 Green Lane* *Washford* *Kent CT4 5BG*				
DATE	LOCATION	TASK COMPLETED	BY WHOM	COST	COMMENTS
17/03/11	*Interior and exterior*	*Carried out annual maintenance check-list inspection.* *Noted area of damp staining at low level on party wall in hallway has spread compared with photo taken last year. Called Building Conservation Services to investigate.* *Mouse droppings in understairs cupboard. Called Pestaway.*	*SELF*	*–*	*Check-list report attached.* *Photographs of damp staining attached.* *Building Conservation Services to visit 10:30am 18/06/11* *Pestaway booked for 9:00am 21/03/11.*
21/03/11	*Interior*	*Inspected for vermin; baited mouse traps provided.*	*Pestaway*	*£ 45.50*	*Invoice paid 21/03/12*
25/04/11	*Gutters & RWPs*	*Inspected rainwater goods during heavy rain.* *Gutter above flank (E) wall overflowing. ?Blockage.* *Contacted roofing contractor to investigate/remedy.*	*SELF*	*–*	
03/05/11	*Gutters & RWPs*	*Cleared blockage in gutter above flank wall.* *Replaced wire balloon to RWP.* *Checked other gutters – all clear.* *Replaced 3 missing slates on front roof slope.*	*Acme Roofing Contractors*	*£ 133.75*	*Invoice paid 07/05/11.*
15/06/11	*Heating & plumbing*	*Annual check-up and gas safety inspection.* *Boiler thermocouple faulty – replaced.* *Cistern in downstairs WC rewashered.* *Gas safety inspection report issued.*	*A1 Plumbing and Heating Services*	*£ 195.23*	*Gas safety inspection enclosed.* *Invoice paid 31/06/11.*

In certain instances the need to provide and maintain records about the building, its services installations and operation may arise from statutory requirements.

REQUIRED DOCUMENTATION	
CONTENTS	REQUIRED WHEN
BUILDING LOG BOOK	
Contains details of newly provided, renovated or upgraded thermal elements, controlled fittings, fixed building services, energy meters and other measures that enable the building owner or manager to monitor and control energy performance	Works carried out where compliance with Building Regulations Part L2 is required For further information, see: • Jones, P. G. (2006); *TM 31 Building a Log Book Toolkit*; London: CIBSE • Jones, P. (n.d.); *Good Practice Guide 348; Building Log Books – A User's Guide*; London: Carbon Trust
HEALTH & SAFETY FILE	
Contains information necessary for ensuring that future construction, maintenance, refurbishment or demolition is carried out safely Prepared by the CDM coordinator and retained by the building owner (and passed on to future owners)	Works notifiable under the Construction (Design and Management) Regulations 2015 [CDM Regulations] Further information available from: *www.hse.gov.uk/construction/cdm/faq/plan.htm*
FIRE SAFETY RISK ASSESSMENT RECORD	
Provides a record of fire hazards identified in the risk assessment and the actions that have been taken to reduce or remove them	Buildings subject to control under the Regulatory Reform (Fire Safety) Order 2005 Further information available from: *www.legislation.gov.uk/uksi/2005/1541/contents/made*

APPENDIX 3

SAMPLE BRIEF FOR A CONDITION SURVEY

Client Aim

To understand in detail the condition of structures on the site, and the need for maintenance and repair during the next 10 years.

Survey Objectives

- Identify the need for emergency repairs to prevent risks to health and safety, and to avoid significant loss to the building fabric in the immediate future.
- Identify the need for further repairs to minimise decay and prevent further loss of the building fabric during the next 5–10 years.
- Identify a long-term approach to sustainable conservation of the site over the next 20–50 years.
- Comment on the success of past repair and maintenance work, and make recommendations for any modification or change in approach.
- Provide a budget for each phase of recommended conservation and repair, and identify if there is any financial benefit in bringing forward future phases of work to form a single project.
- Prepare a specification for conservation and repair methods and materials to assist in obtaining statutory consent for future work.

The Site

Provide a brief description of the site and its history, initially as a functioning building and subsequently as a ruin. The statutory listing or protection of the site should also be identified, and a copy of the listing/scheduling document included.

Background Information

Extant archaeological and condition survey reports should be listed and made available, together with drawings and rectified photographs. If a maintenance regime is in place, details should also be provided.

Survey Method

The type of survey and access methods should be set out. For example, the survey is to be based on close visual inspection from access scaffolding, which will be installed by the client to each building elevation and made available for the surveyors' use for a period of six weeks. If a scaffolding design has been prepared, this should be sent to the surveyor for information.

It is difficult to determine the need for invasive or non-invasive investigation work in advance, but the client brief may request that a provisional sum be allowed. The cost value of the provisional sum should be stated in the brief.

The need for any laboratory-based analysis work should be identified; for example, mortar disaggregation or microscopy. Again, the objectives for this work should be identified. If the need for specialist analysis is unclear, a provisional sum could be given in the brief.

Survey Scope

The scope of the survey work should be identified. This can be summarised in one or two brief paragraphs, followed by a list of building and site elements which are included. For example: *"Carry out a detailed condition survey of all standing masonry ruins within the site confines as defined on the enclosed drawing IC-22024-SPL-01-00. It is the intention of the survey to identify all visible defects to the masonry construction. The survey is to be based on a combination of detailed inspection from scaffolding towers and ground level inspection in all remaining areas where scaffolding is not located. Where ground level inspection is carried out, it is anticipated that the surveyor will use the experience gained from areas of detailed inspection to reasonably predict the probable condition and scope of repair which can be anticipated."*

Access Arrangements

If complex access arrangements are necessary, it is advisable to set these out in a separate section, with a description of the specific type in each location.

Appropriate Skill & Experience

A proven track record of experience should already have been established before the client brief is sent to the tenderers, but it is advisable to request that the individuals who will undertake the work be identified in the tender submission and that a *curriculum vitae* be provided.

Programme

A programme for the work should be identified. This may include the start date, the period for completing the site-based survey and the date for submission of the completed report.

Survey Reporting (Deliverables)

The condition survey is to be presented as an A4 format-bound document with colour photographic illustrations. The report should clearly and concisely summarise any identified defects using numbered paragraphs for ease of reference. The recommendations should include a summary of costs for each recommended phase. Four copies of the bound report should be submitted together with one digital copy in Adobe pdf format. Photographic images should be provided as TIFF files to a 300-dpi resolution.

CONTENTS OF A CONDITION SURVEY REPORT

1.0		REPORT SUMMARY
	1.1	Schedule of the structures included in the survey
	1.2	Description of the structures
	1.3	Construction and materials
	1.4	Inaccessible parts
	1.5	Summary of general condition
	1.6	Recommendations for further investigation
	1.7	Recommendations for further specialist advice
2.0		DESCRIPTION OF DEFECTS: STRUCTURES
	2.1	Masonry walls
	2.2	Floors
	2.3	Surface finishes
	2.4	Protective structures and coverings
	2.5	Past repairs
3.0		DESCRIPTION OF DEFECTS: THE SITE
	3.1	Landscape
	3.2	Invasive plant growth
	3.3	Site drainage
	3.4	Visitor impact
4.0		RECOMMENDATIONS
	4.1	For immediate action
	4.2	For completion within two years
	4.3	For completion within five years
	4.4	For completion within 10 years
	4.5	Monitoring work
	4.6	Adequacy of routine maintenance
5.0		SCHEDULE OF DEFECTS
6.0		PHOTOGRAPHIC ILLUSTRATIONS
7.0		ANNOTATED SURVEY DRAWINGS

SAMPLE CONTACT LIST & EMERGENCY STRATEGY DOCUMENT

CONTACT LIST & EMERGENCY STRATEGY DOCUMENT			
ADDRESS OF PREMISES: Full address of property, including the postcode **Grid Reference:** 12-figure grid reference			**DATE:** Date of completion Dates of any subsequent amendments
PROVISION		**DESCRIPTION**	**IMPORTANT FACTORS**
1	Significant Historic Features	Describe the premises & identify what features are historically important *This is a Grade 1 listed Georgian country house approximately 40 m x 56 m with ground and 3 floors plus a basement. The ceilings and fire places in the ground floor lounge and first floor library are particularly fine, and merit every effort to preserve them.*	Detail any important relevant factors *The ornate ceilings are in a fragile state, and contact with water may weaken or destroy them.*
2	Significant Contents	State whether there is a salvage plan and its location Also identify rooms which contain items that are particularly important *The first floor library contains very rare books, which should be saved as a priority*	If objects are fixed or too heavy to move, detail the precautions to mitigate damage to them
3	Fire Risks	Detail any special risks that cannot be eliminated, such as the location of storage of highly flammable materials, or flammable building construction (for example, matchboard partitions in the loft, thatched roofs, and so on)	Detail any mitigating factors such as whether a suppression system has been provided, or if the level of combustible storage is kept low
4	Flood and Other Risks	Detail any known risks, such as flooding from a tidal or flood-prone river, run off from hills or roofs, internal water systems, mining or any other subsidence problems, approach to airport runways etc.	Detail any precautions taken to mitigate damage, such as precautionary sandbags or leak-detection systems
5	Access for Fire Engines	Consult with local fire-and-rescue service as necessary Briefly detail the location of fire stations on the initial attendance, and the travelling time they are away from the premises: What route will fire engines need to take to get from the nearest main road to the premises, and how much of the perimeter can they reach? Detail any problem areas such as weak bridges, narrow gates, or known parking problems	Check with local fire and rescue service for the pre-determined attendance of fire engines and any special appliances such as water bowsers or turntable ladders
6	Water Supplies	Where are the nearest fire hydrants, identified by yellow marker plate, and what size main are they on? This figure is found in the H on the marker plate, together with the distance to the hydrant. Are there open water supplies such as tanks, lakes, or rivers, and how can they be accessed by fire-and-rescue service pumps?	If open water supplies are relied upon, the water needs to be at least 500 mm deep, and the bank should be substantial enough to support the weight of the pump
7	Compartmentation	Describe and show on a plan the fire-compartment lines, or (if applicable) state that there are no compartments	The roof spaces are the highest risk areas and the compartment separation needs to be in good condition

CONTACT LIST & EMERGENCY STRATEGY DOCUMENT

8	Means of Escape	Detail the number and width of exit doors, and whether they are inward or outward opening *There are four single outward opening doors and one double inward opening door.*	
9	Fire Alarm and Detection	Describe what has been provided (for example, a fire alarm and detection system conforming to *BS5839 Part 1 Fire detection and fire alarm systems for buildings. Code of practice for system design installation, commissioning and maintenance*) and whether the system is connected to a call-receiving centre	Explain how deaf persons or others who may have difficulty responding to a fire alarm be catered for (by a staff sweep when the fire alarm activates, for example)
10	Emergency Lighting	Detail the areas covered by emergency lighting *Emergency lighting of 3-hours duration has been provided in all staircases, outside exit doors and public areas.*	If emergency lighting has not been provided, detail the reasons *The site is closed at dusk, so emergency lighting has not been provided.*
11	Fire-Fighting Equipment	Detail how many fire extinguishers of each type are provided, and exactly where they are sited	State reasons if extinguishers are not mounted, etc.
12	Signs and Notices	Detail the size and positions of all signs *Exit signs with pictograms are provided above all final exit doors, except the patio doors, which obviously lead to the outside area.* *Fire procedure notices are provided adjacent to fire alarm break-glass call point.* *Fire doors are signed with 'fire door keep shut' notices.*	Detail reasons when signs are not in position *Extinguishers are in prominent positions so do not require signage.*

PROCEDURES

A	Evacuation Procedure	Describe the evacuation procedures *All persons evacuate simultaneously on discovery of a fire or on hearing the alarm. Staff will sweep the building to ensure complete evacuation and report to the assembly point on the front lawn. Members of the public will be free to leave the premises or to assemble on the lawn if they wish to continue their visit. Exits to ground level are available on both levels for disabled visitors or staff.*
B	Training	Describe the training undertaken *All staff are:* • *familiar with the layout of the building* • *aware of the location of exit doors* • *trained in the evacuation procedures including a roll call* • *trained in the safe use of first aid firefighting equipment* • *aware of how to call the fire brigade.*
C	Record of Tests	A statement should be made whether a logbook exists, in which all statutory tests on the fire alarm, emergency lighting and extinguishers are recorded, together with staff training and evacuation drills
D	Fire Risk Assessment	A statement should be made regarding the fire risk assessment *A fire risk assessment on an approved format is completed initially, when changes occur and then at regular intervals.*

GLOSSARY

Accredited
An AABC-registered conservation architect, RICS Building Conservation accredited surveyor, ICON accredited conservator/restorer, or CARE accredited engineer. The Institute of Historic Building Conservation and the RIBA Conservation Register are not officially accreditation schemes, but membership is an indicator of recognised conservation expertise.

Adaptation
Change or modification to a place made to accommodate new uses or requirements.

Aesthetic
See Value

Alteration
Work intended to change the function or appearance of a place.
See also Harm

Amenity society
A charitable or unincorporated community association with a stated interest in local social, environmental or development issues. The 'national amenity societies' in England, which must be notified of proposals to demolish listed buildings, are the Ancient Monuments Society, Council for British Archaeology, Society for the Protection of Ancient Buildings, Garden History Society, the Georgian Group, the Victorian Society and the Twentieth Century Society.

Ancient monument
See Scheduled Ancient Monument

Antiquarian
An attitude to historic buildings and remains, prevalent from the mid-18th to the late-19th century, that stresses their value as documentary records of the past. The attitude was associated especially with the Society of Antiquaries.

Approved documents
The practical guidance, published as a series of individual 'Parts' by the Department for Housing, Communities and Local Government, on ways to comply with the functional requirements in the *Building Regulations 2010*. Some of the *Approved Documents* (notably *Approved Document B (Fire Safety)* and *Approved Document L (Conservation of Fuel and Power)*, recognise that historic buildings may require alternative approaches to meet the requirements.

Archaeological building analysis
Systematic analytical process for historic buildings combining visual analysis and interpretation of the fabric (setting, uses, plan form, layout and construction, phasing of development, alterations, and so on), with supporting annotated photographs and drawings, and measured survey data. It is often carried out as part of the preparation of a Conservation Plan for complex assets so as to gain understanding about their history, evolution and significance.

'Archaeological' investigation
An approach to the study and recording of historic buildings and above-ground remains that attempts to understand building evolution using archaeological techniques, in particular stratigraphic techniques. Early efforts relied on scholarship aimed at sorting the evidence according to a pre-established sequence of styles or periods. Modern approaches use both relative and absolute dating techniques (stratigraphic and radiometric) to record historical changes.

Architectural paint analysis
The study of architectural paints and painted schemes based on the analysis of minute samples of paint stratigraphy, strategically chosen after careful consideration of the documented history and/or a condition survey of the painting, and embedded in resin for examination under a microscope.

Article 4 direction

A restriction of permitted development rights under Article 4 of the *Town and Country Planning Act (General Permitted Development) Order 1995 (GPDO)*. Article 4 directions are imposed by a local authority where it deems the exercise of permitted development rights causes, or may cause, a specific local problem for the character or significance of an asset (usually a Conservation Area).

Asbestos

A naturally occurring, fibrous silicate mineral able to strongly resist heat and chemical damage. Used as a fire-retardant or insulator, and as a component in sheeting, tiles, siding, roofing felts, shingles, plasters, cements and textured coatings, as well as for general heat and sound insulation. Serious illnesses (such as cancer and asbestosis) are caused by the inhalation of asbestos fibres.

Asset

See Heritage asset

Asset Management Plan

A strategic management system for the systematic and holistic maintenance and management of complex assets or estates. Based on a system of quadrennial or quinquennial surveys, the plan should integrate survey and assessment of individual sites with a data management system, to control repair and maintenance planning, procurement, and delivery.

Authenticity

Those characteristics that most truthfully reflect and embody the cultural heritage values of a place.

Basic survey

See Preliminary survey

Beneficial use

Any use conducive to public benefit. For a heritage asset, this may mean re-use for its original purpose, or for another purpose consistent with sustaining its significance. It presupposes the asset, after repair or adaptation, is able to generate sufficient market value to secure long-term preservation.

Bill of quantities

A document containing a numbered, itemised list of work items, and quantities of labour, materials, services and other sums required for a project. Used as the basis for tendering and cost management. On a large project it may be prepared by a cost manager/quantity surveyor. For a small project a non-quantitative schedule of works might be produced instead.

Breathable

Descriptive term for any material, system or component that is permeable to water vapour, and/or able to absorb and desorb moisture with changes in relative humidity.

Budget

In cost management, the financial plan for allocating funds for specific activities during the life of a project.

Building fabric survey

Condition survey used to obtain a detailed, comprehensive assessment of fabric condition as a whole. Such surveys are undertaken at close quarters (from scaffolding or a mobile elevated platform), and need to be both descriptive and analytical, diagnosing defects, and setting out the amount and scope of future conservation and repair work, with prioritised recommendations for maintenance and repair over specific time periods. Also known as a full or detailed condition survey, it is usually carried out by a chartered building surveyor or architect.

Building file, Building record
Any dataset compiled and maintained for the purposes of control or management of an individual asset. Property records maintained by local authorities as part of their asset management plan, which include information such as drawings, maps, condition survey and use information, fire performance data, maintenance schedules, valuations, and so on, are an example.

Building fire performance method
In building fire engineering, a methodology for organising the complex and interactive network of fire dynamics, potential active and passive fire safety systems, manual intervention and timescales into a logical framework, taking into account all the relevant factors involved in fire spread.

Building Log Book
A statutory building management document issued to the owner or manager of a building under the *Building Regulations Part L2*. It contains details of newly provided, renovated or upgraded thermal elements, controlled fittings, fixed building services, energy meters and other measures that enable energy consumption to be controlled and monitored. Not to be confused with Maintenance logbook.

Building Preservation Order
A statutory notice served by a local planning authority under the provisions of various *Town and Country Planning Acts*, preventing demolition, alteration or extension without local authority consent. Building Preservation Orders were discontinued with the introduction of the *Town and Country Planning Act* of 1968, when Listed Building Consent was introduced.

Building preservation trust [BPT]
A non-profit organisation established to preserve buildings of architectural or historic importance whose survival is threatened, and for which an economically viable solution is beyond the reach of both the original owner and the normal operation of the market. BPTs are usually constituted as companies limited by guarantee and have charitable status.

Capacitance meter
Electronic device for measuring capacitance (the ability of a material or system to store an electric charge). 'Pinless' devices use radio signals to measure the dielectric properties of the material and obtain a surface moisture content reading.

Commemorative
See Value

Condition survey
General term for any qualitative survey aiming to gather the information needed to understand the construction and condition of a heritage asset, and to identify how its significance is vulnerable to decay. Such surveys vary in scope and detail, and range from periodic or routine inspections, to building fabric surveys, structural survey (or structural appraisal), and further specialist investigations or diagnostics focused on particular features or elements of an asset.

Consent
See Listed Building Consent, Scheduled Monument Consent

Conservation
1. The process of managing change to a significant place in its setting in ways that will best sustain its heritage values, while recognising opportunities to reveal or reinforce those values for present and future generations.
2. Repair 'as found', using techniques and strategies designed to maximise the preservation *in situ* of existing material, and minimise restoration or replacement.
See also Minimum intervention

Conservation area

An area of special architectural or historic interest, the character and appearance of which it is desirable to preserve or enhance. Conservation areas are heritage assets designated under the *Planning (Listed Buildings and Conservation Areas) Act 1990*, primarily by local authorities.

Conservation management plan

A form of Conservation Plan providing a greater level of detail on management of assets, especially complex heritage assets, including detailed maintenance and management guidance for translating policies into appropriate actions. Conservation management plans are normally required as part of the application process for large grants from funding bodies.

Conservation planning, Conservation Plan

A logical, formalised process that entails a systematic approach to understanding significance, identifying conservation issues, and devising policies or strategies to sustain the significance of a heritage asset. The Conservation Plan is a written statement, of varying complexity, that sets out the way in which significance has been understood, the issues identified and the strategies proposed. It includes a careful analysis of fabric and its significance, and serves as both a record of the way judgements were reached, and an operational decision-making tool to guide maintenance, repair, alteration, use and development.

Conservation statement

A synthetic overview of a given heritage asset, which includes an appraisal of its heritage values, a summary of its significance and an initial assessment of the issues likely to affect its future management.

Consolidation

A term generally used to describe the stabilisation or strengthening of historic material or fabric with minimal replacement, such as re-bedding/repointing of loose masonry, re-fixing of detached layers or fragments, or use of adhesives and consolidants.

Construction management contract

Contract in which a construction manager (for example, a managing contractor) is engaged to manage on behalf of the client, for a fee, the project planning and procurement. Rarely suited to conservation projects.

Contingency funds

Funds set aside to cover any unexpected costs arising during the life of a project.

Contract documentation

Complete set of drawings, schedules of work, specifications, cost plan and programme.

Control of Substances Hazardous to Health [COSHH] Regulations

Primary legislation introduced in November 2002 setting out the requirements to be followed by employers to control substances hazardous to health in the workplace, using preventive controls such as risk assessment, exposure control, health surveillance and emergency planning.

Controlled waste

Household, industrial and commercial waste. The definitions for each of these are given in the *Environmental Protection Act 1990* and subsequent environment legislation instruments.

Cultural heritage

Inherited assets which people identify and value as a reflection and expression of their evolving knowledge, beliefs, and traditions, and of their understanding of the beliefs and traditions of others.

Curtilage structures

Buildings or other structures, such as boundary walls or gates, historically associated with the principal listed building in an entry on the statutory list, but generally ancillary to it.

Dendrochronology
Technique of dating historic wooden objects by determining the felling date of the trees from which they were created, using tree-ring growth patterns as revealed by core sampling.

Design-and-build
A method for procuring construction services in which the contractor is responsible for design as well as construction.

Designation
The inclusion of an asset, at international, national or local level, in an inventory, list or register, in accordance with the criteria governing identification and protection. Designation is an indicator of the importance of the particular value(s) and significance of a place, and forms the basis of the statutory system of heritage protection in England, though not all designations are statutory and not all places of significance will meet current criteria for designation.

Detailed design
In architectural projects, a work stage consisting of design drawings, specifications, schedules of work, bills of quantities and any other information necessary for the work to be implemented. It follows the feasibility and design development stages, and precedes the tendering and construction phases.

Development
A legally imprecise term usually interpreted according to the *Town and Country Planning Act 1990*, which defines it as the carrying out of building operations, engineering operations, mining operations, and other operations in, on, over and under land, or the making of any material change of use to any buildings or other land.

Digital terrain model [DTM]
A digitally-generated 3-D representation of terrain, with surface features (such as buildings, vegetation, and so on) removed.

Direct measurement
Obtaining metric data using calibrated measuring devices (such as a tape measure or total station theodolite) which record the actual dimensions of the subject being measured.

Documentary evidence
Written or visual records (such as accounts, deeds, maps, illustrations, historical photographs, and so on) useful in assessing the history, development or alteration of an asset.

Dutyholder
Any person or organisation holding a legal duty. In Health and Safety law, the dutyholder is normally the employer. But in the *Control of Asbestos Regulations 2006*, dutyholder means anyone who has an obligation of any extent in relation to the maintenance or repair of non-domestic premises (and may in some circumstances mean the owner, leaseholder or other responsible party).

Ecclesiastical exemption
The measures adopted in planning law and practice that make ecclesiastical buildings of the five main Christian denominations in England, used for ecclesiastical purposes, exempt from Listed Building Consent. Provisions are currently set out in the *Ecclesiastical Exemption (Listed Buildings and Conservation Areas) (England) Order 2010*.
See also Faculty

Embodied energy
The sum of energy required (measured in megajoules per kilogramme, MJ/kg) for any building material or full product life cycle, from growth or extraction, to manufacture or processing, packaging, transport, installation, de-installation, and disposal. Mass concrete and concrete blocks tend to have a low embodied-energy rating; plastics, and highly refined metals such as copper and aluminium, a high rating.

Endoscopy
Non-destructive technique for localised visual examination of small hollows or cavities using an endoscope, which consists of a flexible tube attached to a light source and viewing system (a video monitor or eyepiece), through which images are transmitted via fibre-optic cables.

Enhancement
Positive measures deemed to improve the character or appearance of a conservation area, or the setting of a designated asset, formulated by a local planning authority.
See also Harm

Environmental impact assessment [EIA]
Assessment and analysis of the potential impact of various forms of human activity on the environment. It is a European Community requirement for major development projects and relevant measures are incorporated into the development consent procedure.

Evidential
See Value

Exemplar
On-site specimen or specimens (for example, mortar mixes, joint treatments, paint removal, cleaning, tooling, and so on), undertaken on a limited scale (often based on suitable trials) to demonstrate the materials or procedures required for specific items of work. Used for testing the method statements included in a specification, setting standards for quality control and acting as a reference during remedial works.

Fabric
The material substance of which places are formed, including their geology, archaeological deposits, structures and buildings, and flora. Fabric is seen as an important aspect of significance of an asset, and the major repository of some or all of its values.

Faculty
Formal permission granted, upon application, by a Diocesan Chancellor on recommendation from the relevant Diocesan Advisory Committee, for works to a building covered by the faculty jurisdiction (that is, Church of England places of worship, excluding cathedrals).

Feasibility study
A process for exploring the viability of a project and the options available. Such a study would include background research, condition survey, options appraisal and financial projections.

Fire safety engineering
Fire mitigation that uses alternative, less disruptive, strategies for meeting the requirements of the Building Regulations 2000 Approved Document B, so as to minimise the adverse effects of fire controls to buildings of special architectural or historic interest. They take into account factors such as fire probability, severity, resistance to fire-spread, and so on.
See Building fire performance method

Fire Safety Risk Assessment Record
A statutory building management document that provides a record of fire hazards identified in the risk assessment and the actions that have been taken to reduce or remove them. It is prepared by the 'responsible person' for buildings that are subject to control under the *Regulatory Reform (Fire Safety) Order 2005*.

Fixed price contract
See Lump sum contract

Gas-carbide meters

Device for measuring free moisture content of drilled masonry or soil samples. The powder sample is weighed, then placed in a sealed flask with a measured amount of calcium carbide powder. The flask is shaken vigorously, and the moisture in the sample reacts with the calcium carbide to produce acetylene gas, which exerts a pressure on the flask that can be read on an external gauge.

Gravimetric analysis

Chemical analysis of a material by separation of the constituents and their estimation by weight. Used for determining particle size distribution by weight, or to calculate percentage moisture content by weight of components of a soil or mortar sample.

Ground Penetrating Radar (GPR)

See Impulse radar

Half-cell potential corrosion survey

Analytical technique, used in combination with other techniques for investigating corrosion in concrete. Using an electrode linked to a high-impedance voltmeter (known as a half cell), potential difference between the reinforcement steel and a stable reference electrode can be measured on the concrete surface. A steep potential gradient would normally signal an area of appreciable corrosion. Corrosion potential is mapped by moving the reference electrode, and the results are entered into a contour-mapping programme, or evaluated using a potential wheel. Further non-destructive analytical methods (for example, corrosion rate mapping) are used for corrosion prognosis.

Harm, Substantial harm

Change for the worse, primarily referring to the effect of inappropriate interventions on the heritage values of a place.

Health and Safety File

A statutory building management document containing information necessary for ensuring that future construction, maintenance, refurbishment or demolition is carried out safely. It is prepared by the CDM coordinator for works notifiable under the *CDM Regulations* and is retained by the building owner (and is passed on to future owners).

Heritage

All inherited resources which people value for reasons beyond mere utility.

Heritage assets

All types of historic building, monument, site, place, area or landscape positively identified as having a degree of significance meriting consideration in planning decisions (regardless of whether or not the asset fulfils existing designation criteria).

Heritage Environment Records [HERs]

A public, map-based data set, primarily intended to inform the management of the historic environment.

Heritage-led regeneration

The improvement of disadvantaged people or places through the delivery of a heritage-focused project.

Heritage values

See Value

Historic area

Any more or less extensive geographical entity viewed from a historical standpoint, or regarded integrally as a heritage asset. It may be a place, settlement, neighbourhood, landscape or otherwise defined entity, and encompass both built fabric and topographical setting.

Historic environment
All aspects of the environment resulting from the interaction between people and places through time, including all surviving physical remains of past human activity, whether visible or buried, and deliberately planted or managed flora.

Historic paint analysis
See Architectural paint analysis

Impact-echo techniques
Non-destructive investigation using low-frequency sound waves generated by mechanical impact (such as small steel balls). The waves are pulsed through homogenous materials and the frequency of reflected signals analysed at the surface using a transducer. Used to reveal discontinuities such as cracks and voids in concrete and masonry, and to assess mechanical properties such as integrity and strength. Also known as dynamic impedance.

Impulse radar
Highly specialised, non-destructive analytical technique using pulsed radio energy, transmitted from an antenna held against the wall surface and reflected back to another antenna. Different materials absorb and reflect differently, so impulse radar can be useful for locating voids, embedded wood, metal and plastic. Also referred to as ground penetrating radar.

Indirect measurement
In surveying, a metric record made by means of techniques (such as rectified photography, photogrammetry, orthophotography and laser scanning) which capture metric data in a non-selective and undifferentiated form. The captured data has to be subsequently processed to produce the required outputs.

Infrared thermography
Specialised non-destructive analytical technique using cameras fitted with infrared detectors and filters, to record variations in the infrared radiation emitted by building materials and features. Can be used to detect voids, concealed or embedded materials, moisture penetration, and to measure thermal performance.

Intangible cultural heritage
Term used in international conservation discourse as a counterpart to tangible heritage. It refers to the practices, representations, expressions, knowledge and skills (including instruments, objects, artefacts, cultural spaces) that communities, groups and, in some cases, individuals recognise as part of their cultural heritage. It includes oral traditions and expressions, language, performing arts, rituals and festive events, knowledge, and practices concerning nature and the universe, traditional craftsmanship, and so on.

Integrity
Wholeness, honesty. In conservation, the term is used both in the material sense, to characterise the identity of assets (or parts of them) as whole entities, and in the moral sense, referring for instance to interventions that seek to safeguard the essential character and significance of the fabric.

Inventory
An official record, in whatever format, of designated heritage assets. In England, the main inventories are the statutory 'lists' of listed buildings and conservation areas protected under the relevant Act, and the non-statutory 'registers' (*Register of Parks and Gardens of Special Historic Interest*, *Register of Historic Battlefields*). The latter do not provide any additional statutory protection for the designated asset, but are taken into account in local planning policy.

Joint contracts tribunal [JCT]
An organisation (now a limited company) formed in 1931 by a number of interest groups in the construction sector for establishing standard forms of contract for use in the construction industry.

Landscape
An area, as perceived by people, whose character is the result of the action and interaction of natural and/or human factors.

Laser scanner

An automated range-finding device that rapidly captures a mass of three-dimensional data by emitting laser light and detecting its reflection from the surface of the subject. Hundreds or thousands of discrete points are generated to create a 'point cloud' of 3-D coordinates. Colour information can also be collected. This is then processed to produce a 3-D model of the scanned surface.

Lead-based paint

Traditional oil paint containing lead carbonate or lead sulphate. Supply and use of these compounds (or any substance containing them) for paint are currently prohibited under the *Environmental Protection (Controls on Injurious Substances) Regulations 1992*, but a derogation established under the *European Marketing and Use Directive 1989* allows for such paints to be used for *"the restoration and maintenance of works of art and historic buildings and their interiors"* subject to certain conditions.

'Lead professional'

Term used in modern construction management for the professional (usually a suitably qualified architect or chartered building surveyor) appointed to advise the client on what can be achieved, how works should be implemented, and what statutory consents and permissions are required. He or she may also be the lead designer and/or project coordinator.

Lichen

A symbiotic organism consisting of a fungus, and an algae or a cynobacterium. In general, they do little harm to the surface of masonry on which they grow and can even slow down weathering.

'Like-for-like' repair

Conservative repair using original, identical or matching materials.

List, listing

See Listed building

Listed building

A building, object or structure judged to be of national significance, and included in the statutory *List of Buildings of Special Architectural or Historic Interest*, drawn up under the *Planning (Listed Buildings and Conservation Areas) Act 1990*. Listed buildings are designated in three (non-statutory) grades: Grade I, Grade II* ('two star') and Grade II (the latter accounting for the majority – roughly 92 % – of listed buildings as a whole).

Listed Building Consent

Consent obtained from the local planning authority for works for the demolition of a listed building, or for its alteration or extension in any manner which would affect its character as a building of special architectural or historic interest. Consent applications are evaluated based on the extent to which significance of the asset in question has been properly understood, and harm eliminated or minimised in the proposals for change.

Listed buildings in ecclesiastical use

See Ecclesiastical exemption

Lump sum contract

A contract in which a buyer agrees to pay a contractor a specified amount for completing a given programme of work. Used for larger, planned packages of repair or maintenance work, such as repainting external joinery and rainwater goods. Also known as a fixed price contract.

Lump sum contract with bill of quantities

Lump sum (or fixed price) contract where the contractor has the complete package of information needed (drawings, specifications, schedule of works and bill of quantities) to price and undertake the project. A 'lump sum without bill of quantities' is a contract based on drawings and specifications only, and is often used in smaller projects or where work cannot be precisely measured.

Maintenance

Routine work necessary to keep the fabric of a building, monument or designed landscape in good order. It is largely preventive, aimed at limiting deterioration, unlike the remedial actions involved in repair. As such it is a cornerstone of conservation practice.

Maintenance logbook

A building management document recording maintenance inspections and actions taken. It is prepared and maintained by the building owner or manager. Not to be confused with Building Log Book.

Maintenance plan

A strategy for managing the routine inspection and maintenance of an asset, organised and carried out with forethought, control and the use of records, to a predetermined plan based on the results of previous condition survey. It can range from a simple inspection schedule to a comprehensive asset management plan.

Measured survey, metric survey

A record of a heritage asset, made by direct or indirect measurement, aiming to produce a dimensionally accurate representation or representations. The metric data and their representation may vary in scope or form depending on the end use of the survey, the techniques used and degree of precision required, practical and technical considerations, timescale, and cost.

Measured term contract

A contract in which a buyer agrees to pay a contractor for repairs or maintenance to a given asset or assets over a fixed period of time. Work is instructed from time to time, and measured and valued on the basis of an agreed schedule of rates provided by the contractor. Used where buyers have a regular flow of maintenance and minor works to procure. Also known as cost reimbursement contracts.

Microdrilling

Portable, low-invasivity technique for assessing the condition of *in-situ* timber. It consists of a high-speed probe drill bit (either 1-mm or 3-mm diameter) inserted into timber at controlled pressure. The rate of penetration is recorded graphically, with sudden or significant changes in the speed of entry indicating decay or other discontinuities within the timber.

Microwave attenuation, microwave absorption

Highly specialised analytical techniques that measure the absorption of electromagnetic energy by water molecules in a sample exposed to microwave radiation.

Minimum intervention

A philosophical tenet of conservation practice which states that the most appropriate interventions to heritage assets are those that entail minimum change to its material character. The principle of minimum intervention (or 'repair as found') emerged at a time when aesthetic and poetic values prevailed, and may be less strictly applied today for many (though not all) types of asset.

Monitoring

The systematic observation, recording and evaluation of dynamic processes, in a selective fashion, either continuously or on a periodic basis, for as long as is necessary to capture all the important cycles, trends or permutations; used to track structural behaviour, characterise building environment, identify risks, record rates of decay, and so on.

Monument

1. A Scheduled Ancient Monument or other designated monument.
2. Any building, sculpture, marker or memorial chiefly significant for its historical or communal associations or values.

Named sub-contractor

A sub-contractor hired directly by a main contractor. Tender documents may include the names of potential named sub-contractors, but evaluation and selection of the sub-contractor is handled by the main contractor.

Nimbyism

A term referring to the resistance shown by residents or other local stakeholders to certain types of land use development (from the acronym NIMBY: "*not in my back yard*"). Originally used in the USA in relation to the siting of industrial facilities in residential or other sensitive areas, it became a popular pejorative term in the UK for local objection to development generally, especially in relation to the historic environment.

Nominated sub-contractor

A sub-contractor hired by a contractor, but nominated by the client or the client's representative in the tender documents.

Nuclear magnetic resonance spectroscopy

Highly specialised analytical technique that measures the magnetic properties of certain atomic nuclei to characterise the atoms and molecules present in a material.

Operative

See Prescriptive

Orthophotograph

A digital photographic image that has been corrected for scaling errors due to tilt and depth displacement. Each pixel is individually scaled and shifted in order to produce an orthographic projection. It can be used as a scaleable, stand-alone rectified image, or imported into CAD and combined with photogrammetry, or manipulated in space using 3-D modelling or visualisation software.

Patina

A change in the surface appearance of a material or materials resulting from natural ageing, wear or use. A patina generally only causes alteration in the colour, sheen or texture of the surface without visibly affecting deterioration, but biological and microbiological activity associated with patinas may potentially influence decay and weathering. In assessing or sustaining significance, patinas may themselves be important to aesthetic or other heritage values.

Performance

The way in which a system, component or structure functions. In work specification, a performance-based approach sets out functional criteria for completing work, and defines the minimum standards to be achieved. It is focused on end results rather than unwieldy prescription, so affording designers and contractors some freedom in their selection of technique and procedure. This is the approach used in the *UK Building Regulations* since the 1980s (with *Approved Documents* providing more detail on compliance in certain performance areas).

Permitted development

Certain types of work to unlisted domestic buildings, which are exempt from the need to seek planning permission (except where a local authority restricts permitted development rights; for example, with an Article 4 direction in a conservation area).

Periodic renewal

Replacement or substitution of elements that have reached the end of their life cycle, or that are incapable of fulfilling their intended functions through more limited intervention. It differs from maintenance in that it occurs on a longer cycle, is usually more drastic in nature and often has a greater visual impact. It involves the temporary loss of certain heritage values, such as aesthetic value, but is desirable unless any harm caused to heritage values as a result will not be lessened over time.

Periodic survey

Condition survey primarily used for maintenance planning and management. Also known as property management survey, it is usually undertaken on a four- or five-year cycle.

Pesticide

Chemical substance and/or certain micro-organisms prepared or used to destroy pests. The general term includes herbicides, fungicides, insecticides and masonry biocides.

Photogrammetry
Survey technique in which precise measurements and representations are produced from two overlapping stereo-images (taken from slightly different positions). The traditional product is a scaled outline drawing ('wire frame'). 'Metric' cameras that have little or no lens distortion, and contain a mechanism for ensuring film flatness, are used to acquire the images. Digital and analytical photogrammetric systems enable accurate drawings to be produced using CAD from very heavily tilted photographs; for example, a whole building façade captured from the street.

Place
Any part of the historic environment, of any scale, that has a distinctive identity perceived by people.

Planned preventive maintenance [PPM]
A programme of maintenance carried out at predetermined intervals to reduce the likelihood of a building or its elements not meeting an acceptable condition in terms of function or significance.

Preliminaries
In contract management or tendering, the initial part of the tender document giving details of the works as a whole, and general conditions and requirements for their execution; for example, sub-contracting, management structure, documents, quality management, and so on. They include general cost items for the contractor such as plant, site staff, facilities, site services, and other items not included in the schedule of rates.

Preliminary survey
A form of survey aimed at providing an initial, broad overview of the construction, and condition of a building, structure or site. It is usually preliminary to a wider programme of survey and analysis, and is the starting point for understanding how significance is vulnerable. Normally carried out from ground level.

Prescriptive
Adjective applied to any statement that gives specific instructions or directions. In a work specification, a prescriptive approach describes materials, work methods, procedures and equipment at length, and in detail. Such specifications are not intended to be open to interpretation. By contrast, an operative or 'performance'-based approach sets out working criteria more loosely, and gives minimum standards to be achieved in terms of performance and quality.

Preservation
A term that can be interpreted in numerous ways (protecting, maintaining in safety, maintaining unharmed or unaltered, safeguarding from decay). The 'presumption in favour of preservation' used in *Planning Policy Guidance (PPG) 15 Planning and the Historic Environment* has been a cherished concept in UK legislation. However, in keeping with a values-based approach, the term 'conservation' is increasingly preferred (*Planning Policy Statement (PPS) 5, Planning for the Historic Environment*).

Primary legislation
Laws and statutes established by Acts of Parliament in the UK (or by founding treaties in the EU).

Prime cost contract
Contract in which a contractor is reimbursed at the end of a project for his cost outlay in materials, labour and plant, plus a fixed fee (or a percentage of the above-mentioned costs). Sometimes used where work has to be carried out urgently and the full scope is not known.

Procurement strategy
The strategy used to define, assess, procure and coordinate all the resources and services needed to deliver a project according to a client's objectives. The strategy covers procurement options in terms of time, cost and quality criteria, tendering and contract negotiation for design and construction stages, risk appraisal, delivery, and cost/performance monitoring.

Project
Resources and funds allocated to achieving a specific set of objectives within a specific time-frame.

Public interest
Common public welfare or wellbeing, as recognised and safeguarded by government and its institutions. Heritage values and significance represent aspects of public interest in the historic environment which enjoy some protection under law. Decisions about change to significant places must, however, be balanced with other aspects of public and private interest.

Quadrennial inspection
A thorough condition survey (periodic survey) of building fabric undertaken on a four-year cycle by a suitably-qualified professional.

Quinquennial inspection
A thorough condition survey (periodic survey) of building fabric undertaken on a five-year cycle by a suitably-qualified professional.

Radiocarbon dating
A radiometric method of determining the age of organic (carbonaceous) material based on the known rate of decay of the carbon isotope 14C (Carbon-14).

Radiography
Non-destructive analysis of materials or components using X-ray or Gamma-ray radiation to investigate interior condition.

Radon
A clear, odourless radioactive gas produced by the decay of uranium, which is found in low concentrations in most soils, especially over granite and limestone bedrock. It produces radioactive dust in the air and increases the risk of lung cancer.

Ramsar sites
Wetlands of international importance designated under the terms of the *Ramsar Convention* of 1971, and the *Control of Habitats and Species Regulations 2010*.

Reactive maintenance
Maintenance carried out in response to a given failure or defect; for example, in response to the findings of an occasional inspection or notification by building users. Less urgent defects may be logged and remedial work programmed as part of a cyclical maintenance programme.

Rectified photography
Photographic survey technique in which a single photograph is taken with the image plane of the camera approximately parallel to the principal plane of the object. Scale errors due to tilt, but not those due to relief, are optically corrected, and the image printed to scale. Unlike an orthophotograph, rectified photographs are not scaleable in multiple planes.

Registered park or garden, Registered battlefield, Registered landscape
A park, garden, battlefield or landscape included in the non-statutory Registers (*Register of Historic Parks and Gardens of Special Historic Interest in England*, *Register of Historic Battlefields*), which are compiled and maintained by Historic England.

Registered waste carrier
A waste carrier registered with the Environment Agency to transport controlled waste. There is a two-tier system for waste registration based on level of risk.

Reinstatement
Putting back or re-introducing missing, damaged or decayed elements (either by repairing the original and putting it back in place, or by restoration or substitution).

Repair
1. Work beyond the scope of maintenance, to remedy defects caused by decay, damage or use, including minor adaptation to achieve a sustainable outcome, but not involving alteration or restoration.
2. Works to remedy defects, damage or decay by patching, mending or refixing *in situ*, rather than by complete or extensive replacement.
See also Conservation; Like-for-like repair

Replacement
Substitution of materials or components after they have completed their life cycle. In contrast to repair (2), it entails a comparatively high loss of existing fabric.

Resistance meter
Electronic device for measuring electrical resistance (by measuring the flow of an electric current between electrodes inserted into the sample). In porous materials, electrical resistance changes according to moisture content and resistance meter readings can provide an indication of moisture levels within a certain range (NB they do not measure absolute moisture content).

Restoration
Intervention made to a place with the deliberate intention of revealing or recovering a known element of heritage value that has been eroded, obscured or previously removed. Such measures are acceptable provided they meet certain criteria: heritage values of the elements that would be restored must be shown to decisively outweigh the values of those that would be lost; there must be compelling evidence of the evolution of the place; relationships between significance and evolution must be fully understood; previous forms of the place must be respected; and maintenance and management of the restored place must be shown to be sustainable.

Retreatability
The quality of being retreatable. It is a key consideration in pragmatic approaches to conserving heritage assets using interventions that do not produce permanent negative consequences that could restrict future interventions (for instance, if more appropriate techniques are developed in the future). The concept has gained ground in view of the practical difficulties of making many types of intervention fully 'reversible'.
See Reversibility

Re-use
Conservation strategy based on developing a beneficial use for a heritage asset and carrying out interventions sufficient to enable its rehabilitation without adversely affecting significance. The strategy is also seen as contributing to the sustainable use of energy and material resources.

Reversibility
The quality of being reversible. It is used in reference to interventions that can subsequently be undone (for instance, if shown to be unsuitable) without damage or harm to significant original fabric. In practice, few interventions are fully reversible, and the term is a relative, rather than absolute, criterion for judging the long-term implications of repair, restoration, refurbishment and other interventions.

Risk
The likelihood that a hazard (or threat) will cause harm. Risk assessment is the process of identifying hazards or threats, assessing likelihood of injury, harm or other negative impact, and deciding on the control measures needed to keep the risk at an acceptable level.

Salvage
Recuperation of heritage assets, in whole or in part, to prevent permanent loss, particularly in a disaster situation. To be effective, salvage has to be handled systematically on the basis of a comprehensive emergency strategy and salvage plan.

Scheduled Ancient Monument

A site, usually an archaeological monument, designated under the provisions of the *Ancient Monuments and Archaeological Areas Act 1979*, and afforded statutory protection under the Act. The category includes buildings as well as below-ground sites, such as caves and excavations, some submarine structures and the remains of some vehicles, aircraft ,and so on.

Scheduled Monument Consent [SMC]

Authorisation, obtained from the Secretary of State for Digital, Culture, Media and Sport, to carry out works to a Scheduled Ancient Monument. Consent applications are handled by Historic England and assessed on the basis of the impact that the proposed works would have upon the significance of the monument(s) concerned, as specified in Department policy. For some works (existing agricultural and gardening operations carried out lawfully in the same place in previous years) consent is not usually required.

Schedule of rates

A list of unit-prices for work items in a tender document.

Schedule of works

A concise list of items of work, used in conjunction with the drawings and the specifications, for pricing by the contractor. Schedules of works do not include quantities and are often used by smaller firms for smaller projects. They merely list the individual items of work to be carried out in a given location.

Secondary thickening

Increase in girth of the stems and roots of a woody plant due to the formation of secondary tissue in the vascular cambium.

Setting

The surroundings in which a place is experienced, its local context, embracing present and past relationships to the adjacent landscape.

Significance

The sum of the cultural and natural heritage values that people associate as an asset.
See Values

Siliceous abrasives

Blasting aggregate containing siliceous materials such as sand, flint, opal and quartzite. The silica dust is linked to the lung disease silicosis and the use of such aggregates is all but impossible under the *Control of Substances Hazardous to Health [COSHH] Regulations*.

Sites of special scientific interest [SSSI]

Sites of special ecological or geological significance designated by *Natural England under the Wildlife and Countryside Act 1981 (as amended)*.

Soft capping

The technique of using soil, turf and vegetation (principally grasses) to protect exposed wall tops. Known to protect the covered masonry from deterioration by freeze/thaw weathering and reduce the potential damage to underlying stone from the thermal expansion of exposed wall tops.

Special areas of conservation [SACs]

Designated areas which have been given special protection under the *Conservation of Habitats and Species Regulations 2010*. They provide increased protection to a variety of wild animals, plants and habitats in line with EU legislation to conserve biodiversity. They are also designated nationally as Sites of Special Scientific Interest.

Special protection areas [SPAs]
Designated areas under the *Conservation of Habitats and Species Regulations 2010* and *EC Habitats Directive*, identified as being of international importance for the breeding, feeding, wintering or migration of rare and vulnerable species of birds. They are also designated nationally as Sites of Special Scientific Interest.

Strategic environmental assessment [SEA]
An overarching strategic assessment and consultation at regional level of the potential impact of various forms of land use in terms of sustainability (including heritage impact). It must be undertaken, in line with European Community requirements, ahead of any corresponding Environmental Impact Assessments.

Structural survey
A specialist condition survey to investigate structural condition and performance (for example, cracking, bulging or movement). Provided in addition to building fabric surveys and undertaken by a qualified structural engineer with experience in working with historic buildings. Also referred to as a structural appraisal or structural assessment.

Sustainability, sustainable
Capable of meeting present needs without compromising ability to meet future needs.

Symbolic
See Value

Tangible heritage, Tangible cultural heritage
Term used in international conservation discourse, usually understood as referring to the physical manifestations of culture and heritage (in particular, monuments, groups of buildings, sites, objects and artefacts) considered worthy of preservation. The term 'cultural property' is also used. The terms 'natural heritage' or 'natural property' have been preferred for environments, habitats, ecosystems and other types of physical landscape, though these can also be both cultural and tangible (or, indeed, intangible).

Thermal mass
Property of a material relating to its ability to absorb, store and release heat energy. Thick-walled traditional masonry and earth building will absorb and transmit heat slowly and store it effectively, owing to the buffering effects of mass, density, specific heat capacity and conductivity.

Time domain reflectometry
Highly specialised technique for measuring electrical conductivity of porous materials by measuring the velocity of electrical pulses passed through the material via electrical cables to an embedded probe.

Toluene
See Volatile organic compounds

Toponymy
Place names or the study of place names.

Total station theodolite [TST]
Electronic surveying instrument that combines precise electromagnetic distance measurement [EDM] and angle measurement (theodolite) with a datalogging capacity to record 3-D points at ranges from approximately 0.25 m to 2000 m. A beam (usually infrared) is emitted, and distances are measured by recording the signal reflected from the selected target. Using the distance measurement, and the horizontal and vertical angles between the instrument and the target, three-dimensional coordinates are calculated using trigonometry.

Tree Preservation Order

An order made by a local planning authority (*Town & Country Planning Act 1971*) to prohibit the cutting down, uprooting, topping, lopping or wilful damage/destruction of trees without the authority's consent. Trees in Conservation Areas have additional protection as a result of Conservation Area designation.

Trials

Tests to explore materials, production methods or finishing techniques. They may be undertaken on a limited scale off-site or on-site, and are used to refine the specification or method statement for specific work items. Techniques and outcomes agreed on the basis of trials are used to create exemplars for quality control.

Ultrasound pulse velocimetry

Highly specialised non-destructive analytical technique, using high-frequency sound waves pulsed through building materials to reveal cracks, voids, fracture planes and variations in density.

Value

Any of the attributes, which are not fixed but fluid and changing over time, ascribed by people to a place and forming the basis of its significance. Values range from material evidence of past cultures or human activity (evidential), associative with or illustrative value connected with past events and people (historical), role or importance in collective memory, commemoration, spiritual life or shared experience (communal), natural, environmental or artistic (aesthetic) character.

Values-based assessment

An assessment that reflects the values of the person or group making the assessment.

Vernacular Architecture

Common domestic or indigenous architecture, built with local materials according to local environmental and cultural contexts.

Volatile organic compounds [VOCs]

A group of organic chemicals that easily form vapours at normal temperature. Known air pollutants and hazards to human health, they include a number of solvents used in modern paints, adhesives, sealants and coatings (benzene, styrene, xylene, toluene, formaldehyde, and so on).

Weathering

Physical deterioration and chemical decomposition of materials on exposure to atmospheric agents such as wind, water, soluble salts, acidic gases, temperature and radiation.

Wire frame

See Photogrammetry

Works specification

In contract management and tendering, the part of the tender document containing design information, drawings, specifications for materials, suppliers, manufacturers, and methods of work.

BIBLIOGRAPHY

Anderson, J. M. (1888); *Conservation of Ancient Monuments and Remains: General Advice to the Promoters of the Restoration of Ancient Buildings* (revised and enlarged edition); London: Royal Institute of British Architects

Ashurst, J., Ashurst, N. (1988); *Practical Building Conservation: English Heritage Technical Handbook: Vols.1–5* (series); Aldershot: Gower Technical Press

Australia ICOMOS (1999); *The Burra Charter: The Australia ICOMOS Charter of Places of Cultural Significance*; available from *http://australia.icomos.org/publications/charters/*

Betjeman, J. (ed.) (1934–67); *Shell Guides (A series of guides to the counties of England and Wales)*; London: Architectural Press

Brereton, C. (1995); *The Repair of Historic Buildings: Advice on Principles and Methods* (2nd edition); London: English Heritage

British Property Federation, Historic England, Royal Institution of Chartered Surveyors (2017); *Heritage Works: A Toolkit of Best Practice in Heritage Regeneration*; Swindon: Historic England; available from *www.historicengland.org.uk/images-books/publications/heritage-works/*

British Standards Institution (1998); *BS 7913:1998 Guide to the principles of the conservation of historic buildings*; London: BSI

Brown, G. B. (1905); *The Care of Ancient Monuments*; Cambridge: University Press

Byron, R. (1937); *How we Celebrate the Coronation (On the demolition of historic buildings in London)* (Reprinted from the Architectural Review); London: Architectural Press

Carter, J. (1804); 'Westminster Abbey church and Henry the Eighth's chapel'; in *The Gentleman's Magazine*; Vol.74, No.2 (August), pp.738–40

Charter of George II incorporating the Society of Antiquaries of London (London, 1751); available from *www.sal.org.uk/about-us/governance/*

Colvin, H. (1978); *A Biographical Dictionary of British Architects, 1600–1840* (New edition); London: J. Murray

Cooper, T. P. (1911); *History of the Castle of York from its Foundation to the Present Day*; London: E. Stock

Council of Europe (1985); *Convention for the Protection of the Architectural Heritage of Europe (Granada, 3/10/1985)*; available from *conventions.coe.int/Treaty/en/Treaties/html/121.htm*

Council of Europe (2000); *European Landscape Convention CETS No.: 176 (Florence Convention; Florence, 20/10/2000)*; available from *conventions.coe.int/Treaty/Commun/QueVoulezVous.asp?NT=176&CM=8&CL=ENG*

Council of Europe (2005); *Council of Europe Framework Convention on the Value of Cultural Heritage for Society (Treaty of Lisbon; Faro, 27/10/2005)*; available from *conventions.coe.int/Treaty/en/Treaties/Html/199.htm*

Croad, S. (1992) 'The National Building Record: The early years'; in *Transactions of the Ancient Monuments Society*; Vol.36, pp.79–98

Department for Communities and Local Government (1990); *Planning Policy Guidance 16: Archaeology and Planning (PPG 16)* (Cancelled)

Department for Communities and Local Government (2011); *Draft National Planning Policy Framework*; available from *webarchive.nationalarchives.gov.uk/20120919182957/http://www.communities.gov.uk/documents/planningandbuilding/pdf/1951811.pdf*

Department for Communities and Local Government, Department for Culture, Media and Sport, English Heritage (2010); *PPS5: Planning for the Historic Environment: Historic Environment Planning Practice Guide*; London: The Stationery Office; available from *webarchive.nationalarchives.gov.uk/20120920011334/http://www.communities.gov.uk/documents/planningandbuilding/pdf/1514132.pdf*

Department for Culture, Media and Sport (2001); *The Historic Environment: A Force for Our Future*; London: DCMS

Department for Culture, Media and Sport (2003); *Protecting our Historic Environment: Making the System Work Better*; London: DCMS

Department for Culture, Media and Sport (2004); *Review of Heritage Protection: The Way Forward*; London: DCMS

Department of the Environment, Department of National Heritage (1994); *Planning Policy Guidance: Planning and the Historic Environment (Planning Policy Guidance; PPG 15)*; London: HMSO

Dobrée, B., Webb, G. (eds.) (1927); *The Complete Works of Sir John Vanbrugh: Volume 4*; London: The Nonesuch Press

Ela Palmer Heritage (2008); *The Social Impacts of Heritage-led Regeneration*; London: The Agencies Coordinating Group

English Heritage (2000); *Power of Place: The Future of the Historic Environment*; London: English Heritage

English Heritage (2002); *State of the Historic Environment Report (Heritage Counts 2002)*; London: English Heritage

European Commission (1985); *Environmental Assessment Council Directive of 27 June 1985 on the Assessment of the Effects of Certain Public and Private Projects on the Environment 85/337/EEC*; available from *eur-lex.europa.eu/legal-content/EN/TXT/PDF/?uri=CELEX:31985L0337&from=EN*

European Parliament, Council of the European Union (2001); *Directive 2001/42/EC of the European Parliament and of the Council of 27 June 2001, on the Assessment of the Effects of Certain Plans and Programmes on the Environment*; available from *eur-lex.europa.eu/LexUriServ/LexUriServ.do?uri=OJ:L:2001:197:0030:0037:EN:PDF*

Evans, J. (1956); *A History of the Society of Antiquaries*; Oxford: University Press

Feilden, B. M. (1982); *Conservation of Historic Buildings*; London: Butterworth Scientific

Fergusson, A. (1973); *The Sack of Bath: a Record and an Indictment*; Salisbury: Compton Russell

Gillon, J. (1996); 'Conservation charters and standards'; in *Context*; Vol.51, p.20; available from *www.ihbconline.co.uk/context/51/#20*

Great Britain, Ministry of Housing and Local Government (1968); *Chichester: A Study in Conservation*; London: HMSO

Great Britain, Ministry of Housing and Local Government (1968); *Chester: A Study in Conservation*; London: HMSO

Great Britain, Ministry of Housing and Local Government (1968); *Bath: A Study in Conservation*; London: HMSO

Great Britain, Ministry of Housing and Local Government (1968); *York: A Study in Conservation*; London: HMSO

Harvey, J. H. (1993) 'The origin of listed buildings'; in *Transactions of the Ancient Monuments Society*; Vol.37; pp.1–20

Harvey, J. H. (1994); 'Listing as I knew it in 1949'; in *Transactions of the Ancient Monuments Society*; Vol.38; pp.97–104

Historic England (2008); *Conservation Principles, Policies and Guidance for the Sustainable Management of the Historic Environment*; London: Historic England; available from *www.historicengland.org.uk/images-books/publications/conservation-principles-sustainable-management-historic-environment/*

Hope, W. H. St John (1913); *Windsor Castle: An Architectural History* (3 volumes); London: Country Life

Insall, D. W. (1973); *The Care of Old Buildings Today*; London: The Architectural Press

Jokilehto, J. (1999); *A History of Architectural Conservation*; Oxford: Butterworth-Heinemann

Lemaire, R., Stovel, H. (eds.) (1994); *The Nara Document on Authenticity*; ICOMOS; available from *www.icomos.org/charters/nara-e.pdf*

Miele, C. (ed.) (2005); *From William Morris: Building Conservation and the Arts and Crafts Cult of Authenticity 1877–1939* (Studies in British Art 14); New Haven and London: Yale University Press

Ministry of Housing, Communities and Local Government (2019); *National Planning Policy Framework*; available from *www.gov.uk/government/publications/national-planning-policy-framework--2*

Morris, W. (1877); *Manifesto*; June 1877; London: SPAB archive

Morris , W. (1890); *News from Nowhere*; Boston: Roberts Bros.

Mynors, C. (2006); *Listed Buildings, Conservation Areas and Monuments* (4th edition); London: Sweet & Maxwell

Orbasli, A. (2008); *Architectural Conservation: Principles and Practice*; Oxford; Blackwell Science

Pendlebury, J. R. (2009); *Conservation in the Age of Consensus*; Abingdon: Routledge

Powys, A. R. (1929); *Repair of Ancient Buildings*; London: J. M. Dent

Pugin, A. W. N. (1836); *Contrasts, or a Parallel between the Noble Edifices of the Fourteenth and Fifteenth Centuries, and Similar Buildings of the Present Day; Shewing the Present Decay of Taste*; London: A. W. N. Pugin

'R. G.' (Richard Gough) (1789); 'Gentleman's Magazine, October 1789 (p874)'; in Jokilehto, J. (1999); *A History of Architectural Conservation*; Oxford: Butterworth Heinemann

Rickman, T. (1825); *An Attempt to Discriminate the Styles of Architecture in England from the Conquest to the Reformation: with Notices of Above Three Thousand British Edifices, Preceeded by a Sketch of the Grecian and Roman Orders* (3rd edition); London: Longman, Hurst, Rees, Orme

Robertson, M. (ed.) (1993); 'Listed Buildings: The National Resurvey of England'; in *Transactions of the Ancient Monuments Society*; Vol.37, pp.21–94

Ruskin, J. (2008); *The Lamp of Memory* (Reprint); London: Penguin

Stanley Price, N. *et al.* (eds.) (1996); *Historical and Philosophical Issues in the Conservation of Cultural Heritage*; Los Angeles: The Getty Conservation Institute

Stubbs, J. H. (2009); *Time Honored: A Global View of Architectural Conservation: Parameters, Theory, & Evolution of an Ethos*; Hoboken (New Jersey): John Wiley and Sons

Summerson, J. (1949); 'The past in the future'; in *Heavenly Mansions (and Other Essays on Architecture)* (Reprinted 1998); New York and London: W. W. Norton & Co.; pp.21–42

Thompson, M. (2006); *Ruins Reused: Changing Attitudes to Ruins since the Late Eighteenth Century*; King's Lynn: Heritage Marketing & Publishing

United Nations Economic Commission for Europe [UNECE] (1998); *Convention on Access to Information, Public Participation in Decision-Making and Access to Justice in Environmental Matters (Aarhus Convention; Aarhus, Denmark, 25 June 1998)*; available from *www.unece.org/fileadmin/DAM/env/pp/documents/cep43e.pdf*

Watkin, D. (1980); *The Rise of Architectural History*; London: Architectural Press

Williams-Ellis, C. (1928); *England and the Octopus*; London: Geoffrey Bles

Williams-Ellis, C. (ed.) (1937); *Britain and the Beast*; London: J. M. Dent & Sons

CURRENT LAW, POLICY & GUIDANCE

British Standards Institution (1998); *BS 7913:1998 Guide to the principles of the conservation of historic buildings*; London: BSI

Department for Communities and Local Government (2006); *Regulatory Reform (Fire Safety) Order 2005: A Short Guide to Making your Premises Safe from Fire*; Wetherby: DCLG Publications; available from *www.gov.uk/government/publications/making-your-premises-safe-from-fire*

Department for Communities and Local Government (2011); *Draft National Planning Policy Framework*; available from *webarchive.nationalarchives.gov.uk/20120919182957/http://www.communities.gov.uk/documents/planningandbuilding/pdf/1951811.pdf*

Department for Culture, Media and Sport (2010); *The Operation of the Ecclesiastical Exemption: Guidance*; available from *www.gov.uk/government/publications/the-operation-of-the-ecclesiastical-exemption-and-related-planning-matters-for-places-of-worship-in-england-guidance*

Department for Digital, Culture, Media and Sport (2018); *Principles of Selection for Listing Buildings* (web document); available from *www.gov.uk/government/publications/principles-of-selection-for-listing-buildings*

Europa (2010); *Reducing the emissions of volatile organic compounds [VOCs]* (web page); available from *europa.eu/legislation_summaries/environment/air_pollution/l28029b_en.htm#amendingact*

European Community (2002); *Consolidated Version of the Treaty Establishing the European Community*; available from *eur-lex.europa.eu/legal-content/EN/TXT/PDF/?uri=CELEX:12002E/TXT&from=EN*

Health and Safety Executive (2006); *Health and Safety in Construction* (3rd edition); Sudbury: HSE Books; available from *www.hse.gov.uk/pubns/books/hsg150.htm*

Health and Safety Executive (2012); *Revision of the Biocidal Products Directive (98/8/EC)* (web page); available from *www.hse.gov.uk/aboutus/europe/euronews/dossiers/biocide.htm*

Health and Safety Executive (2013); *Workplace Health, Safety and Welfare: Workplace (Health, Safety and Welfare) Regulations 1992: Approved Code of Practice and Guidance*; Sudbury: HSE Books; available from *www.hseni.gov.uk/publications/l24-workplace-health-safety-and-welfare-gb-acop-approved-use-ni*

Health and Safety Executive (2015); *Managing Health and Safety in Construction: Construction (Design and Management) Regulations 2015: Guidance on Regulations*; Sudbury: HSE Books; available from *www.hse.gov.uk/pUbns/priced/l153.pdf*

Health and Safety Executive (n.d.); *Guidance* (web page; a comprehensive range of guidance leaflets); available from *www.hse.gov.uk/guidance/index.htm*

Historic England (2008); *Conservation Principles, Policies and Guidance for the Sustainable Management of the Historic Environment*; London: Historic England; available from *www.historicengland.org.uk/images-books/publications/conservation-principles-sustainable-management-historic-environment/*

Historic England (2015); *Easy Access to Historic Buildings*; Historic England; available from *www.historicengland.org.uk/images-books/publications/easy-access-to-historic-buildings/*

Historic England (2017); *Listing Selection Guides* series (web document); available from *www.historicengland.org.uk/listing/selection-criteria/listing-selection/*

Historic England (2017); *Energy Efficiency and Historic Buildings: Application of Part L of the Building Regulations to Historic and Traditionally Constructed Buildings*; Historic England; available from *www.historicengland.org.uk/images-books/publications/energy-efficiency-historic-buildings-ptl/*

Historic England (2017); *Understanding Place: Historic Area Assessments*; Historic England; available from *www.historicengland.org.uk/images-books/publications/understanding-place-historic-area-assessments/*

Historic England (2019); *Conservation Area Appraisal, Designation and Management Second Edition, Historic England Advice Note 1*; Swindon: Historic England; available from *www.historicengland.org.uk/images-books/publications/conservation-area-appraisal-designation-management-advice-note-1/*

Historic England (n.d.); *Listed Buildings* (web page); available from *www.historicengland.org.uk/listing/what-is-designation/listed-buildings/*

HM Government (2010); *The Building Regulations 2010: Structure: Approved Document A* (2004 edition incorporating 2004, 2010, 2013 amendments); available from *www.planningportal.co.uk/info/200135/approved_documents/62/part_a_-_structure*

HM Government (2010); *The Building Regulations 2010: Fire Safety: Approved Document B: Volume 1 – Dwellinghouses* (2019 edition); available from *www.gov.uk/government/publications/fire-safety-approved-document-b*

HM Government (2010); *The Building Regulations 2010: Fire Safety: Approved Document B: Volume 2 – Buildings Other than Dwellinghouses* (2019 edition); available from *www.gov.uk/government/publications/fire-safety-approved-document-b*

HM Government (2010); *The Building Regulations 2010: Resistance to the Passage of Sound: Approved Document E* (2003 edition incorporating 2004, 2010 and 2013 amendments); available from *www.gov.uk/government/publications/resistance-to-sound-approved-document-e*

HM Government (2010); *The Building Regulations 2010: Ventilation: Approved Document F* (2010 edition incorporating 2010 and 2013 amendments); available from *www.gov.uk/government/publications/ventilation-approved-document-f*

HM Government (2010); *The Building Regulations 2010: Conservation of Fuel and Power in Existing Dwellings: Approved Document L1B* (2010 edition incorporating 2010, 2011, 2013, 2016 and 2018 amendments); available from *www.planningportal.co.uk/info/200135/approved_documents/74/part_l_-_conservation_of_fuel_and_power/2*

HM Government (2010); *The Building Regulations 2010: Conservation of Fuel and Power in Existing Buildings Other than Dwellings: Approved Document L2B* (2010 edition incorporating 2010, 2011, 2013 and 2016 amendments); available from *www.planningportal.co.uk/info/200135/approved_documents/74/part_l_-_conservation_of_fuel_and_power/4*

HM Government (2010); *The Building Regulations 2010: Access to and Use of Buildings: Approved Document M* (2015 edition incorporating 2016 amendments); available from *www.gov.uk/government/publications/access-to-and-use-of-buildings-approved-document-m*

HM Government (2010); *The Building Regulations 2000: Approved Document K – Protection from Falling, Collision and Impact*; (2013 edition); available from *www.planningportal.co.uk/info/200135/approved_documents/73/part_k_-_protection_from_falling_collision_and_impact*

Mynors, C. (2006); *Listed Buildings, Conservation Areas and Monuments* (4th edition); London: Sweet & Maxwell

National Archives (1995); *The Town and Country Planning (General Permitted Development) Order 1995* (web page); available from *www.legislation.gov.uk/uksi/1995/418/contents/made*

National Archives (2005); *The Regulatory Reform (Fire Safety) Order 2005* (web page); available from *www.legislation.gov.uk/uksi/2005/1541/contents*

National Archives (2010); *The Ecclesiastical Exemption (Listed Buildings and Conservation Areas) (England) Order 2010* (web page); available from *www.legislation.gov.uk/uksi/2010/1176/contents/made*

National Archives (2005); *Care of Cathedrals (Amendment) Measure 2005* (web page); available from *www.legislation.gov.uk/ukcm/2005/2/contents*

National Archives (2010); *The Building Regulations 2010* (web page); available from *www.legislation.gov.uk/uksi/2010/2214/contents*

Paul Drury Partnership (PDP), The Environmental Project Consulting Group (2006); *Streamlining Listed Building Consent: Lessons from the Use of Management Agreements: A Research Report*; London: English Heritage; available from *webarchive. nationalarchives.gov.uk/20120920000903/http://www.communities.gov.uk/publications/planningandbuilding/ streamlininglistedbuilding*

Planning Portal (2012); *Building Regulations* (web page); available from *www.gov.uk/building-regulations-approval*

Scheduled Monuments: Identifying, Protecting, Conserving and Investigating Nationally Important Archaeological Sites under the Ancient Monuments and Archaeological Areas Act 1979; London: DCMS; available from *www.gov.uk/government/publications/ scheduled-monuments-policy-statement*

Town and Country Planning Act 1990: Section 215: Best Practice Guidance; Wetherby: ODPM Publications

www.parliament.uk (1997); *Judgments – Shimizu (U.K.) Ltd. v. Westminster City Council* (web page); available from *www.publications.parliament.uk/pa/ld199697/ldjudgmt/jd970206/shimiz01.htm*

CONSERVATION PLANNING FOR MAINTENANCE & REPAIR

Anon. (1840); *Reform Club Archives – Minutes of the Building Committee May 1840*; London: Reform Club Library

Barry, C. (1838); *Transcripts of Two Specifications of Works by Sir Charles Barry for the Reform Club, Pall Mall, London: The First for Constructing the Club House, The original Document Dated April 1838* (28 p., ms.) a*nd the Second for Completing the Painting and Decoration of a Portion of the Interior* (7 p., ms.); Augustus Tanner Papers (Collection); TaA/1/1; London: RIBA Library

Brereton, C. (1995); *The Repair of Historic Buildings: Advice on Principles and Methods* (2nd edition); London: English Heritage

Clark, K. (ed.) (1999); *Conservation Plans in Action: Proceedings of the Oxford Conference;* London: English Heritage

English Heritage (2002); *Conservation Bulletin 43: The Value of Historic Places*; available from *www.historicengland.org.uk/ images-books/publications/conservation-bulletin-43*

English Heritage (2007); *Conservation Bulletin 55: Heritage: Broadening Access;* available from *www.historicengland.org.uk/ images-books/publications/conservation-bulletin-55*

English Heritage (2010); *Conservation Bulletin 63: People Engaging with Places;* available from *www.historicengland.org.uk/ images-books/publications/conservation-bulletin-63*

English Heritage (2011); *Knowing Your Place: Heritage and Community-Led Planning in the Countryside* (web document); available from *www.stratford.gov.uk/doc/173665/name/English%20Heritage%20Knowing%20your%20Place.pdf/*

Fagan, L. (1887); *The Reform Club: Its Founders and Architect*; London: Bernard Quaritch

Feilden, B. M., Jokilehto, J. (1998); *Management Guidelines for World Cultural Heritage Sites* (2nd edition); Rome: ICCROM

Historic England (2008); *Conservation Principles, Policies and Guidance for the Sustainable Management of the Historic Environment*; London: Historic England; available from *www.historicengland.org.uk/images-books/publications/conservation-principles- sustainable-management-historic-environment/*

Historic England (2008); *Understanding Historic Buildings: Policy and Guidance for Local Planning Authorities* (web document); available from *www.historicengland.org.uk/images-books/publications/understanding-historic-buildings-policy-and-guidance/*

Historic England (2016); *Historic England Advisory Note on the Reconstruction of Heritage Assets*; available from *www. historicengland.org.uk/content/docs/guidance/draft-reconstruction-of-heritage-assets-apr16-pdf/*

Historic Scotland (2000); *Conservation Plans: A Guide to the Preparation of Conservation Plans* (web document); available from *www.historicenvironment.scot/media/2786/conservation-plans.pdf*

Insall, D. W. (2008); *Living Buildings: Architectural Conservation, Philosophy, Principles and Practice*; Mulgrave (Australia): The Images Publishing Group

Jones, L. S. (1971); 'St Michael and All Angels, Thornhill: A catalogue of the medieval glass contained in seven windows of the church together with material relating to the history and restoration of the same'; B.Phil. dissertation, University of York

Kerr, J. S. (1996); *The Conservation Plan: A Guide to the Preparation of Conservation Plans for Places of European Cultural Significance* (4th edition); Sydney: The National Trust of Australia

National Lottery Heritage Fund; *Conservation Planning Guidance*; (web page); available from *www.heritagefund.org.uk/publications/conservation-planning-guidance*

Paul Drury Partnership (PDP) (2007); *Forty Hall and Estate Enfield: Conservation Management Plan: Executive Summary: Introduction*; available from *governance.enfield.gov.uk/documents/s6190/*

SURVEY & INVESTIGATION METHODS

Bristow, I. C. (1996); *Architectural Colour in British Interiors, 1615–1840;* New Haven, London: Yale University Press

Clarke, K. (2001); *Informed Conservation: Understanding Historic Buildings and their Landscapes for Conservation;* London: English Heritage

Curteis, T. (2008); 'The Survey and Identification of Environmental Deterioration'; in *The Building Conservation Directory 2008*; available from *www.buildingconservation.com/articles/envdet/environment.html*

D'Ayala, D., Smars, P. (2003*); Minimum Requirements for Metric Use of Non-metric Photographic Documentation;* available from *smars.yuntech.edu.tw/papers/eh_report.pdf*

D'Ayala, D., Fodde, E. (2008); *Structural Analysis of Historic Construction: Preserving Safety and Significance: Proceedings of the VI International Conference on Structural Analysis of Historic Construction, SAHC08, 2–4 July 2008, Bath, United Kingdom* (Volumes 1&2); London: CRC Press

Dallas, R. (ed.) (2003); *Guide for Practitioners 4: Measured Survey and Building Recording for Historic Buildings and Structures*; Edinburgh: Historic Scotland; available from *www.engineshed.scot/publications/publication/?publicationId=7610cb30-bf3e-4da9-8b5e-a5ad00a569dc*

Doubleday, H. A. (and subsequently others) (ed.) (1901–2011); *Victoria History of the Counties of England* (series); London: Archibald Constable (1901–14), London: St Catherine's Press (1923–38), Oxford: Oxford University Press (1936–2002), Woodbridge: Boydell and Brewer (2002–11); available from *www.british-history.ac.uk/catalogue.aspx*

Eppich, R., Chabbi, A. (eds.) (2011); *Recording, Documentation, and Information Management for the Conservation of Heritage Places: Illustrated Examples*; Shaftesbury: Donhead

GBG (GB Geotechnics Ltd) (2001); *Non-destructive Investigation of Standing Structures: Technical Advice Note 23*; Edinburgh: Historic Scotland

Historic England (n.d.); *The National Heritage List for England* (web page, searchable archive of historic photographs of England); available from *www.historicengland.org.uk/listing/the-list/*

Historic England (2016); *Understanding Historic Buildings: A Guide to Good Recording Practice*; Historic England; available from *www.historicengland.org.uk/images-books/publications/understanding-historic-buildings/*

Historic England (2019); *PastScape* (web page, searchable database of records held in the national historic environment record); available from *www.pastscape.org.uk*

Historic England (2019); *https://archive.historicengland.org.uk*

Historic England, IHBC, ALGAO-England (2012); *Heritage Gateway* (web page, national and local records of England's historic sites and buildings, including images of listed buildings); available from *www.heritagegateway.org.uk/gateway/*

Hughes, H. (ed.) (2002); *Layers of Understanding: Setting Standards for Architectural Paint Research* (Papers taken from the proceedings of English Heritage's national seminar held in London on 28th April 2000); Shaftesbury: Donhead

Laxton, R. R., Litton, C. D. and Howard, R. E. (eds.) (2001); *English Heritage Research Transactions: Research and Case Studies in Architectural Conservation, Vol. 7, Timber: Dendrochronology of Roof Timbers at Lincoln Cathedral*; London: James & James

Letellier, R., Schmid, W., LeBlanc, F. (2011); *Recording, Documentation, and Information Management for the Conservation of Heritage Places: Guiding Principles*; Shaftesbury: Donhead

Marshall, D., Worthing, D., Heath, R. (2009); *Understanding Housing Defects* (3d edition); London: Routledge

Pevsner, N. (1951–2011); *Buildings of England* (series); London: Penguin (1951–59), New Haven: Yale University Press (2002–11)

Pinchin, S. (2008); 'Techniques for monitoring moisture in walls'; in *Reviews in Conservation*; London: IIC; pp.33–45

Robson, P. (2005); *Structural Appraisal of Traditional Buildings* (2nd revised edition); Shaftesbury: Donhead

Royal Commission on the Historical Monuments for England [RCHME] (1910–85); *RCHME Inventories*; London: HMSO

Swallow, P., Dallas, R., Jackson, S., Watt, D. (2004); *Measurement and Recording of Historic Buildings*; Shaftesbury: Donhead

Taylor, J. (1999); 'The appointment of professionals for quinquennial inspections: An introduction to accreditation and approval systems'; in *The Conservation and Repair of Ecclesiastical Buildings*; available from *www.buildingconservation.com/articles/quinap/ quinap.htm*

Urquhart, D. (2007); *Technical Advice Note 31: Stone Masonry Materials and Skills: A Methodology to Survey Sandstone Building Façades*; Edinburgh: Historic Scotland; available from *www.engineshed.scot/publications/publication/?publicationId=1ded2862- 0599-44cb-81d4-a5c300a2bb9e*

Watt, D. (2007); *Building Pathology: Principles and Practice*; Oxford: Blackwell

Watt, D. (2011); *Surveying Historic Buildings* (2nd edition); Shaftesbury: Donhead

Wood, J. (1996); 'Record making and the historic environment'; in *The Building Conservation Directory 1996*; available from *www.buildingconservation.com/articles/records/records.htm*

ECOLOGICAL CONSIDERATIONS

Altringham, J. D. (1996); *Bats: Biology and Behaviour*; Oxford: Oxford University Press

Baker, J., Beebee, T., Buckley, J., Gent, A., Orchard, D. (2011); *Amphibian Habitat Management Handbook*; Bournemouth: Amphibian and Reptile Conservation; available from *www.arc-trust.org/habitat-management-handbooks*

Cameron, S., Urquhart, D., Wakefield, R., Young, M. (1997); *Biological Growths on Sandstone Buildings: Control and Treatment (Historic Scotland Technical Advice Notes No.10)*; Edinburgh: Historic Scotland

Darlington, A. (1981); *Ecology of Walls*; London: Heinemann Educational

Department for Environment, Food and Rural Affairs [Defra] (2002); *MAGIC* [Multi-Agency Geographical Information for the Countryside] (web page; updated 06/03/12); available from *www.magic.gov.uk/*

Department for Environment, Food and Rural Affairs [Defra] (2007); *Code of Practice on How to Prevent the Spread of Ragwort* (Revised edition); London: Defra; available from *assets.publishing.service.gov.uk/government/uploads/system/uploads/attachment_ data/file/801153/code-of-practice-on-how-to-prevent-the-spread-of-ragwort.pdf*

Dunwell, A. J., Trout, R. C. (1998); *Burrowing Animals and Archaeology (Historic Scotland Technical Advice Note No.16)*; Edinburgh: Historic Scotland

Edgar, P., Foster, J., Baker, J. (2010); *Reptile Habitat Management Handbook;* Bournemouth: Amphibian and Reptile Conservation; available from *www.arc-trust.org/habitat-management-handbooks*

English Heritage, National Trust, Natural England (2009); *Bats in Traditional Buildings*; Swindon: English Heritage; available from *www.historicengland.org.uk/images-books/publications/bats-in-traditional-buildings/*

Gilbert, O. L. (1992); *Rooted in Stone: The Natural Flora of Urban Walls*; Peterborough: English Nature

Hundt, L. (2012); *Bat Surveys: Good Practice Guidelines* (2nd edition); London: Bat Conservation Trust

Mayle, B., Ferryman, M., Pepper, H. (2007); *Controlling Grey Squirrel Damage to Woodlands (Forestry Commission Practice Note August 2007)*; Edinburgh: Forestry Commission; available from *www.forestresearch.gov.uk/research/controlling-grey-squirrel- damage-to-woodlands-2/*

Mitchell-Jones, A. J. (2004); *Bat Mitigation Guidelines (IN136)*; Peterborough: English Nature; available from *webarchive. nationalarchives.gov.uk/20140605171643/http://publications.naturalengland.org.uk/publication/69046?category=31008*

National Archives (1997); *The Control of Pesticides (Amendment) Regulations 1997* (web page); available from *www.legislation.gov.uk/ uksi/1997/188/contents/made*

Natural England (2012); *Advisory Leaflets and Guidance Notes* (web documents; Advice on wildlife, problems that may occur and solutions); available from *webarchive.nationalarchives.gov.uk/20130301211844/http://www.naturalengland.org.uk/ourwork/regulation/wildlife/advice/advisoryleaflets.aspx*

National Tree Safety Group (2011); *Common Sense Risk Management of Trees: Guidance on Trees and Public Safety in the UK for Owners, Managers and Advisers*; Edinburgh: Forestry Commission; available from *www.forestresearch.gov.uk/research/common-sense-risk-management-of-trees*

Office of the Deputy Prime Minister [ODPM] (2005); *Planning Policy Statement 9: Biodiversity and Geological Conservation*; London: ODPM; available from *webarchive.nationalarchives.gov.uk/20120920030656/http://www.communities.gov.uk/documents/planningandbuilding/pdf/147408.pdf*

Paine, S. (1993); 'The effects of bat excreta on wall paintings'; in *The Conservator*; Vol.17, pp.3–10

Parsons, K. *et al.* (eds.) (2007); *Bat Surveys: Good Practice Guidelines*; London: Bat Conservation Trust

Pepper, H. (1998); *The Prevention of Rabbit Damage to Trees in Woodlands (Forestry Practice Note July 1998)*; Edinburgh: The Forestry Authority (Forestry Commission); available from *www.forestresearch.gov.uk/research/the-prevention-of-rabbit-damage-to-trees-in-woodland/*

Quy, R., Poole, D. (2004); *A Review of the Methods Used within the European Union to Control the European Mole* (Talpa Europaea); London: Defra; available from *webarchive.nationalarchives.gov.uk/20130301224839/http://www.naturalengland.org.uk/Images/molereview_tcm6-4393.pdf*

Richardson, D. H. S. (1992); *Pollution Monitoring with Lichens (Naturalists' handbooks No.19)*; Slough: Richmond Publishing Co.

Segal, S. (1969); *Ecological Notes on Wall Vegetation*; The Hague: W. Junk

Stebbings, R. E. (1986); *Which Bat is it?: A Guide to Bat Identification in Great Britain and Ireland*; London: The Mammal Society and The Vincent Wildlife Trust

Watkins, J., Wright, T. (eds.) (2007); *The Management and Maintenance of Historic Parks, Gardens & Landscapes: The English Heritage Handbook*; London: Frances Lincoln

Wood, C., Cathersides, A., Viles, H. (2018); *Soft Capping on Ruined Masonry Walls*; available from *www.historicengland.org.uk/advice/technical-advice/parks-gardens-and-landscapes/landscape-management-of-monuments/*

MANAGING MAINTENANCE & REPAIR

Ashurst, J. (ed.) (2007); *Conservation of Ruins*; Oxford: Butterworth-Heinemann

Ashurst, N. (1994); *Cleaning Historic Buildings* (Volumes 1&2); Shaftesbury: Donhead

Brereton, C. (1995); *The Repair of Historic Buildings: Advice on Principles and Methods* (2nd edition); London: English Heritage

British Standards Institution (1986); *BS 8210:1986 Guide to building maintenance management*; London: BSI

Cocroft, W. *et al.* (2004); *Military Wall Art: Guidelines on its Significance, Conservation and Management*; London: English Heritage; available from *www.historicengland.org.uk/images-books/publications/military-wall-art-guidelines/militarywallartguidelines/*

Collings, J. (2002); *Old House Care and Repair*; Shaftesbury: Donhead

Cox, J. (2000); *English Heritage Research Transactions: Research and Case Studies in Architectural Conservation: Volume 6, Thatch: Thatching in England 1940–1994*; London: James & James

Daniels, M. (2006); 'Health and safety'; in *The National Trust Manual of Housekeeping: The Care of Collections in Historic Houses Open to the Public*; Oxford: Butterworth-Heinemann; pp.650–59

Earl, J. (2003); *Building Conservation Philosophy*; Shaftesbury: Donhead

English Heritage (2000); *Thatch and Thatching: A Guidance Note*; available from *www.periodproperty.co.uk/docs/English%20Heritage%20-%20Thatch%20and%20Thatching%20Guidance%20Note.pdf*

English Heritage (2010); *Caring for Places of Worship*; available from *www.historicengland.org.uk/images-books/publications/caring-for-places-of-worship-report/*

Fidler, J. (ed.) (2002); *English Heritage Research Transactions: Research and Case Studies in Architectural Conservation, Vol.2, Stone, Stone Building Materials, Construction and Associated Component Systems: Their Decay and Treatment*; London: James & James

Gibbons, P., Newsom, S., Whitfield, E. (2004); *Technical Advice Note 26: Care and Conservation of 17th Century Plasterwork in Scotland*; Edinburgh: Historic Scotland; available from *www.engineshed.scot/publications/publication/?publicationId=184ddba3-d6db-49b1-bd44-a5c30099c8ac*

Gowing, R., Heritage, A. (2003); *Conserving the Painted Past: Developing Approaches to Wall Painting Conservation: Post-prints of a Conference Organised by English Heritage, London 2–4 December, 1999*; London: James & James

Harrison, R. (1999); *English Heritage Research Transactions: Research and Case Studies in Architectural Conservation: Volume 3, Earth: The Conservation and Repair of Bowhill, Exeter: Working with Cob*; London: James and James

Heritage, A., Gowing, R. (2002); *Temporary Protection of Wall Paintings During Building Works: Practical Information Leaflet 2*; London: English Heritage; available from *www.lakenheathwallpaintings.co.uk/wallpaintprotect.pdf*

Historic England (2016); *Energy Efficiency and Historic Buildings: Secondary Glazing for Windows*; Historic England; available from *www.historicengland.org.uk/images-books/publications/eehb-secondary-glazing-windows/*

Historic England (2016); *Energy Efficiency and Historic Buildings: Draught-proofing Windows and Doors*; Historic England; available from *www.historicengland.org.uk/images-books/publications/eehb-draught-proofing-windows-doors/*

Historic England (2016); *Sourcing Stone for Historic Building Repair*; Historic England; available from *www.historicengland.org.uk/images-books/publications/sourcing-stone-for-historic-building-repair/*

Historic England (2016); *Stopping the Rot: A Guide to Enforcement Action to Save Historic Buildings*; available from *www.historicengland.org.uk/images-books/publications/stoppingtherot/*

Historic England (2017); *Adapting Traditional Farm Buildings: Best Practice Guidelines for Adaptive Reuse*; Historic England; available from *www.historicengland.org.uk/images-books/publications/adapting-traditional-farm-buildings/*

Historic England (2017); *The Maintenance and Repair of Traditional Farm Buildings: A Guide to Good Practice*; Historic England; available from *www.historicengland.org.uk/images-books/publications/maintenance-repair-trad-farm-buildings/*

Historic England (2018); *Insuring Historic Buildings and other Heritage Assets*; Swindon: Historic England; available from *www.historicengland.org.uk/images-books/publications/insuring-historic-buildings-and-other-heritage-assets/*

Historic England (2018); *Vacant Historic Buildings: Guidelines on Managing Risks*; Swindon: Historic England; available from *www.historicengland.org.uk/images-books/publications/vacanthistoricbuildings/*

Historic England (2018); *Wall Paintings: Anticipating and Responding to their Discovery*; available from *www.historicengland.org.uk/images-books/publications/wall-paintings/*

Historic England (n.d.); *Caring for Historic Cemetery and Graveyard Monuments*; Historic England (web page); available from *www.historicengland.org.uk/advice/caring-for-heritage/cemeteries-and-burial-grounds/monuments/*

Hunt, R., Suhr, M., The Society for the Protection of Ancient Buildings [SPAB] (2008); *Old House Handbook: A Practical Guide to Care and Repair*; London: Frances Lincoln

Insall, D. W. (2008); *Living Buildings: Architectural Conservation, Philosophy, Principles and Practice*; Mulgrave (Australia): The Images Publishing Group

Kent, D. (2011); 'Conservative repair'; in *The Building Conservation Directory 2011*; available from *www.buildingconservation.com/articles/conservative-repair/conservative-repair.htm*

Klemisch, J. (2011); *Maintenance of Historic Buildings: A Practical Handbook*; Shaftesbury: Donhead

Lithgow, K. (2006); 'Building work: planning and protection'; in *The National Trust Manual of Housekeeping: The Care of Collections in Historic Houses Open to the Public*; Oxford: Butterworth-Heinemann; pp.710–21

Macdonald, S. (2001); *Preserving Post-war Heritage: The Care and Conservation of Mid-Twentieth Century Architecture*; Shaftesbury: Donhead

Macdonald, S., Normandin, K., Kindred, B. (eds.) (2007); *Conservation of Modern Architecture*; Shaftesbury: Donhead

Matulionis, R. C., Freitag, J. C. (1991); *Preventive Maintenance of Buildings*; New York: Van Nostrand Reinhold

Moir, J. (1999); *English Heritage Research Transactions: Research and Case Studies in Architectural Conservation: Volume 5, Thatch: Thatching in England 1790–1940*; London: James and James

National Archives (1998); *The Control of Lead at Work Regulations 1998* (web page); available from *www.legislation.gov.uk/all?title=Control%20of%20Lead%20at%20Work%20Regulations%201998*

National Heritage Training Group (n.d.); *NHTG Initiatives* (web page); available from *www.nhtg.org.uk/*

Normandin, K. C., Slaton, D. (2005); *Cleaning Techniques in Conservation Practice*; Shaftesbury: Donhead

Ridout, B. V. (ed.) (2001), *English Heritage Research Transactions: Research and Case Studies in Architectural Conservation, Vol. 4, Timber: The EC Woodcare Project: Studies of the Behaviour, Interrelationships and Management of Deathwatch Beetles in Historic Buildings*; London: James & James

Robson, P. (1999); *Structural Repair of Traditional Buildings*; Shaftesbury: Donhead

Seeley, N. (2006); 'Commissioning conservation work'; in *The National Trust Manual of Housekeeping: The Care of Collections in Historic Houses Open to the Public*; Oxford: Butterworth-Heinemann; pp.774–83

Taylor, J. (2009); 'Site protection'; in *The Building Conservation Directory 2009*; available from *www.buildingconservation.com/articles/siteprotection/siteprotection.htm*

Teutonico, J. M. (ed.) (1998); *English Heritage Research Transactions: Research and Case Studies in Architectural Conservation, Vol.1: Metals*; London: English Heritage/James & James

Teutonico, J. M., Fidler, J. (eds.) (2001); *Monuments and the Millennium: Proceedings of a Joint Conference Organised by English Heritage and the United Kingdom Institute for Conservation*; London: James & James

Wood, C. (ed.) (2003); *English Heritage Research Transactions: Research and Case Studies in Architectural Conservation Research Transactions, Vol.9, Stone Roofing: Conserving the Materials and Practice of Traditional Stone Slate Roofing in England*; London: James & James

Wrightson, D. (2002); *A Stitch in Time: Maintaining Your Property Makes Good Sense and Saves Money*; Tisbury: The Institute of Historic Building Conservation in Association with the Society for the Protection of Ancient Buildings; available from *www.ihbc.org.uk/stitch/Stitch%20in%20Time.pdf*

Young, M. E., Ball, J., Laing, R. A., Urquhart, D. C. M. (2003); *Technical Advice Note 25: Maintenance and Repair of Cleaned Stone Buildings*; Edinburgh: Historic Scotland; available from *www.historicenvironment.scot/archives-and-research/publications/publication/?publicationid=7e1f71ec-a9e7-47f0-bf8f-a5c2010fb88d*

EMERGENCY PLANNING

Forsyth, M. (ed.) (2007); *Structures and Construction in Historic Building Conservation (Historic Building Conservation series)*; Oxford: Blackwell

Historic England (2015); *Flooding and Historic Buildings: Revised 2nd Edition*; Historic England; available from *www.historicengland.org.uk/images-books/publications/flooding-and-historic-buildings-2ednrev/*

Historic England (2017); *Metal Theft from Historic Buildings: Prevention, Response and Recovery*; available from *www.historicengland.org.uk/images-books/publications/metal-theft-from-historic-buildings/*

Historic England (2019); *Lightning Protection: Design and Installation for Historic Buildings*; Swindon: Historic England; available from *www.historicengland.org.uk/images-books/publications/lightning-protection/*

Ingham, N. (2004); *Practical Conservation Guidelines for Successful Hospitality Events in Historic Houses*; available from *www.english-heritage.org.uk/learn/conservation/collections-advice-and-guidance/*

Kidd, S. (ed.) (2006); *Heritage Under Fire: A Guide to the Protection of Historic Buildings* (2nd edition); London: Fire Protection Association

Kidd, S. (2010); *Guide for Practitioners 7: Fire Safety Management in Traditional Buildings: Part 1: Principles and Practice*; Edinburgh: Historic Scotland

Kidd, S. (2010); *Guide for Practitioners 7: Fire Safety Management in Traditional Buildings: Part 2: Technical Applications and Management Solutions*; Edinburgh: Historic Scotland

Matthews, G., Smith, Y., Knowles, G. (2009); *Disaster Management in Archives, Libraries and Museums*; Burlington: Ashgate

Maxwell, I. (ed.) (2008); *COST Action C17: Fire Loss to Historic Buildings Built Heritage* (Conference Proceedings /Historic Scotland Technical Conservation, Research and education Group; 4 vols); Edinburgh: Historic Scotland

O'Farrell, G. (2007); *External Lighting for Historic Buildings*; London: English Heritage; available from *www.historicengland.org.uk/images-books/publications/external-lighting-for-historic-buildings/*

Roberts, B. (2009); *Historic Building Engineering Systems & Equipment: Heating & Ventilation*; London: English Heritage; available from *www.historicengland.org.uk/images-books/publications/heating-ventilation*

Rowell, C., Martin, J. (1996); *Uppark Restored;* London: National Trust

Stovel, H. (1998); *Risk Preparedness: A Management Manual for World Cultural Heritage*; Rome: ICCROM; available from *www.iccrom.org/sites/default/files/ICCROM_17_RiskPreparedness_en.pdf*

Xavier-Rowe, A. *et al.* (2010); *Historic House Collections: Drawing up a Collections Management Plan: Guidance Notes*; available from *www.english-heritage.org.uk/siteassets/home/learn/conservation/collections-advice--guidance/drawing-up-collections-management-plan.pdf*

INDEX

ACKNOWLEDGEMENTS
& PICTURE CREDITS

ACKNOWLEDGEMENTS

Project Management: Sally Embree

Picture Editors: Sarah Pinchin and John Stewart

Picture Research: Celia Dearing and Suzanne Williams

Design & Layout: Robyn Pender and Tracy Manning

Editing: Elizabeth Gold

Thanks are also due to HoWoCo for Design Consultancy, 4word for copy editing, and Altaimage for colour processing and print preparation. Most of the diagrams for this series of Practical Building Conservation were produced by Iain McCaig, who also provided many of the diagrams in the first series. The glossaries for all volumes in the series were produced by David Mason.

All British Standards extracts reproduced with permission from the British Standards Institution (BSI – _www.bsigroup.com_) © British Standards Institution. All Rights Reserved.

Historic England and the volume editor would like to thank the peer reviewers of this book, Robert Gowing, Mike Harlow, Alan Johnson, Matthew Slocombe and David Watt. Thanks are also extended to The London Library for their help. We would also like to express our gratitude for the generous advice offered on parts of the text by Christopher Brookes, David Drewe, Sally Embree, Simon Houghton, Geraldine O'Farrell, John Renshaw and Joy Russell. Thanks are also extended to the many individuals, both within Historic England and without, who gave support and assistance, and without whom this book could never have been produced.

Ashgate Publishing and Historic England would like to thank the following people and organisations for their kind permission to reproduce their pictures.

Abbreviations key: t=top, b=bottom, l=left, r=right, c=centre, ca=centre above, cb=centre below

Unless otherwise stated, all diagrams are: Iain McCaig © Historic England•

Cover picture: Iain McCaig © Historic England•

3: Arrival of Cardinal Francesco Gonzaga; detail of the background showing an ideal city, from the *Camera degli Sposi* or *Camera Picta*, 1465–74, fresco by Andrea Mantegna in Palazzo Ducale, Mantua, Italy. The Bridgeman Art Library•

4: The Bodleian Library, University of Oxford. Gough Maps. 231, fol.5•

5: Plan A, from John Aubrey's Monumenta Britannica. The Bodleian Library, University of Oxford. MS TOP.c.24, Plan A 39v–40•

6: The Bridgeman Art Library/ Private Collection•

7: The Bridgeman Art Library/ Ashmolean Museum, University of Oxford•

8: © Historic England•

9: © Drury McPherson Partnership•

11: The Bridgeman Art Library/ Private Collection•

12: Plate from Vol. III of Monasticon Anglicanum, by William Dugdale, 1673 © The London Library•

13: RIBA Library Drawings & Archives Collections•

15: The Bridgeman Art Library/ Private Collection/ The Stapleton Collection•

16: Frontispiece of *News From Nowhere*, by William Morris, 1892, wood engraving by W.H. Hooper, 1892. The Bridgeman Art Library/ Private Collection•

19: City of London, London Metropolitan Archives•

21: Tintern Abbey, watercolour on paper by Thomas Girtin (1775–1802). The Bridgeman Art Library © Norwich Castle Museum and Art Gallery (l); Alamy/ © James Jagger (r)•

22: Black Fryers in Glocester, Aug 24, 1721, Plate 22 from *Itinerarium Curiosum*, by William Stukeley, 1776 © The London Library•

23: © Historic England•

24: Copyright John Ashurst Estate•

27: Courtesy SPAB•

28: © Historic England•

31: © Historic England•

33: Alamy/ © Janine Wiedel Photolibrary•

35: Getty Images / Hulton Archive•

38: Alamy/ © VIEW Pictures Ltd.•

41: Rivenhall Church, the south wall of the chancel, from W.J. and K.A. Rodwell, *Rivenhall: Investigations of a Villa, Church and Village 1950-77*, CBA Research

Report 55 (1985), fig 90. By kind permission of Professor Warwick Rodwell (t); Digital photography and rectification by historic building analyst Barry Hillman-Crouch 2008 and annotation by Drury McPherson Partnership. Reproduced with the kind permission of the National Trust (b)•

43: Iain McCaig © Historic England•

45: Front cover of *Chester, a Study in Conservation*, by Donald W. Insall and Associates, 1968•

47: © Historic England•

51: Alamy/ © Bob Richardson•

56: Alamy/ © Andrew Michael•

57: Alamy/ © Dorling Kindersley•

59: Rex Features / Times Newspapers•

63: © Historic England•

66: © Historic England (tl, tr, bl, br); Iain McCaig © Historic England (cr)•

67: Iain McCaig © Historic England•

68: © Historic England•

71: © Historic England (t); Iain McCaig © Historic England (bl); Alamy/ © Skyscan Photolibrary (br)•

72: Iain McCaig © Historic England•

79: Steve Emery © Historic England•

83: © Historic England•

85: Claire Craig © Historic England (l); Iain McCaig © Historic England (r)•

86: © Historic England•

91: © Historic England•

92: Oxford Castle site © Oxford Archaeology Ltd.•

93: Iain McCaig © Historic England (t); Alamy/ © Vic Pigula (b)•

94: © Historic England (t); Iain McCaig © Historic England (b)•

95: © Historic England (t); Getty Images (bl); sculpture by Fritz Koenig, *Great Spherical Caryatid, 1971*, relocated to Battery Park, New York. Photo Iain McCaig © Historic England (br)•

100: Iain McCaig © Historic England•

101: © Linda Hall (l); © Colin Burns (r)•

102: Iain McCaig © Historic England (t); © Historic England (bl); © Colin Burns (br)•

103: © Colin Burns (l); © Historic England (r)•

104: © Historic England (t); © Patrick Stow (c); Iain McCaig © Historic England (b)•

105: © Historic England•

106: © Historic England•

107: © Historic England•

110: © Colin Burns•

113: Iain McCaig © Historic England•

114: Iain McCaig © Historic England•

115: © Historic England•

116: Uppark, West Sussex, after the fire on 30 August 1989. © National Trust Images/ David Bradfield•

119: © Colin Burns•

120: © Drury McPherson Partnership•

121: © Drury McPherson Partnership, drawing by Richard Peats•

122: © Drury McPherson Partnership•

123: © Historic England, drawing by the late George Langlands Wilson•

124: Chris Wood © Historic England•

125: The Bridgeman Art Library/ Guildhall Library, City of London•

126: Rick Mather Architects•

127: Kathryn Davies © Historic England•

128: John Neale © Historic England•

129: John Neale © Historic England•

130: John Neale © Historic England•

131: John Neale © Historic England•

133: © Historic England (l); drawing by Burlison & Grylls of a panel (Panel 1c CVMA) from the Savile Chapel east window. Deposited in the West Yorkshire Archives at Wakefield, reproduced by kind permission of the parish church of St Michael and All Angels, Thornhill, West Yorkshire. Photo © Jonathan & Ruth Cooke (tr); Reproduced by kind permission of the parish church of St Michael and All Angels, Thornhill, West Yorkshire. Photo © Jonathan & Ruth Cooke (cr); Photo © Jonathan & Ruth Cooke (br)•

134: Iain McCaig © Historic England•

135: Architects: Communion Design. Photography: Infinity Unlimited•

139: Iain McCaig © Historic England (from John Stewart, 'An approach to the management of mosaics through preliminary condition and risk assessment', *Managing Archaeological Sites with Mosaics: From Real Problems to Practical Solutions, The 11th Conference of the International Committee for the Conservation of Mosaics, Meknes, Morocco, October 24–27 2011*, in press)•

141: © Richard Oxley, Oxley Conservation Ltd.•

142: © Historic England•

143: © Historic England•

144: © Historic England•

145: © Historic England•

146: © Historic England•

147: © Historic England•

148: James O. Davies © Historic England•

149: © Historic England•

150: © Historic England•

154: Tithe map showing Little Sutton, ref EDT 245/2, Cheshire Record Office, reproduced with permission from Cheshire Archives and Local Studies (l); Cheshire Record Office, reproduced with permission from Cheshire Archives and Local Studies (r)•

155: *The Builder's Compleat Assistant*, Vol. II Plate LIV., by Batty Langley, 1738 (t); *Building Construction*, Plate VI., by Henry Adams, 1906 (b)•

156: © Shropshire Archives (tl); Courtesy of M. & L. V. Bate Properties (tr); © Historic England (cl); Whitchurch History and Archaeological Group (c); © Crown copyright. HE. (cr); Iain McCaig © Historic England (b)•

157: © Bath in Time – Bath Central Library•

158: Courtesy of the Brooking Collection / photo Raymond Smith, Environmental Historian•

159: Iain McCaig © Historic England•

160: © Historic England•

161: Drawn by Peter Ferguson DipArch (UCL) RIBA for Hugh Harrison Conservation (t); Survey by James Brennan Associates, annotation by Drury McPherson Partnership for a conservation plan commissioned by Manx National Heritage, 2011 (b)•

162: Hugh Harrison (l); J. Dobie / Historic England (r)•

164: Jerry Sampson © Caroe & Partners Architects•

165: Iain McCaig © Historic England•

166: Iain McCaig © Historic England•

167: © Bath & North East Somerset Council•

173: Photogrammetric Unit, IoAAS, University of York Copyright © Historic England (l); Sarah Pinchin © Historic England (r)•

174: © Historic England•

176: © Graham Abrey (t); Drawn by Peter Ferguson DipArch (UCL) RIBA for Hugh Harrison Conservation (bl); Sarah Pinchin © Historic England (br)•

180: © Crown copyright. HE. (tl); © Historic England (tr, cl, bl); © Crown Copyright. HE. (br)•

184: © Robert Demaus/ Demaus Building Diagnostics Ltd.•

185: © Sean Wheatley – Plastering Specialist•

186: © Robert Demaus/ Demaus Building Diagnostics Ltd.•

187: © Sandberg (l, l inset); Iain McCaig © Historic England (r)•

188: © Stratascan Ltd. (t); © Stratascan Ltd. Courtesy Stainburn Taylor and Michael Reardon (b)•

189: Hugh Harrison (t); Mobile Radiographic Services Ltd. (b)•

191: Brian Ridout (t); © Odgers Conservation Consultants (bl); Sarah Pinchin © Historic England (br)•

193: © Tobit Curteis Associates (t, br); © The Wall Paintings Workshop (bl)•

198: Iain McCaig © John Renshaw Architects•

199: Iain McCaig © John Renshaw Architects (t); Triage plan by Iain McCaig © John Renshaw Architects and original site map DP004709 © Derived from information compiled by and/or copyright of RCAHMS (b)•

205: © Historic England•

208: Science Photo Library/ Adrian Bicker•

209: © Historic England•

211: Iain McCaig © Historic England•

212: Iain McCaig © Historic England•

213: Alamy/ © Rick & Nora Bowers (t); PestFix, www.pestfix.co.uk (b)•

214: © Historic England•

215: Alan Cathersides © Historic England•

216: Alamy/ © David Chapman•

217: Tim Allen © Historic England•

219: © Historic England (t); © Alan Cathersides (b)•

221: Science Photo Library/ John Devries (t); © Catherine Woolfitt (cl); Alan Cathersides © Historic England (bl, br)•

223: © Alan Cathersides (tl, tr); Alan Cathersides © Historic England (cl); © Historic England (bl); Alan Cathersides © Historic England, with kind permission of Historic Royal Palaces (br)•

224: © Alan Cathersides•

225: Iain McCaig © Historic England (l); © Alan Cathersides (r)•

227: Iain McCaig © Historic England (tl, br); © Alan Cathersides (tr); Alan Cathersides © Historic England (bl)•

228: Alamy/ © Emmanuel Lattes (t); Alan Cathersides © Historic England (b)•

229: Alan Cathersides © Historic England•

230: Alan Cathersides © Historic England•

231: Iain McCaig © Historic England (t); Alan Cathersides © Historic England (b)•

232: © Alan Cathersides•

234: Alan Cathersides © Historic England•

235: © Alan Cathersides (tl, bl); Alan Cathersides © Historic England (r)•

237: John Ashurst © Historic England (t); Chris Wood © Historic England (bl, br)•

241: Iain McCaig © Historic England•

242: Mary Evans Picture Library•

243: © Tobit Curteis Associates•

244: Iain McCaig © Historic England•

245: John Ashurst © Historic England•

246: Courtesy of Out of the Gutter – Property Cleaning and Maintenance Service•

248: Iain McCaig © Historic England•

256: © Graham Abrey (l); © Sally Strachey Historic Conservation (r)•

258: © Historic England•

259: © Historic England•

262: Iain McCaig © Historic England•

263: John Stewart © Historic England•

268: Iain McCaig © Historic England, with thanks to West Dean College•

273: Iain McCaig © Historic England (tl); © Robert

Demaus/ Demaus Building Diagnostics Ltd. (tc); Survey by James Brennan Associates, annotation by Drury McPherson Partnership for a conservation plan commissioned by Manx National Heritage, 2011 (r); Drawn by Peter Ferguson DipArch (UCL) RIBA for Hugh Harrison Conservation (bl)•

279: Iain McCaig © Historic England (t); Chris Wood © Historic England (c); Bovingdon Brickworks Ltd. (b)•

280: © Alexander Holton (l); © Graham Abrey (tr, br)•

281: © Richard Ireland (l); © Alison Henry (r)•

282: © Julie Haddow, Lime Repair Ltd.•

290: © Graham Abrey•

291: Iain McCaig © John Renshaw Architects•

297: © Graham Abrey•

298: © Banx•

299: © Tribal Education Limited 2012. All rights reserved.•

304: Rex Features / Action Press•

306: The Banqueting Room, Guildhall © Bath and North East Somerset Council (tl); Alamy/ © Image Source (tr); Steve Emery © Historic England (bl, br)•

309: Courtesy of Oxfordshire Fire & Rescue Service•

310: Courtesy of Oxfordshire Fire & Rescue Service•

312: Steve Emery © Historic England•

314: © Douglas Kent•

315: Steve Emery © Historic England (t); Courtesy of Oxfordshire Fire & Rescue Service (b)•

317: © Historic England (tl, bl); Steve Emery © Historic England (tr, br)•

318: © Historic England•

319: Simon Revill © Historic England with thanks to Steve Emery (l, r); Steve Emery © Historic England (inset)•

320: Courtesy of Oxfordshire Fire & Rescue Service•

321: © Historic England•

322: Elizabeth Ridout © Ridout Associates•

323: Brian Ridout (l); Elizabeth Ridout © Ridout Associates (r)

T - #0487 - 071024 - C394 - 240/220/18 - PB - 9781032609157 - Gloss Lamination